Global Oil and Gas Industry Development Report: 2023

Xingkun Qian · Ruquan Lu · Liangcai Luo ·
Mouyuan Wu · Qin Liao

Global Oil and Gas Industry Development Report: 2023

石油工业出版社
PETROLEUM INDUSTRY PRESS

Springer

Xingkun Qian
CNPC Economics and Technology
Research Institute
Beijing, China

Ruquan Lu
CNPC Economics and Technology
Research Institute
Beijing, China

Liangcai Luo
CNPC Economics and Technology
Research Institute
Beijing, China

Mouyuan Wu
CNPC Economics and Technology
Research Institute
Beijing, China

Qin Liao
CNPC Economics and Technology
Research Institute
Beijing, China

ISBN 978-981-95-2374-0 ISBN 978-981-95-2375-7 (eBook)
https://doi.org/10.1007/978-981-95-2375-7

Jointly published with Petroleum Industry Press
The print edition is not for sale in China (Mainland). Customers from China (Mainland) please order the print book from: Petroleum Industry Press.
ISBN of the Co-Publisher's edition: 978-7-5183-6562-3

This Springer imprint is published by the registered company Springer Nature Singapore Pte Ltd.
The registered company address is: 152 Beach Road, #21-01/04 Gateway East, Singapore 189721, Singapore

Preface

In 2023, the century-long changes are accelerating, the world economy lacks growth momentum, and multiple factors such as the Ukrainian crisis, Israeli-Palestinian conflict, unilateral sanctions, trade barriers, climate change, and extreme weather have led to a mismatch in global energy supply and demand and intensified market fluctuations. The global political and economic development shows characteristics of polarization and regionalization. The competition among major powers intensifies the decoupling of the energy industry chain and supply chain, bringing serious shocks to the stability of the international energy market and the safety of the production and supply chain. The global energy development environment has undergone profound changes, and energy security has become a priority issue for energy development in various countries. Countries around the world are actively promoting energy transformation and upgrading, promoting the rapid development of clean energy, low-carbon technology, and energy efficiency improvement. In the face of changes in the world, the era, and history, China's energy structure is accelerating its adjustment. The industry is actively implementing the important instructions and directives of General Secretary Xi Jinping on vigorously enhancing domestic oil and gas exploration and development, and the energy bowl must be held in our own hands. Speed up the improvement of the resilience of the energy system, promote the complementarity of traditional energy and new energy, enhance energy reserves and system backup, promote marketization and intelligence, coordinate the energy governance system, fully launch and implement a new round of mineral exploration breakthrough strategic actions, and the national energy supply and demand is generally stable and orderly, and the foundation for energy supply is further consolidated.

Since 2008, the Economic and Technical Research Institute of China National Petroleum Corporation (hereinafter referred to as CNPC ETRI) has organized dozens of experts and research backbones to track and analyze the development of the domestic and foreign oil and gas industry every year, analyze major events and hot issues, and form an understanding and judgment of the development trend of the oil and gas industry, and compile and publish the "Report on the Development of the Domestic and Foreign Oil and Gas Industry" to provide references for the country, industry, enterprises, and research institutions to understand the industry situation and grasp the development trend.

The “2023 Report on the Development of the Domestic and Foreign Oil and Gas Industry” is the sixteenth annual analysis report, with a total of 13 chapters and 35 sections. This report highlights the annual characteristics in content, pays more attention to visual and intuitive forms of expression, and emphasizes originality in data analysis. At the same time, this report also looks forward to the development trend of the oil and gas industry in 2024.

The chief editors of this book, Qian Xingkun and Lu Ruquan, organized the framework and manuscript review of the whole book, and the deputy editors Luo Liangcai, Wu Mouyuan, and Liao Qin organized the writing and review work. Leaders and experts such as Yu Guo, Jiang Xuefeng, Xu Yongfa, Liu Keyu, Zhu He, and Zhou Yong guided and reviewed the content of the whole book.

Due to the limitations of the editors, there are inevitably unsatisfactory places in the book, and we sincerely hope to hear your opinions and suggestions.

Beijing, China
February 2024

Xingkun Qian
Ruquan Lu

Editorial Committee—2023 Domestic and International Oil and Gas Industry Development Report

Contents

2023 Development and 2024 Outlook of the Domestic and International Oil and Gas Industry

Global Oil and Gas Industry Development and Outlook

The Global Energy Landscape Continues to Adjust, with a Steady Transition to Green and Low-Carbon

In 2023, the international political and economic landscape continues to change, compounded by the lingering effects of the COVID-19 pandemic, the global economic situation is generally stable with flexibility, and energy consumption maintains slow growth. The global GDP growth rate fell back to 3% in 2023, lower than the 3.5% level in 2022. Energy consumption growth is slow due to the economic downturn, with global primary energy consumption at 146.1 billion tons of oil equivalent, an increase of 1.7% compared to 2022, and a growth rate increase of 0.5 percentage points.

The global energy landscape continues to undergo profound adjustments, with frequent disturbances in energy supply caused by geopolitical conflicts. The impact of the Ukraine crisis on the international supply and demand pattern of major energy sources such as oil, gas, and coal continues. The Israeli-Palestinian conflict has not yet spread to the main oil-producing countries in the Middle East, but it has once again exacerbated the uncertainty of energy supply. After the Ukraine crisis, the European Union banned the import of all types of coal from Russia, leading to a significant increase in US coal exports to Europe, and Russia redirected its coal originally exported to Europe to Asia–Pacific countries. In terms of natural gas, Norway has replaced Russia as the EU's largest pipeline gas supplier, and LNG supplies from the United States have also increased significantly. In terms of oil, Russia has successfully redirected its oil and petroleum product exports affected by the EU and G7 embargoes to countries in Asia, Africa, Latin America, and the Middle East. The trend of international oil and gas trade flows changing from "counterclockwise" to "clockwise" has intensified.

X. Qian et al., *Global Oil and Gas Industry Development Report: 2023*,
https://doi.org/10.1007/978-981-95-2375-7_1

In 2023, the global energy consumption structure continues to maintain a trend of declining fossil energy and rising non-fossil energy. Global coal demand hit a new historical high, but the growth rate slowed to 0.5%, a significant drop of 5.3 percentage points from the highest point in 2021, with demand growth mainly coming from emerging economies in Asia. Oil demand growth gradually slowed down under the promotion of factors such as the accelerated popularization of electric vehicles and improved energy efficiency, with a year-on-year increase of 2.1%, and growth mainly came from a few non-OECD countries led by China. Natural gas demand reversed the decline of the previous year, with consumption increasing by 0.5%, mainly from North America and the Asia–Pacific region. The overall consumption of fossil energy increased slightly by 1.1%, and the consumption ratio fell by 0.6 percentage points from the previous year, falling below 80% for the first time, reaching 79.7%. Non-fossil energy consumption increased by 4.1%, with the Asia–Pacific region and Europe leading the global non-fossil energy consumption growth. Nuclear power generation increased, with a year-on-year increase of 2.3%. The installed capacity of non-hydro renewable energy generation led by wind and solar power grew rapidly, with an estimated increase of more than 430 million kilowatts in 2023, breaking through 2.5 billion kilowatts, a significant year-on-year increase of 20%. Non-hydro renewable energy generation exceeded 4700 terawatt-hours, basically meeting 70% of the global power generation growth. The global energy transition process is steadily advancing.

Global energy transition-related investment is growing strongly. In 2023, global clean energy investment is estimated to exceed 1.7 trillion US dollars, an increase of more than 20% compared to 2021, and the gap between clean energy and fossil fuel investment is widening. The growth of clean energy investment is mainly concentrated in renewable energy led by solar energy and electric vehicles, as well as areas such as lithium batteries, heat pumps, and nuclear energy. It is estimated that investment in the solar energy sector will reach 380 billion US dollars, surpassing upstream oil investment for the first time. Investment in electric vehicles is expected to reach 130 billion US dollars, more than double the level in 2021.

Looking forward to 2024, the uncertainty of geopolitical conflicts is intensifying, coupled with high inflation and high interest rate pressures, the global economy still faces huge downside risks, energy commodity prices will continue to fluctuate, and the energy industry will become an important field of great power competition. The affordability and security concerns triggered by the global energy crisis will make governments and enterprises pay more attention to the development of clean energy technologies. Hydrogen energy and nuclear energy will become important forces to support the development of clean energy and the process of carbon neutrality. Countries such as the United States and Japan have successively updated their national hydrogen energy strategies, clarifying the strategic position of clean hydrogen energy; new technologies and new directions represented by advanced nuclear energy technology and controllable nuclear fusion will become hotspots; lithium-ion batteries, flow batteries, compressed air energy storage and other energy storage technologies will also welcome important developments. The continuously expanding energy industry will continue to drive global demand for key minerals, and countries such as Europe and the United States will further achieve the localization and diversification of the new energy industry chain through policies, and the global competition for new energy and mineral resources will become more intense.

Weak Demand and Geopolitical Premium Gradually Receding, International Oil Prices Fluctuate and Fall

In 2023, international oil prices fell compared to the previous year. The average annual price of Brent crude oil futures was $82.17 per barrel, a 17% decrease from $99.04 per barrel in 2022. The annual trend was a fluctuating decline under the backdrop of the fading geopolitical premium and the continuous interest rate hikes by the Federal Reserve. In the first half of the year, the continuous interest rate hikes by the Federal Reserve and the turmoil in the European and American banking industry, dominated by concerns about the macroeconomic situation, led to a downward trend in international oil prices. In the third quarter, as the US interest rate hike cycle neared its end, Saudi Arabia and Russia voluntarily extended their production cuts until the end of the year, significantly tightening market supply. Oil inventories fell to low levels, and international oil prices rebounded for a time. In the fourth quarter, although the Federal Reserve stopped this round of interest rate hikes, due to the off-season for oil consumption, the speed of oil inventory reduction significantly slowed down. The production cut agreement reached by "OPEC+" at the end of November did not meet expectations, and international oil prices fell again. The international oil market supply and demand tended to be relaxed. The production of non-OPEC oil-producing countries such as the United States recovered quickly, offsetting the production cut strategy implemented by "OPEC+". Global oil supply increased by 1.7 million barrels/day to 101.8 million barrels/day. After the transition of China's COVID-19 prevention and control, residents' compensatory travel grew rapidly, driving the recovery of consumption. Global oil consumption increased by 2.1 million barrels/day year-on-year, to 101.7 million barrels/day, with an annual oversupply of 150,000 barrels/day. In the second half of the year, the oil market entered a rapid inventory reduction stage, and oil inventories were at a low level of the 5-year average. In 2024, the international oil market was generally relaxed, and international oil prices continued to decline, but they were still at a medium-to-high level. The post-epidemic global oil consumption rebound phase will end and return to normal. China's oil consumption growth rate will significantly slow down. The rapid development of new energy vehicles will have a greater impact on oil consumption. It is expected that the global oil demand growth rate will slow down, with a year-on-year increase of only 950,000 barrels/day, to 102.7 million barrels/day. "OPEC+" will still restrain production, and it is expected that the global oil supply growth will be lower than the previous year, with a year-on-year increase of 1.3 million barrels/day, and the oversupply will further expand to 500,000 barrels/day. Oil inventories will still remain low, providing some support for oil prices. It is expected that international oil prices may fluctuate widely at high frequencies, and the central price will be lower than in 2023. The average price of Brent crude oil will be \$75–80 per barrel. In 2024, due to the change of governments in the United States, Russia, India and other countries, the spillover effects of the Ukraine crisis and the Israeli-Palestinian conflict, uncertainties in Sino-US relations, and the still unclear global economic outlook, it is very likely to trigger "black swan" and "grey rhino" events, which may make future oil price fluctuations the norm, and the global oil trade pattern will continue to adjust deeply.

The Natural Gas Market Continues to Adjust, and International Gas Prices Fall from High Levels

In 2023, the geopolitical premium gradually faded, international gas prices fell sharply, the natural gas market slowly recovered, and supply and demand tended to a surplus. Global consumption growth turned positive, with an estimated consumption of 3.96 trillion cubic meters, a year-on-year increase of 0.5%. Asia–Pacific consumption steadily recovered, with a growth rate of about 1.8%, driven primarily by China and India. In North America, increased renewable energy adoption slowed gas consumption growth to 1.2%. European gas consumption fell by about 6.2% year-on-year, falling for two consecutive years. It is estimated that the global natural gas production in 2023 will be 4.28 trillion cubic meters, a year-on-year increase of 0.8%. The trend of "US rise and Russia fall" continued, with US production reaching a new historical high of 1.15 trillion cubic meters, a year-on-year increase of 3.3%; Russian production continued to decline to 642 billion cubic meters, a year-on-year decrease of 4.5%; production in the Middle East region steadily increased to 738.8 billion cubic meters, a year-on-year increase of 0.8%.

Gas price fell sharply from historical highs. Price in the three major markets have all fallen back to the level before the Ukraine crisis. The average prices of European TTF spot, Northeast Asia LNG spot, and US HH spot were \$12.9/MMBtu, \$15.8/MMBtu, and \$2.54/MMBtu, respectively, a year-on-year decrease of 65.8%, 54.4%, and 60.5%.

The evolution of the international natural gas trade pattern further deepened, with more Russian pipeline gas exiting the European market and Asia driving the growth of LNG trade. Global gas trade volume was 1.25 trillion cubic meters, a year-on-year decrease of 1.0%. Among them, the pipeline gas trade volume was 694.4 billion cubic meters, a year-on-year decrease of 3.3%. Russia exported 49 billion cubic meters of pipeline gas to Europe, a year-on-year decrease of 36.4 billion cubic meters. LNG trade volume was 411 million tons, a year-on-year increase of 2.0%. European LNG import volume was 126 million tones, a year-on-year decrease of 1.8%; Asian import volume was 264 million tones, a year-on-year increase of 3.5%. U.S. LNG exports surged 9.5%, whereas Qatar's fell 1.5%. U.S. exports to Europe and Asia rose 6.8% and 17.5%, respectively, while Qatar's exports to these regions dropped 25.3% and grew 5.3%. The global LNG liquefaction capacity was 457 million tons/year, a year-on-year increase of 0.9%, the lowest growth rate in a decade, while regasification capacity expanded 6.2% to 1.042 billion tons/year. The total amount of new long-term LNG contracts signed globally fell back, and the proportion of new long-term contracts linked to oil prices rose to 50.0%, with the average link slope rising to 13–14%, and the way of linking to gas prices became more diversified.

In 2024, the global natural gas market will continue to maintain a fragile balance, with increasing uncertainty risks. Natural gas consumption is expected to be 4.02 trillion cubic meters, with a growth rate of 1.5%. The main hotspots are in Asia, continuing to be driven by China, India, and emerging markets. Europe's natural gas consumption continues to decline, with imports of pipeline gas from Russia remaining low. Upstream oil and gas investment is increasing, with global natural gas production expected to be 4.37 trillion cubic meters, a growth rate of 2.0%. North America and the Middle East continue to increase production, and Russia's output is expected to rebound. The global LNG market is generally loose, with periodic tension. It is expected that the actual increase in global liquefaction capacity will be 7.71 million tons/year, and LNG demand will increase by 4.7 million tons/year. Prices across the three key markets will show "two declines, one rise" with heightened volatility: The average spot price of European TTF is $11.0–12.0/MMBtu, a year-on-year decrease of 6.7–14.5%; the average spot price of Northeast Asia LNG is $12.0–13.0/MMBtu, a year-on-year decrease of 17.4–23.8%; the average spot price of US HH is $2.6–3.0 USD/MMBtu, a year-on-year increase of 2.5–18.2%.

Oil and Gas Exploration and Development Investment Continues to Grow, and Oil and Gas Production Continues to Increase

After a significant increase in global oil and gas exploration and development investment in 2022, oil companies maintained a cautiously optimistic attitude towards exploration and development investment in 2023, mainly considering factors such as the Ukraine crisis, the uneven global economic recovery, changes in the pace of energy transition, and future oil demand expectations. Global oil and gas exploration and development investment was 568.7 billion USD, a year-on-year increase of 10.6%, a lower increase than the previous year. Among them, investment in the Russia-Central Asia region showed negative growth, a year-on-year decrease of 4.5%. In 2023, the global engineering technology service market size was about 300.4 billion USD, a year-on-year increase of 10.2%. The oilfield measurement and testing services and drilling and completion services sectors performed better than other sectors, and the market size of the oilfield production services and geophysical equipment and services sectors increased by 7.5% and 8.6%, respectively, lower than the average level. In 2023, a total of 213 oil and gas discoveries were made globally, with discovered oil and gas reserves of 11.24 billion barrels of oil equivalent, and the average size of discoveries reached 52.8 million barrels of oil equivalent; compared with the previous year, the number of oil and gas discoveries decreased, the discovered reserves decreased, and the average size of new discoveries significantly decreased, setting a new low in nearly five years. In 2023, there were four discoveries exceeding 500 million barrels of oil equivalent, three of which were natural gas and one was oil, two of which were located on land in Iran, one in deep water in Indonesia, and one in deep water in Guyana.

In 2023, global oil and gas production continued to increase, with oil production of 4.45 billion tons, a year-on-year increase of 70.9 million tons, an increase of 1.6%. Among them, the North American region had the largest increase (66.86 million tons), followed by the Latin American region (30 million tons). Countries with production growth exceeding 10 million tons were the United States (64.9 million tons) and Brazil (19.19 million tons), and Russia reduced production by 12 million tons. Natural gas production reached 4.28 trillion cubic meters, a year-on-year increase of 0.8%. The North American region had the largest increase (500 billion cubic meters), followed by the Middle East region (180 billion cubic meters), with increases of 3.8% and 2.4% respectively. Under the stimulus of strong LNG demand in Europe and Asia, the growth of US LNG exports drove the growth of natural gas production, with US natural gas production at 1.15 trillion cubic meters. Due to the reduction in Russia-Europe pipeline natural gas trade, Russia's natural gas production continued to decline, decreasing by 300 billion cubic meters. The top three countries in global natural gas production are the United States, Russia, and Iran.

In 2024, it is expected that global exploration and development investment will increase by 2.1% year-on-year, which will continue to drive the expansion of the global engineering technology service market size. Global oil and gas production will also increase slightly, with the Americas continuing to maintain growth, leading other regions, while Russia's oil and gas production will continue to decline, but the decline will further narrow. According to IEA forecasts, global oil production will grow by about 1.67% in 2024, a lower growth rate than in 2023; natural gas production will grow by about 2%, a faster growth rate.

Refining Ethylene Capacity Continues to Grow, Overall Gross Profit Declines

In 2023, global refining capacity rose to 5.18 billion tons/year, a net increase of 59.425 million tons/year, slightly lower than the increase in 2022 (62.1 million tons/year), with new capacity mainly concentrated in the Middle East, North America, and China. The total crude oil processing volume was about 82.4 million barrels/day, an increase of about 1.7 million barrels/day compared to 2022, and the total processing volume has returned to the level before the COVID-19 pandemic in 2019. The overall utilization rate of refineries rose from 78.6% in 2022 to 80%. The utilization rate of refineries in OECD countries declined to varying degrees, while the utilization rate of refineries in Asia and the Middle East increased significantly. Refining gross profit fell from the 2022 high, with the gross profit of US Gulf medium sulfur, Northwest Europe light low sulfur, and Singapore light low sulfur crude oil cracking falling to 29.16 USD/barrel, 12.11 USD/barrel, and 8.28 USD/barrel, respectively, a year-on-year decrease of 18%, 34%, and 36%. The global yield of gasoline, diesel, and fuel oil continues to decline, while the yield of aviation kerosene (abbreviated as aviation coal) continues to increase.

The growth rate of ethylene production capacity has significantly slowed down, with the total ethylene production capacity reaching 228 million tons/year, an increase in capacity of 5.323 million tons/year, a year-on-year increase of 2.2%. The average operating rate of ethylene units has slightly increased, rising from 83.1 to 83.8%. The ethylene output is about 189 million tons, a year-on-year increase of 5.5%. Ethylene prices remain depressed, with the average spot prices of ethylene in North America, Northwestern Europe, and Northeast Asia falling by 29%, 14.6%, and 18% respectively compared to the previous year. The gross profit of ethylene in various major regions has increased and decreased. The average gross profit of ethylene produced by ethane cracking in North America is $410/ton, a year-on-year increase of only 1.7%; the average gross profit of ethylene produced by naphtha cracking in Northwestern Europe and Northeast Asia is $283/ton and $3.84/ton respectively, a year-on-year decrease of 37.8% and 90% respectively.

Looking forward to 2024, the global refining capacity is expected to increase significantly, with an additional capacity of over 100 million tons/year, mainly from India, China, Thailand, Nigeria, Mexico, etc. Nigeria's largest Dangote refinery with a capacity of 32 million tons/year is expected to be finally completed and put into operation. The refinery's crude oil processing volume is expected to increase by about 1 million barrels/day year-on-year, and the refinery's utilization rate is expected to remain around 80%. The tight refining capacity in Europe and America is expected to be resolved. The number of new ethylene projects will significantly decrease, and the process of ethylene expansion will begin to slow down. The total ethylene production capacity is expected to reach 232 million tons/year, with only an additional capacity of 4.33 million tons/year. Ethylene demand will continue to grow, the operating rate of the unit will still remain at a low level, the gross profit of ethylene will rebound, and the green and low-carbon development of the ethylene industry will continue to advance.

The Overlay of New and Old Hotspots of Geopolitical Conflict Makes the Overseas Investment Environment More Complex

In 2023, the geopolitical situation continues to evolve, the confrontation between the West and Russia triggered by the Ukraine crisis continues, the outbreak of the Israeli-Palestinian conflict has intensified the intensity of the great power game and camp confrontation, the risk of great power confrontation has reached its highest since the "Cold War", the trend of "camping" in the world political pattern has intensified, and the global trade may be affected as the Red Sea attack incident continues to ferment.

The political, economic, and security environment of major overseas oil and gas cooperation areas has changed significantly. The political situation in Central Asia is generally stable, the economy is gradually recovering, and the economic growth is expected to be 5.7%. Russia, under the joint pressure of the West, has high political risks, and although economic growth has turned from negative to positive, risks such as energy export restrictions, foreign capital withdrawal, and ruble exchange rate fluctuations are still threatening Russia's economic recovery. The geopolitical risks in the Middle East are rising, economic growth is slowing down, and economic growth is estimated to be 2.0%, lower than the growth rate of 5.3% in 2022. The new round of Israeli-Palestinian conflict has increased the uncertainty of the investment environment in the Middle East. Political and social risks in Africa are still important factors affecting the investment environment. The situation in various countries continues to be turbulent. Niger, Gabon, and Guinea have experienced military coups or attempted coups. Large-scale armed conflicts continue in Sudan, terrorist attacks occur frequently in many countries, and the number of deaths from terrorist attacks in Somalia, Burkina Faso, Mali, and Niger accounts for 43% of the total number of deaths worldwide, and the security situation continues to deteriorate. Latin America still faces high political and economic risks, the game between the left and right wings is intensifying, economic growth continues to be sluggish, inflation remains high, economic growth is estimated to be 1.7%, and the average inflation rate is 4.9%. The ruling party faces difficulties and challenges in governing the country. The United States temporarily relaxed sanctions on Venezuela, easing its domestic situation. Brazil, Guyana, Suriname, and other countries held bidding for offshore oil and gas blocks, and international oil companies increased their investment in deep-sea Latin America, increasing opportunities for deep-sea oil and gas cooperation. The risk in the Asia–Pacific region is relatively low, the regional economic growth momentum is obvious, and the economic growth is estimated to be 4.6%. The political and security situation in Myanmar continues to be complex, the social situation continues to be chaotic, and it may face the risk of the West increasing energy sector sanctions in the future.

In 2024, the uncertainty of the global geopolitical situation has increased. The situation in Russia in Central Asia remains severe, the Ukrainian crisis continues to ferment, the Western sanctions against Russia have become normalized, and Russia's oil and gas exports and economic growth will continue to decline. The spillover of the Israeli-Palestinian conflict in the Middle East has escalated regional tensions, and the relationship between Iran and the United States faces significant uncertainty, with geopolitical risks becoming long-term. The security situation in Africa remains bleak, with the Sudanese conflict becoming protracted, terrorist organizations active, and Chad and Mali facing political instability risks due to transitional government elections. The power struggle between left and right-wing regimes in Latin America has intensified, with countries such as Mexico, Peru, and Venezuela facing elections and political uncertainty. Venezuela also faces the risk of US sanctions. Geopolitical risks in the Asia–Pacific region continue to rise due to the US rallying regional countries to contain China, and the conflict between the Myanmar military government and local ethnic armed groups (EAOs) is difficult to resolve in the short term, and the situation remains unstable.

International Corporations Re-emphasize Upstream Oil and Gas, Balancing Carbon Reduction and Profitability

In 2023, affected by the downturn in international oil prices, the operating performance of major international oil companies generally declined. In the first three quarters, the total operating income and net profit of the five major international oil companies decreased by 16% and 10% respectively year-on-year, and their investment and capital expansion intentions significantly increased, with total capital expenditure increasing by 14% year-on-year. Crude oil production saw its first increase since 2019, up 3.12% year-on-year; affected by a sharp drop in gas prices and reduced gas demand, natural gas production fell 7.84% year-on-year.

Major international oil companies have adjusted their strategic goals and development paths, emphasizing both carbon reduction and profitability, re-emphasizing upstream operations, increasing oil and gas investment and production targets, and oil and gas asset mergers and acquisitions are active. BP has increased its oil and gas production targets for 2025 and 2030 by 300,000 barrels/day and 500,000 barrels/day respectively, adjusting them to 2.3 million barrels/day and 2 million barrels/day. Shell has abandoned its original goal of reducing oil and gas production by 1–2% annually. ExxonMobil and Chevron have made large-scale acquisitions of upstream assets, totaling $125.2 billion, making it their largest acquisition case since the twenty-first century. Among them, ExxonMobil acquired Pioneer Natural Resources for $59.5 billion, and Chevron acquired Hess Corporation for $53 billion. After the acquisition, the production of ExxonMobil and Chevron increased by 700,000 barrels/day (to 4.4 million barrels/day) and 400,000 barrels/day (to 3.4 million barrels/day) respectively, widening the production gap with their peers, increasing the reserve replacement rate, and showing diverse transition paths.

Major international oil companies are actively exploring new models of integration and synergy between low-carbon and oil and gas businesses. First, they pay more attention to profitability, reassess and adjust the scope of shrinking low-carbon industries, and strategically review renewable energy businesses, abandoning areas with unclear development prospects or where they lack competitive advantages. Shell has divested its low-carbon domestic electricity business, which had lower returns than expected, and ExxonMobil has exited its algae biofuel project, which it has been operating for 14 years. Second, they prioritize the development of low-carbon businesses that are highly integrated with oil and gas businesses, with carbon capture, utilization, and storage (CCUS), biomass energy, and hydrogen energy project acquisitions accounting for 60% of low-carbon business acquisitions. Lithium battery business has become a new favorite for investment, with ExxonMobil spending $100 million to purchase 120,000 acres of lithium mining rights from Galvanic Energy. Third, they are committed to creating industrial clusters that integrate and synergize the development of oil and gas and low-carbon businesses, integrating oil and gas, renewable energy power generation, hydrogen energy, energy storage, and CCUS, etc., to achieve point-to-chain, chain-to-group development. They are actively opening up new low-carbon businesses with high growth potential. Fourth, they are creating innovative incentive mechanisms to promote transformation and development, such as Shell's implementation of the "Shell Energy Transition Road" evaluation system, and Total Energy's formulation of the "Green Allowance" plan, etc.

Consensus on Carbon Reduction Further Solidified, and Influence of Carbon Market Continues to Grow

In 2023, global fossil fuel-related CO_2 emissions reached 36.8 billion tons, an increase of 1.1% from the previous year. During the 28th Conference of the Parties (COP28) to the United Nations Framework Convention on Climate Change, a consensus was reached on several topics such as the first global stocktake and just transition. To align with the 1.5 °C global warming target, global greenhouse gas emissions must be reduced by 43% from 2019 levels by 2030, by 60% by 2035, and achieve net-zero emissions by 2050. Countries such as France and Australia, as well as the international maritime sector, have proposed more ambitious emission reduction targets.

The EU Carbon Border Adjustment Mechanism (CBAM) officially entered the transition phase, initially covering six major sectors including steel, aluminum, cement, fertilizers, electricity, and hydrogen. The UK planned to implement CBAM from 2027. The United States and Australia have been also accelerating the process of carbon tariffs, aiming to enhance the competitiveness of domestic manufacturing and mitigate carbon leakage risks.

By the end of 2023, 29 emission trading markets have been established in different countries and regions around the world, with new markets launched in Washington State, USA, and Indonesia. The EU carbon market remained one of the most active globally. Driven by factors such as market reform and energy price fluctuations, the EU carbon price historically broke through 100 euros/ton for the first time in February. Throughout the year, prices fluctuated at high levels, with an average of 92.21 dollars/ton, a year-on-year increase of 7.6%. The price range was 72.28–106.82 dollars/ton, with the price deviation narrowing by 27.1% compared to the previous year. In the United States, the Regional Greenhouse Gas Initiative (RGGI), which covers the electricity sector, was operated relatively stable, with an annual average of 13.49 dollars/ton, remaining basically unchanged year-on-year. The price of Western Climate Initiative (WCI) California-Quebec reached a historical high at 38.73 dollars/ton, with an annual average of 33.03 dollars/ton, a year-on-year increase of 16.1%, mainly driven by California's plan to reduce allowance budgets between 2025 and 2030 and the Canadian federal government's more ambitious carbon price target. The average annual prices in the UK, South Korea, and New Zealand fell sharply, with year-on-year declines of 29.2%, 52.7%, and 24.0% respectively.

As the scale of the international carbon market continues to expand in the future, the role of carbon pricing mechanisms is expected to become more prominent. The established carbon markets are likely to include more sectors, while the development of carbon markets in countries such as Thailand, India, Brazil, and Chile is accelerating. For the oil and gas industry, the EU carbon market and WCI have already included refining and oil and gas supply in their coverage respectively. Some countries and regions have introduced carbon taxes to regulate carbon emissions from local oil and gas production and consumption. The EU also plans to control methane emission intensity of imported oil and gas from 2030. Overall, the oil and gas industry will face increasing pressure to reduce emissions in the coming years, compelling companies to take effective measures to reduce greenhouse gas emissions.

Development and Outlook of the Domestic Oil and Gas Industry

1. Based on Safe and Clean Development, the Energy Industry Policy System is Being Improved Rapidly 2023 is the beginning of the full implementation of the spirit of the 20th National Congress of the Communist Party of China. Focusing on ensuring energy security, building a new energy system, steadily advancing carbon peaking and neutrality, and promoting high-quality development, China has successively introduced or revised a series of energy industry policies and reform opinions. Substantial progress has been made in the reform of three major areas. In July 2023, the second meeting of the Central Committee for Comprehensive Deepening Reform approved the "Opinions on Promoting the Gradual Transition of Energy Consumption Dual Control to Carbon Emission Dual Control", "Implementation Opinions on Further Deepening the Reform of the Oil and Natural Gas Market System to Enhance National Oil and Gas Security", and "Guiding Opinions on Deepening Electricity System Reform and Accelerating the Construction of a New Electricity System", clarifying the overall direction of China's energy system reform. First, in terms of steadily promoting the policy of transitioning from dual control of energy consumption to dual control of carbon emissions, the system design has been improved, arrangements have been made for full coverage of renewable energy green electricity certificates, efforts have been made to restart the China Certified Emission Reduction (CCER) trading market, and 100 cities and parks with typical representativeness have been selected nationwide to carry out carbon peaking pilot construction. Second, in terms of orderly advancing the market-oriented reform of the oil and gas industry, the "Opinions on Several Issues Concerning Deepening the Reform of Mineral Resources Management" by the Ministry of Natural Resources, the "Notice on Further Improving the

Registration Management of Mineral Resources Exploration and Mining", and the "Method for Levying Mining Rights Transfer Income" have been issued; the exploration right retention period has been extended, the exploration right deduction area ratio has been reduced, and the mining right approval procedure has been simplified, among other encouraging policies; guiding opinions have been put forward on establishing and improving the natural gas upstream and downstream price linkage mechanism, and the natural gas utilization policy has been adjusted and improved. Third, in terms of deepening the reform of the electricity system, the construction of the electricity market has been promoted, the safe and stable supply of electricity has been ensured, the construction of the spot market has been accelerated, the ability to ensure the safe operation of the power grid and power load management has been strengthened; efforts have been made to actively absorb renewable energy electricity; a coal-fired power capacity price mechanism has been established, a two-part tariff has been implemented for coal-fired power, industry expectations have been stabilized, and the transformation and development of coal-fired power has been promoted. Energy supply and price stability are given top priority. The self-sufficiency rate of energy is steadily increased, the investment in energy mineral resources development is increased, the exploration, development, and increase of reserves and production of important energy and mineral resources are strengthened domestically, the construction of overseas energy resource supply bases is promoted, and the capacity reserve of key links is arranged. Relevant guiding policies for the steady growth of the chemical industry and the green innovation and high-quality development of the refining industry have been issued. A batch of "oil reduction and chemical increase" projects have been started, and efforts have been made to cultivate a batch of world-class energy and chemical enterprises. By 2025, China's refining capacity will be controlled within 1 billion tons/year, the industry concentration will be increased, and the proportion of refining capacity of 10 million tons will be about 55%. By 2030, the ability of the refining industry to guarantee raw materials and special products for the chemical industry will be greatly improved, green hydrogen refining and chemical industry, CCUS and other technologies will complete industrial and large-scale demonstration verification, and green innovation and high-quality development will be achieved. Improve the construction of the energy sector and "dual carbon" standard system. The construction of the carbon peak and carbon neutrality standard system is fully deployed, and the "Quality Power Construction Outline", "2023 National Standard Project Guide", "Carbon Peak Carbon Neutrality Standard System Construction Guide" and "2023 Energy Industry Standard Plan Project Guide" are successively released. A "dual carbon" standard construction policy system with clear goals, a complete framework, and both long and short-term considerations is established. The implementation plan for the new industry standardization leading project before 2035 is clarified, focusing on eight major fields such as new generation information technology, new energy, new materials, high-end equipment, new energy vehicles, green environmental protection, civil aviation, ship and marine engineering equipment, as well as bio-manufacturing, new energy storage and other fields. With enterprises as the main body, government guidance, and an open and integrated standardization work system, the level of new industry standardization is improved.

Looking forward to 2024, China will continue to promote the implementation of the new energy security strategy, guided by three key documents, breaking through after establishing, focusing on further implementing the transition from dual control of energy consumption to dual control of carbon, deepening market-oriented reforms of oil and gas and power system, introducing more policy measures, promoting energy development and emission reduction, improving oil and gas security capabilities, building a new power system overall goal, promoting the transformation and upgrading of traditional high-energy-consuming industries, improving carbon trading, green certificates and the three major markets of electricity, and steadily advancing the implementation of the "Fourteenth Five-Year Plan" goals.

The Overall Energy Supply and Demand is Stable, and the Construction of the New Energy System is Orderly

In 2023, China's green and low-carbon energy transition is steadily advancing, and the proportion of renewable energy consumption continues to increase. It is estimated that the total primary energy consumption of the country is 5.72 billion tons of standard coal, an increase of 5.7% year-on-year. Among them, the proportion of non-fossil energy in total energy consumption increased to 17.2%; the proportions of coal, oil, and natural gas consumption were 55.3%, 18.3%, and 8.7%, respectively, decreasing by 0.9 percentage points, increasing by 0.4 percentage points, and 0.2 percentage points. The electrification of the whole society continues to accelerate, and it is estimated that the social electricity consumption is 92 trillion kilowatt-hours, an increase of 6.5% year-on-year, faster than the GDP growth rate. The electrification substitution in the terminal energy field is accelerating, and the domestic sales of new energy vehicles in 2023 reached 9.495 million, an increase of 37.9% over the previous year, and the market penetration rate reached 31.6%.

The diversified energy supply system is more perfect, and the diversified energy supply guarantee system that is comprehensively coordinated by clean coal, oil and gas, renewable energy and nuclear energy is gradually clear. In 2023, China's coal, crude oil, natural gas and other major energy products maintained growth, and the energy supply guarantee capability steadily improved. The coal market supply and demand is basically balanced, with raw coal output of 4.66 billion tons, a year-on-year increase of 2.9%, and the growth rate has slowed down. Oil and gas reserves have increased, with annual crude oil production stabilizing at 200 million tons, natural gas production exceeding 230 billion cubic meters, and an increase of more than 10 billion cubic meters for seven consecutive years. Renewable energy power generation is 30 trillion kilowatt-hours, accounting for about 1/3 of the total social electricity consumption. The installed capacity of renewable energy power generation exceeded 1.47 billion kilowatts, accounting for more than half of the total installed capacity.

Among them, the installed capacity of wind power and photovoltaic power exceeded 1.05 billion kilowatts, accounting for more than 1/3 of the total installed capacity, an increase of 6.3 percentage points year-on-year.

In 2024, with the recovery and improvement of China's economy and the solid advancement of high-quality development, it will further drive the growth of energy consumption. The green and low-carbon transformation of energy is accelerating, the construction of the new energy system is accelerating, and maintaining a stable supply of energy security is still an important mission of the energy industry. The supply and demand of coal will continue to be stable, the exploration and development of oil and gas and the increase of reserves will continue to exert force, the installed capacity and power generation of renewable energy such as wind power and photovoltaic power will continue to grow, and the proportion of non-fossil energy consumption will further increase.

The Consumption of Refined Oil is Accelerating, and the Apparent Consumption of Oil Hits a New High

In 2023, China's oil market achieved a simultaneous recovery of supply and demand, and many indicators hit historical records. The domestic and international markets operated smoothly, and the industry developed in an orderly manner. The demand for oil and oil products rebounded sharply, with annual oil consumption of about 756 million tons and crude oil processing volume of 738 million tons, both hitting historical highs, with year-on-year increases of 11.5% and 8.6% respectively; the dependence on foreign oil rebounded, still maintaining around 72%; the consumption of refined oil was 399 million tons, a year-on-year increase of 9.5%, recovering to 98.6% of the same period in 2019. Among them, the consumption of gasoline, diesel, and kerosene increased by 9.7, 3.3, and 65.7% year-on-year, respectively, recovering to 99.4, 98.7, and 94.6% of 2019. Domestic crude oil production has made steady progress, reaching 209 million tons, a year-on-year increase of 2.1%, and crude oil imports hit a record high of 564 million tons, a year-on-year increase of 11%. The production of refined oil was 444 million tons, a year-on-year increase of 11.1%, and the supply and demand were loose. The export of refined oil ended the trend of continuous decline for three consecutive years, with annual exports reaching 41.9 million tons, a year-on-year increase of more than 20%. The industry's regulatory policies have been further refined and implemented, promoting the refined oil market to continue to develop in an orderly, standardized, and fair manner. The volume of off-balance-sheet gasoline and diesel resources plummeted to 14.69 million tons, a year-on-year contraction of 55.3%.

In 2024, the national policy for steady growth will continue to exert its force, and the economic growth trend will continue to rebound and improve. The annual economic growth rate is expected to be around 5.0%. The demand for oil and oil products will continue to grow, but the recovery rebound phase has ended, and the growth rate of demand will slow down. It is estimated that the domestic oil consumption in 2024 will be 764 million tons, an increase of 1.0% year-on-year; the consumption of refined oil will be 400 million tons, an increase of 0.3% year-on-year. Among them, although gasoline consumption is affected by the accelerated replacement of new energy vehicles, the growth rate is expected to reach 1.3%; diesel consumption will decrease by 2.8% due to the continuous transformation and upgrading of the economic structure and the expansion of the LNG substitution effect; kerosene consumption will benefit from the recovery of the tourism industry and is expected to increase by 13%. With the official commissioning of Zhenhai Refining and Chemical and Yulong Island Petrochemical, the domestic crude oil processing volume will increase, and the annual crude oil processing volume is expected to be 752 million tons, an increase of 1.8% year-on-year; the output of refined oil will be 451 million tons, an increase of 1.6% year-on-year, with production exceeding demand by at least 50 million tons, which still needs to be adjusted by exports to balance domestic supply and demand. In 2024, the oil industry will continue to exert its strength in high-quality development and orderly market operation.

Natural Gas Consumption Improved, "One National Network" Accelerated

In 2023, China's natural gas consumption was 391.7 billion cubic meters, an increase of 24.2 billion cubic meters, a year-on-year increase of 6.6%. It was mainly affected by three factors: first, the macroeconomic recovery supported the demand for gas. The prosperity of the commercial service industry rapidly increased, industrial production gradually recovered, driving the growth of commercial service and industrial gas demand. Second, domestic LNG prices fell, and the economy of natural gas improved. The decline in international gas prices led to a year-on-year decrease of 28% in domestic LNG market prices, and the price ratio of calorific value LNG to diesel reached a nearly three-year low. The economic advantage of LNG for vehicles was obvious, and the sales volume of LNG vehicles increased by more than 300% year-on-year, and the demand for gas for transportation grew rapidly. Third, hydropower did not meet expectations, and the demand for gas power increased. On the demand side, the total electricity consumption of the whole society increased by 6.7% year-on-year, an increase of 3.1 percentage points from the previous year; on the supply side, coal power increased by 5.8% year-on-year, wind power and photovoltaic power generation increased by 12.3% and 17.2% respectively, but hydropower decreased by 5.1% year-on-year, and the demand for gas power grew rapidly.

The annual domestic gas production was 235.3 billion cubic meters (including shale gas, coalbed methane, and coal gas), an increase of 5.7%, an increase of 12.6 billion cubic meters, and an increase of more than 10 billion cubic meters for seven consecutive years. Affected by the decline in international oil and gas prices, the growth of imported natural gas resumed, with a total import volume of 165.6 billion cubic meters, an increase of 9.5%. The degree of foreign dependence was 42.3%, an increase of 1.1 percentage points from the previous year. Among them, the import of pipeline gas was driven by the increase in the China-Russia East Line, reaching 67.1 billion cubic meters, a year-on-year increase of 6.4%; imported LNG was 98.4 billion cubic meters, a year-on-year increase of 11.7%.

Infrastructure construction has made positive progress, and the "one national network" is rapidly improving. The Hubei section of the West–East Gas Pipeline III, the first phase of the Mongolian West Pipeline, and other pipelines were put into operation. The China-Russia East Line, the West–East Gas Pipeline I, and other natural gas trunk pipelines were interconnected. By the end of 2023, the total mileage of natural gas long-distance pipelines was 87,000 km. With the accelerated construction of the West–East Gas Pipeline IV, the completion of the Sichuan Gas Eastward Transmission II Line and the Nenjiang and Jining branches of the China-Russia East Line, the total mileage of natural gas long-distance pipelines will further increase. Provincial natural gas pipelines are being built at an accelerated pace, with the Shandong Province Ring Network South Main Line officially supplying gas, and the Guiyang to Linwu gas pipeline in Hunan Province put into operation. The gas storage and peak-shaving capacity has reached a new level, with the Leilongwan gas storage depot of PetroChina Changqing Oilfield, the Wenba gas storage depot of Tuha Oilfield, and the Wen 24 gas storage depot of Sinopec successively put into operation. By the end of 2023, China had 29 operating gas storage depots (groups), forming a gas storage and peak-shaving capacity of 23 billion cubic meters, a year-on-year increase of 19.8%, accounting for 5.9% of domestic natural gas consumption. The national LNG receiving capacity exceeded 100 million tons for the first time. Six LNG terminals such as Xintian Tangshan and Tianjin Nangang were put into operation (including expansion), adding a total of 18.8 million tons/year of LNG receiving capacity, setting a historical record. By the end of 2023, the total receiving capacity of the country was 116 million tons/year.

Looking forward to 2024, affected by the steady growth of the macro economy, the decline in international gas prices, and the increase in the supply of alternative energy, it is expected that natural gas consumption will be 415.7 billion cubic meters, an increase of 6.1%. The growth rates of city gas, industrial fuel, power generation gas, and chemical gas are 6.1%, 6.8%, 7.5%, and 1.0% respectively. More than 50% of the new domestic natural gas supply in 2024 will come from imported gas. The construction of domestic natural gas production capacity is steadily advancing, and the output will reach 245.8 billion cubic meters, an increase of 4.5%. It is expected that the import volume of natural gas will be 179.1 billion cubic meters, an increase of 8.2%, and the degree of foreign dependence will increase to 43.1%. In terms of pipeline gas, the import capacity of the China-Russia East Line continues to increase, and it is expected that the annual import of pipeline gas will be 72.6 billion cubic meters, an increase of 8.2%. In terms of LNG, affected by factors such as the performance of long-term contracts, the commissioning of terminals, and the fall in international gas prices, it is expected that the annual import of LNG will be 106.4 billion cubic meters, an increase of 8.1%.

Oil and Gas Reserves Continue to Grow, and "Two Deeps and One Non" Become New Areas for Increasing Reserves and Production

In 2023, domestic oil and gas companies continued to increase their exploration and development efforts, strengthening their scale of reserves and efficient development, achieving significant results. The investment in oil and gas exploration and development was about 390 billion yuan, and both exploration and development investments reached historical highs. The newly discovered geological reserves of oil were about 1.3 billion tons, and the newly discovered geological reserves of natural gas were nearly 1 trillion cubic meters. The output of crude oil was 209 million tons, and the output of natural gas was 235.3 billion cubic meters. The progress in the theory, technology, and equipment of oil and gas exploration and development supported significant breakthroughs in deep-water and deep-layer oil and gas exploration and development.

In terms of exploration, excellent results were achieved in the deep-sea, deep-land, and unconventional "two deep and one unconventional" oil and gas resources. Multiple exploration breakthroughs were achieved in new areas and fields, old areas with new depressions, and new strata. Multiple billion-ton oil and gas areas were discovered in basins such as Tarim, Ordos, and Bohai Bay, and multiple trillion-cubic-meter gas areas were discovered in basins such as Junggar, Sichuan, and Ordos. Significant breakthroughs were made in deep-layer coalbed methane exploration, opening up new areas for reserve increases.

In terms of development, deep and ultra-deep layers have become new areas for oil and gas production growth, and multiple deep-layer large oil and gas fields have been efficiently built. Continental shale oil has started a new round of "oil revolution", with production reaching a new high, showing a good prospect for the development of unconventional oil in China. Offshore oil fields have a strong momentum of production, and the increase in oil production accounts for more than half of the national increase.

Domestic oil service companies have seen a significant rebound in business volume, and the expansion of international business has promoted the continuous development of oil service companies. Most oilfield technical service companies have seen a rise in profits, and their operations are in a good state. Oil service companies continue to promote development transformation, and new businesses related to clean energy bring growth points to the oil service market. The progress of core engineering technology and equipment provides important support for the increase in revenue in the oil service industry.

In 2024, oil and gas field companies will continue to exert efforts in the strategic succession areas of deep and ultra-deep layers, deep water, and unconventional oil and gas reserves. The newly discovered reserves of oil and natural gas nationwide will maintain high growth, maintaining the same level as the previous year. The output of crude oil will remain stable at more than 200 million tons, and the output of natural gas will maintain a high growth rate, with production expected to reach 245.8 billion cubic meters. China will lead the development of deep oil and gas exploration and development, making progress in the theory and major engineering technology and equipment of deep and ultra-deep layer oil and gas exploration and development. The oil and gas industry will deeply implement the action plan for the integrated development of oil and gas exploration and development and new energy, explore and form a new model of multi-energy complementarity and integrated development, and accelerate the formation of a new pattern of integration and multi-energy complementarity in the upstream field of oil and gas and new energy industries.

Refining Capacity Continues to Grow Slightly, and Many Production and Operation Indicators Rebound Significantly

In 2023, China's refining capacity continued to grow slightly, increasing by 12 million tons/year, the lowest increase since 2017, and the total refining capacity rose to 936 million tons/year, ranking first in the world. The degree of industry concentration and scale further improved. The number of refineries with a capacity of over 10 million tons increased from 35 in the previous year to 36, with a total refining capacity of 522 million tons/year, accounting for 55.8% of the total domestic refining capacity. Driven by favorable factors such as the rapid rebound in demand for refined oil and the continuous advancement of market rectification, many major production and operation indicators have significantly increased. Among them, the crude oil processing volume was 738 million tons, an increase of 8.6% year-on-year; the average operating rate of refineries was 78.9%, an increase of 5.3 percentage points year-on-year, reaching a historical high; the production of gasoline, diesel, and kerosene was 444 million tons, an increase of 11.1% year-on-year.

Looking forward to 2024, China's refining capacity will grow rapidly again, with an expected net increase in refining capacity of 25 million tons/year. The additional capacity mainly comes from the second phase of Zhenhai Refining and Chemical and Shandong Yulong Island Petrochemical. The total refining capacity will rise to 961 million tons/year. As the projects are all put into operation in the second half of the year, the impact on the growth of crude oil processing volume and refined oil production in the current year is not significant. It is expected that the crude oil processing volume for the whole year will be 752 million tons, an increase of 1.8% year-on-year; the production of refined oil will be 451 million tons, an increase of 1.6% year-on-year. It is worth noting that with the continuous promotion of "reducing oil and increasing chemical" in recent years, the pressure of oversupply of refined oil has eased, but due to the continuous increase in refining capacity in the future, the problem of excess refining capacity still needs to be highly concerned and seriously solved.

The Production Capacity of Bulk Petrochemical Products is Expanding, and the Rise of Private Enterprises is Eye-Catching

In 2023, China's ethylene, propylene, p-xylene and other bulk petrochemical raw materials and three major synthetic materials expanded their capacity for the fifth consecutive year, and their capacities all ranked first in the world.

The total ethylene production capacity reached 51.95 million tons/year, with an increase of 2.42 million tons/year, growing for five consecutive years, which is lower than the average annual growth of 6.05 million tons/year over the past four years. The proportion of ethylene production capacity of private enterprises has risen to 35.8%, and the combined production capacity of the three major companies, China Petroleum, Sinopec, and CNOOC, accounts for 49.7%. Ethylene prices have fallen significantly, further narrowing production profits. The development of ethylene production materials continues to be lighter, more diversified, and integrated. Propylene has seen the largest scale expansion in history, with an increase in production capacity of up to 7.29 million tons/year, and the total production capacity has surged to 63.34 million tons/year. Ninety percent of the increased capacity comes from private enterprises, and the proportion of propylene production capacity of private enterprises has risen to 28%. The proportion of propane dehydrogenation (PDH) capacity accounts for 30%, surpassing the capacity of oil-based steam cracking routes for the first time, becoming the main production process route for domestic propylene. The price of imported propane remains high, the demand for propylene is insufficient, the PDH start-up rate has dropped to below 70%, and the profits are not good. The new capacity of xylene is 5.7 million tons/year, and the total capacity has surged to 42.1 million tons/year, nearly three times that of 2018. The combined capacity of the three major oil companies for xylene is only 39%, and the market is dominated by private enterprises. With the sharp increase in capacity, the dependence on foreign xylene has dropped from 61% in 2018 to 22%. The new capacity of synthetic resin is 13.42 million tons/year, a year-on-year increase of 13.3% (polypropylene and ABS resin account for 64% of the new capacity, and the growth rate of polypropylene is significantly faster than polyethylene), which is the most increased in the past five years. The total capacity has reached 115 million tons/year. Due to insufficient effective demand in domestic and foreign markets, the overall profit of synthetic resin production has declined, and ABS resin and polystyrene have suffered significant losses. The capacity of synthetic rubber continues to grow, with an annual increase in capacity of 350,000 tons/year, and the total capacity reaches 7.202 million tons/year, mainly for butadiene rubber and styrene-butadiene rubber, and the proportion of private enterprise capacity has risen to 52.4%. The profitability of major varieties of synthetic rubber is differentiated, the profitability of styrene-butadiene rubber has improved, and the profits of butadiene rubber and nitrile rubber have narrowed. The new capacity of synthetic fibers is 5.43 million tons/year, and the increase mainly comes from the polyester filament capacity of private enterprises. By 2024, it is expected that the new capacity of ethylene will be 3.2 million tons/

year, the new capacity of propylene will be 6 million tons/year, the capacity of xylene will temporarily stop rapid growth, the start-up rate will rebound, and the price will remain at a medium-to-high level. The capacity of the three major synthetic materials will continue to grow, and the profit level will still remain low. The new capacity of synthetic resin is 15.5 million tons/year, with the fastest increase in polypropylene capacity. The new capacity of synthetic rubber is 785,000 tons/year, mainly for private enterprises. The new capacity of synthetic fibers is 1.85 million tons/year, mainly from private enterprises and leading enterprises under central enterprises. In the next few years, ethylene and propylene will each have more than 20 million tons/year of new capacity, and the capacity of many domestic bulk petrochemical products will tend to be saturated or even surplus, and may repeat the old path of rapid expansion of refining leading to excess capacity, keeping industry profits at a low level, and the petrochemical market competition dominated by central enterprises, private enterprises, and foreign enterprises will intensify. Deepening the integration of refining and chemical, absorbing excess capacity, breaking through technical bottlenecks, extending and expanding the industrial chain, vigorously developing mid-to-high-end products, achieving import substitution and expanding exports, and promoting high-quality green and effective development, from large to strong has become the main task of the petrochemical industry in the future.

Overseas Business Improves Quality and Efficiency, and International Cooperation Continues to Deepen

In 2023, Chinese oil companies will focus on strategy, mutual benefit, and the whole industry chain of overseas oil and gas cooperation will continue to deepen. Exploration and development are steadily advancing, and production is stable and increasing. Chinese oil companies overcome the adverse effects of armed conflicts in some areas, sanctions from the West, etc., aim at the target of equity production, optimize production operations, and the overseas oil and gas equity production equivalent reaches 188 million tons, a year-on-year increase of 1.9%. Among them, crude oil production is 149 million tons, natural gas production is 480 billion cubic meters, mainly due to the production of several major projects in Guyana, Brazil, etc., as well as the high-level operation of major projects in Iraq, Kazakhstan, Turkmenistan, etc. The reserve scale of projects in Guyana, Brazil, Russia, Chad, etc. has further jumped, and the overseas resource base is constantly consolidated. LNG, refining, pipeline and other businesses are steadily advancing. Breaking through the difficulties of service and equipment supply caused by safety and sanctions, PetroChina's Coral floating LNG project in Mozambique's 4th district was put into operation as scheduled. Sinopec has a stake in Kazakhstan's largest polyethylene project. Hengyi Petrochemical's second phase project in Brunei has received preliminary approval from the government. There has been gratifying progress in the downstream field of oil and gas. After several Chinese banks provided financing, Chinese companies have successively signed agreements for the construction of various East African crude oil export pipeline projects with host countries. Accelerate asset disposal, acquire high-quality assets, and optimize overseas asset allocation. The speed of disposal of inefficient and high-cost assets has been accelerated, and a number of significant high-quality assets have been obtained during the same period. PetroChina replaced ExxonMobil as the operator of the West Qurna 1 oil field in Iraq, Sinopec and PetroChina respectively invested in the expansion project of the North Field in Qatar, and Continental Oil and Gas and United Energy Company respectively won the bid for the exploration and development project of the southern oil and gas field in Iraq.

2023 marks the 10th anniversary of the "Belt and Road" initiative. Over the past decade, Chinese oil companies have seized opportunities and continuously promoted deep and solid international energy cooperation. They have completed a number of key cooperation projects such as the Missan oil field in Iraq and the Yanbu refinery in Saudi Arabia; participated in a number of major world projects such as the Arctic in Russia, deep water in Brazil, and LNG in Qatar; deeply participated in dialogues and exchanges under the international energy governance mechanism, and created high-end dialogue and exchange platforms such as the "Belt and Road" oil and gas cooperation roundtable. They have fully utilized the consensus and opportunities provided by the China-Foreign Cooperation Summit to continuously deepen cooperation with Arab countries, Gulf countries, and Central Asian countries in the field of traditional energy. Energy cooperation has extended from upstream oil and gas to the entire industry chain, with increased synergy and integration. Further expanded from traditional energy to renewable energy fields, PetroChina's photovoltaic projects in the upstream of the second phase in Niger and the pilot projects of photovoltaic energy storage in the Nibe pipeline have entered the construction phase. The cooperation methods include both bilateral and multilateral, which are more flexible, diverse, and pragmatic.

In 2024, Chinese oil companies will continue to focus on the opportunities of the "Belt and Road", deepen energy cooperation with core regional countries, expand new projects, optimize asset structure, and achieve the transformation and upgrade of international oil and gas cooperation business. With the implementation of proposed and under-construction oil and gas, pipeline, new energy and other cooperation projects, international energy cooperation across the entire industry chain will reach a new level.

Domestic Companies Focus on "Two Cores", Making Remarkable Progress

In 2023, the three major domestic oil companies comprehensively deepened reforms, focused on improving core competitiveness and enhancing core functions, fully ensured national energy security, actively promoted the construction of world-class enterprises, actively built a new energy security system, and achieved significant results.

Continue to increase the intensity of reserves and production. Strengthen the "two deep and one non-conventional" breakthroughs, and demand oil and gas from deep land and deep sea. Shale oil exploration and development are steadily advancing, and the construction of three national demonstration zones in Jimsar, Xinjiang, Gulong in Daqing, and Jiyang in Shengli, and the Qingcheng shale oil field is accelerating. The multi-well type test in Yantong depression in Subei has achieved commercial breakthroughs, and the shale oil production has broken through 4 million tons to create a new high. Onshore deep and ultra-deep exploration and development continue to make major discoveries, and multiple deep oil fields have been efficiently built. China has become a global leader in the field of onshore oil and gas over 6000 m deep. The production of unconventional natural gas has broken through 96 billion cubic meters, accounting for 43% of the total natural gas production, and has become an important growth pole for natural gas reserves and production. Offshore oil and gas exploration and development continue to exert force, through innovative basin formation and oil and gas accumulation model recognition, billion-ton oil and gas exploration new discoveries have been made in the Bohai Sea and South China Sea deepwater areas, opening up new exploration areas such as deepwater, deep layer, hidden oil and gas reservoirs, and basin edge depressions. In 2023, the domestic oil and gas production equivalent exceeded 390 million tons, maintaining rapid growth of tens of millions of tons for seven consecutive years, with an average annual increase of 11.7 million tons of oil equivalent, forming a new peak period of production growth.

Accelerate the transformation and upgrade of refining and chemical industry and structural adjustment. Actively respond to the adverse effects of oil price fluctuations and weak demand for chemical products, promote product structure adjustment mainly based on "oil transformation and oil to special", strengthen refining integration and production and sales coordination, accelerate the construction of world-class refining integrated bases, and steadily increase the proportion of high-value-added chemical products and new material products. Promote the upgrading and transformation projects of Zhenhai, Hainan, Yueyang, Luoyang and other refining and chemical integration. The 10 million ton Guangdong Petrochemical refining and chemical integration project was successfully put into operation. The second phase of Dushanzi Petrochemical, Jilin, Guangxi and other projects continue to advance. Continue to improve heavy oil catalytic cracking, hydrocracking, catalytic pyrolysis, catalytic gasoline adsorption desulfurization, diesel ultra-deep hydrodesulfurization and other technologies to support the construction of 10 million ton refineries.

Emphasize innovation leadership and technological breakthroughs. With the goal of building a national strategic technological force, we will deeply implement the science and technology innovation strengthening project, accelerate the advancement of key core technology breakthroughs, actively promote the construction of original technology source areas, and actively promote the domestication of major equipment. The drilling of China's first 10,000-m deep earth exploration well—the Tarim Oilfield Deep Earth Ta Ke 1 well marks a new major breakthrough in China's deep earth exploration technology series, and the drilling capability has entered the

"10,000-m era". There have been major breakthroughs in deep earth oil and gas exploration and development technology and related theories. A batch of major deep-sea exploration and development equipment represented by "Deep Sea No. 1" has been put into use one after another, and offshore oil and gas development has achieved a historic leap in ultra-deep water. A series of world-leading full-process refining technologies such as electric fracturing equipment, high-end drill bits, large-scale hydrogenation reactors, and large-scale refining DCS have continued to make breakthroughs. Intelligent technology is widely used in all major links of the oil and gas industry chain, new achievements have been made in the digital transformation of the entire industry, and we are gradually moving towards the stage of intelligent oil and gas fields through data sharing, business collaboration, and intelligent application construction, and new progress has been made in the construction of intelligent manufacturing factories and offshore intelligent oil fields. Accelerate the transformation to a comprehensive energy company. Actively layout new energy business, and develop oil and gas and new energy in a coordinated and integrated manner. Several new companies related to charging, photovoltaic power generation, and hydrogen energy have been established one after another. Projects such as Xinjiang's million-kilowatt photovoltaic power station and Jilin's large-scale centralized wind power have been completed. Construction of China's largest CCUS full-industry demonstration base has started. The country's first offshore carbon dioxide sequestration demonstration project has been officially put into use. The first deep-sea floating wind power platform "CNOOC Guanlan" has successfully connected to the grid for power generation. The layout of the hydrogen energy "production, transportation, storage, and sales" full industry chain is actively promoted. In 2024, facing the complex and changing domestic and foreign environment, domestic oil companies will further play their resource advantages, comprehensively enhance the resilience of the energy system, and accelerate the construction of a mature and reliable energy industry chain supply chain system. Consolidate the resource base, maintain a steady increase in domestic crude oil production, and maintain rapid growth in natural gas production; accelerate the development of new energy business, promote green energy low-carbon transformation, gradually increase the development and utilization of renewable energy, and seek breakthroughs in emerging clean energy fields such as hydrogen energy; increase scientific research investment, strengthen innovation drive, achieve more breakthroughs in key technologies with independent intellectual property rights in the fields of exploration and development, refining and chemical industry, oil and gas pipeline transportation, and enhance the core competitiveness of enterprises; actively participate in the division of labor and cooperation in the global energy industry, accelerate the construction of a new energy cooperation platform mainly based on the "Belt and Road", actively participate in the energy governance of regional and comprehensive international mechanisms such as the G20, APEC, ASEAN, and enhance China's discourse power in dealing with climate change and building a green and low-carbon global energy governance pattern.

The Pace of Carbon Market Construction is Accelerating, and the Oil and Gas Industry Faces Both Opportunities and Challenges

In 2023, the national carbon market trading activity significantly increased, and prices rose significantly. The daily average trading volume of the carbon market was 876,000 tons, 4.2 times that of 2022, with an annual average price of 68.15 yuan/ton, a year-on-year increase of 23.2%. The highest daily transaction price broke through 80 yuan/ton for the first time in more than two years since the operation of the national carbon emissions trading market, reaching 81.67 yuan/ton. The price increase was driven by companies' concentrated performance, and was also affected by factors such as the country's adjustment of carbon emission benchmarks and moderate tightening of carbon quotas.

The pace of national carbon emissions trading market construction is accelerating, and supporting policies are gradually being perfected. The China's domestic voluntary carbon market officially restarted, issued the "The Measures for the Administration of Voluntary Greenhouse Gas Emission Reduction Trading (Trial)", "Voluntary Greenhouse Gas Emission Reduction Project Methodology Afforestation Carbon Sink", "Voluntary Greenhouse Gas Emission Reduction Project Approval and Emission Reduction Verification Implementation Rules" and other documents, laying the institutional foundation for the start of the CCER market. Among them, the new CCER methodology implements a "top-down" management model, which is collected and selected by the Ministry of Ecology and Environment and published, and is no longer filed by each methodology developer. The first batch of methodologies for afforestation carbon sinks, grid-connected solar thermal power generation, grid-connected offshore wind power, and mangrove forest construction projects were released. The restart of CCER will promote industries and enterprises that contribute significantly to carbon reduction and increase in industries such as renewable energy and forestry carbon sinks to participate in greenhouse gas emission reduction actions. The country revised and issued the national carbon market power industry carbon quota allocation plan for 2021 and 2022, implemented annual quota management, introduced balance values as the basis for setting benchmarks, added personalized relief policies such as quota advance, and overall followed the principle of moderately tightening carbon quota benchmarks and promoting energy conservation and carbon reduction in the power industry. The carbon market and the green electricity market have effectively linked. In 2023, Tianjin, Beijing, Shanghai, and Hubei, the four pilot carbon markets, clearly stated that the carbon emissions of green electricity purchased through market means are accounted for as zero.

The Ministry of Ecology and Environment has further refined the emission reporting and verification requirements for industries such as cement and electrolytic aluminum, and the signal for carbon market expansion is becoming increasingly clear. The "Carbon Market-Green Electricity" linkage mechanism will also gradually become clear with the expansion, and the top-level design of carbon market trading management related rules and legislation will further advance. In the future, the coverage of the national carbon market will gradually expand, and the policy system will also be further improved.

Looking ahead to 2024, with the accelerated construction of the national emissions trading carbon market, oil and gas companies are facing increasing pressure to reduce carbon emissions. While promoting energy conservation and emission reduction, strengthening zero-carbon technology research, implementing clean energy substitution, and accelerating industrial transformation and development, attention should be paid to the opportunities brought by the accelerated construction of the carbon market. Utilize potential CCER projects such as CCUS, offshore wind power, and solar thermal power generation for the development and offset mechanism market opportunities, and at the same time, use the carbon power market mutual recognition and interconnection mechanism to tap the energy-saving and carbon-reducing potential of corporate green power projects, and solidly promote green and low-carbon development.

Domestic and Foreign Energy Industry Development

World Energy Industry Development Status

In 2023, global energy consumption growth accelerated compared to the previous year, with the recovery of natural gas consumption driving the growth of fossil energy consumption, and renewable energy consumption continued to maintain a high growth momentum. The turbulence in the global energy price system caused by the Ukraine crisis has basically subsided. In 2023, the average price of European TTF natural gas futures was 10.9 US dollars/million British thermal units, Brent crude oil futures price was 83 US dollars/barrel, Rotterdam coal futures average price was 130 US dollars/ton, and the price ratio of gas, oil, and coal has recovered from 3.2:1.5:1 in 2022 to 1.9:2.5:1, close to the 2019 level. The deep adjustment of the global energy pattern is still ongoing, the aftermath of the Ukraine crisis is not over, the global energy trade "counterclockwise" pattern is further consolidated, but with the escalation of the Israeli-Palestinian conflict in the second half of the year, the uncertainty of energy trade flow direction increases, and countries' emphasis on energy security deepens. Overall, the proportion of fossil energy in the global energy consumption structure will not decrease significantly in the short term, but the process of energy transition is accelerating; in the long term, the trend of diversification, low-carbonization, and digitalization of energy production and consumption will continue to strengthen.

Slight Increase in World Energy Consumption

Global Primary Energy Consumption Slowly Recovers

In 2023, the global primary energy consumption is estimated to be 14.61 billion tons of oil equivalent, an increase of 1.7% compared to 2022, and the growth rate increased

X. Qian et al., *Global Oil and Gas Industry Development Report: 2023*,
https://doi.org/10.1007/978-981-95-2375-7_2

by 0.5 percentage points. The overall consumption of fossil energy increased by 1.1%, of which natural gas consumption increased by 0.5%, and coal and oil consumption increased by 0.5% and 2.1% respectively. The marginal effect of geopolitical factors such as the Ukraine crisis on the international energy market is weakening, the fall in natural gas prices stimulates demand recovery, and global natural gas demand in 2023 reverses the decline of the previous year, rebounds by 3.3 percentage points, and achieves positive growth; the "counterclockwise" flow of natural gas trade intensifies, Russia's natural gas production and exports continue to decline, Europe gradually gets rid of dependence on Russian energy, and turns to other countries such as the United States for LNG imports. Global coal consumption cools down from the high growth momentum under the 2022 Ukraine crisis, the deepening of the US and Europe's coal phase-out action offsets the growth of coal in Asia, the overall consumption volume slightly increases, and still maintains a historic high. Global oil demand growth gradually slows down under the push of factors such as the accelerated popularization of electric vehicles and improved energy efficiency, and the growth rate drops by 0.3 percentage points compared to the previous year. Non-fossil energy overall still maintains a high growth trend, nuclear power output increases in many countries, nuclear energy consumption grows by 2.3%, hydropower output decreases due to drought in many regions, consumption decreases by nearly 3%, non-hydropower renewable energy develops rapidly, growing by 12.6%, non-fossil energy consumption overall grows by about 4.1% (Fig. 1).

In 2023, the global energy structure continues to maintain a trend of declining fossil energy and rising non-fossil energy (Fig. 2). The proportion of fossil energy consumption decreased by 0.6 percentage points from the previous year, falling below 80% for the first time, to 79.7%. Global oil demand remains high, with the proportion of oil consumption slightly increasing by 0.1 percentage points to 29.4%; the growth rate of global coal consumption has significantly slowed down, with

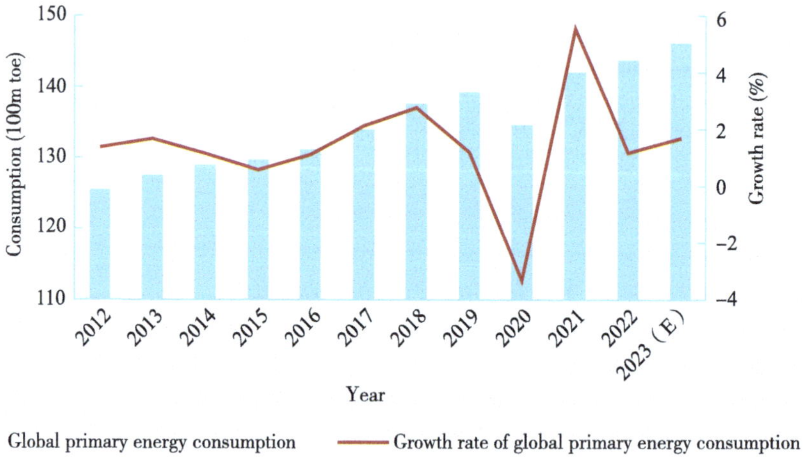

Fig. 1 Global primary energy consumption and growth rate from 2012 to 2023

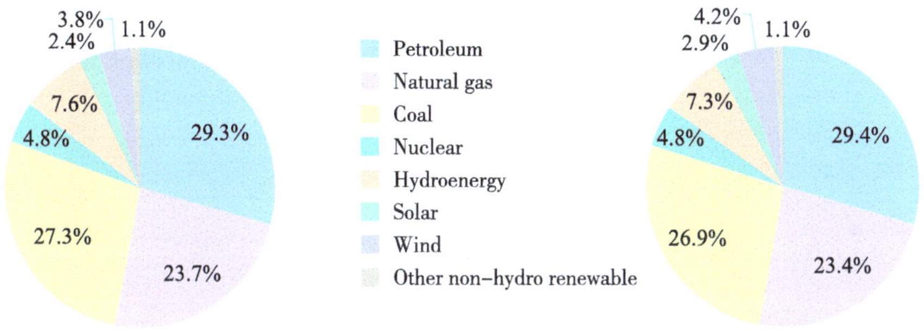

(a) 2022 (b) 2023

Fig. 2 Energy consumption structure of the world in 2022 and 2023

the proportion of coal consumption decreasing by 0.4 percentage points to 26.9%; natural gas demand has resumed growth, but the growth rate is still slow, with the proportion of natural gas consumption decreasing by 0.3 percentage points to 23.4%. The consumption of global non-hydro renewable energy is growing rapidly, the pace of energy transformation is further accelerating, and the proportions of solar and wind energy consumption have increased by 0.5 percentage points (to 2.9%) and 0.4 percentage points (to 4.2%), respectively. The proportion of other non-hydro renewable energy sources, including bioenergy and geothermal energy, remains unchanged at 1.1%; in the first half of the year, droughts in many parts of the world led to a significant reduction in hydroelectric consumption, with the proportion of hydroelectric power decreasing by 0.3 percentage points from the previous year to 7.3%; the completion of maintenance or delayed retirement of some nuclear power units worldwide has driven the growth of nuclear energy consumption, with the proportion of nuclear energy remaining basically the same as the previous year, at 4.8%.

Energy Consumption Growth Slows Down in Many Regions

In 2023, regional energy consumption mainly shows the characteristics of accelerated growth in Asia–Pacific, recovery in Europe and Eurasia, and general slowdown in North America, Central and South America, the Middle East, and Africa. The Asia–Pacific region remains the main driver of global economic and energy growth. In 2023, China's economy grew by 5.2%, India's economy grew by 6.2%, driving regional energy consumption to increase by 2.7%, an increase of 0.4 percentage points from the previous year, and it is the only region where consumption growth has accelerated apart from Europe and Eurasia. Energy consumption in Europe and Eurasia has recovered from the impact of the Ukraine crisis, reversing the decline of the previous year and achieving a growth of 0.7%. North America has a large economic volume, and the economic recovery has slowed down due to high inflation, with energy consumption slightly increasing by 0.4%, a decrease of 2.6 percentage

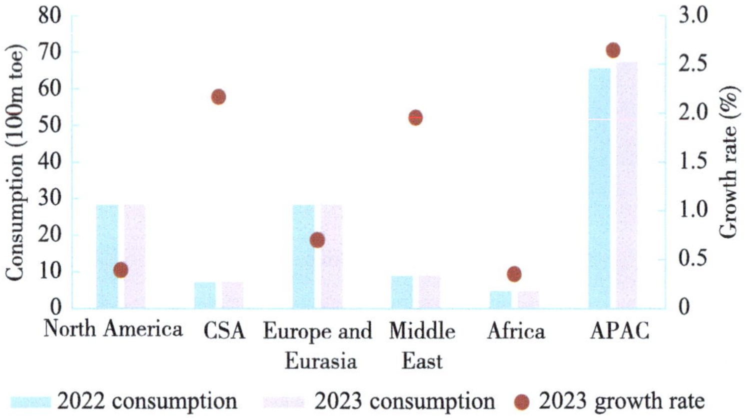

Fig. 3 Primary energy consumption and growth rate by region in 2022 and 2023

points from the previous year. The economies of the Middle East, Africa, and Central and South American oil-producing countries have maintained steady growth under high oil and gas prices, but the growth rate has slowed down compared to the previous year. The energy consumption growth rates of the three regions have decreased by 2.7 percentage points (to 2%), 0.1 percentage points (to 0.4%), and 2 percentage points (to 2.2%) respectively from the previous year (Fig. 3).

The Growth Rate of Coal Consumption Continues to Decline

In 2023, global coal consumption increased slightly by 0.5% year-on-year, reaching 3.93 billion tons of oil equivalent, a decrease of 0.7 percentage points from the previous year and a decrease of 5.3 percentage points from the highest point in 2021, but consumption still remains at a historical high (Fig. 4). Regional consumption shows the characteristics of decline in the US and Europe and growth in Asia–Pacific. The energy transition process in the United States and Europe has further accelerated, and coal demand has declined. In the United States, cheap natural gas has also put pressure on coal demand. In the Asia–Pacific region, China's hydropower generation was very low due to drought in the first half of the year, which increased the dependence on coal-fired power generation. The growth rate of coal demand slowed slightly in the second half of the year, but overall it still remained high. India's strong economic growth has strengthened the demand for coal. Overall, global coal consumption remains at a historical high, but the decline in the US and Europe has offset the growth in the Asia–Pacific region, and the growth rate of consumption has significantly slowed down.

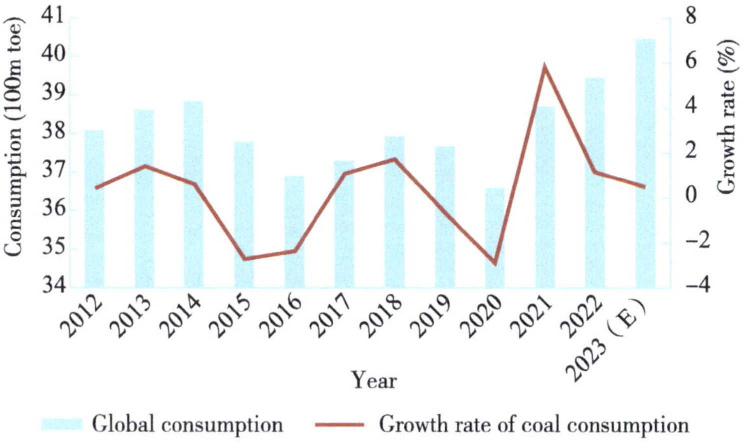

Fig. 4 Global coal consumption and growth rate from 2012 to 2023. *Data source* PetroChina Economic Research Institute

The Process of Energy Consumption Cleaning Continues to Advance

Global geopolitical conflicts are frequent, the impact of the Ukraine crisis is profound, and the Israeli-Palestinian conflict is not over. Governments around the world have deepened their emphasis on the importance of fossil energy in ensuring their own energy security. Although the proportion of global fossil energy is still in a downward trend, the growth rate of fossil energy consumption has slightly accelerated from the previous year, from 0.4% to 1.1%. Non-fossil energy, due to the historic decline in hydropower in 2023, the growth rate of consumption has slightly slowed down from the previous year, to 4.1%, but it is still much higher than the growth rate of fossil energy consumption, with Asia–Pacific and Europe leading the global growth of non-fossil energy consumption (Table 1). Countries' emphasis on renewable energy in the pursuit of energy independence is also further increasing. Wind power, photovoltaic and other projects are accelerating, electric vehicles are becoming more popular, energy efficiency is constantly improving, and the process of global energy consumption cleaning continues to advance. In 2023, the estimated global energy-related carbon emissions increased by 1.1%, an increase of 0.2 percentage points from 2022.

Global Electricity Generation Maintains Recovery Growth

Since the outbreak of the COVID-19 pandemic in 2020, global electricity generation has maintained recovery growth for three consecutive years. In 2022, the demand growth rate fell due to the Ukraine crisis, but in 2023, electricity generation slightly recovered, with the growth rate increasing to 2.5%, an increase of 0.2 percentage

Table 1 Primary energy demand and energy structure by region in 2022 and 2023

Project		North America		Central and South America		Europe and Eurasia		Middle East		Africa		Asia Pacific	
		2022	2023	2022	2023	2022	2023	2022	2023	2022	2023	2022	2023
Energy structure (%)	Coal	8.87	8.49	3.89	3.83	12.74	12.54	1.01	0.96	20.16	19.75	48.07	47.34
	Oil	33.16	33.21	37.32	37.31	28.74	28.64	43.24	43.23	41.34	41.49	24.20	24.53
	Natural gas	33.45	33.91	19.32	18.83	32.13	31.19	54.08	54.13	28.92	29.18	11.74	11.59
	Non-fossil energy	24.52	24.39	39.47	40.03	26.39	27.63	1.67	1.68	9.58	9.58	15.99	16.54
Primary energy consumption (Billion tons of oil equivalent)		28.40	28.51	7.20	7.35	28.34	28.54	8.95	9.13	4.91	4.93	65.86	67.61

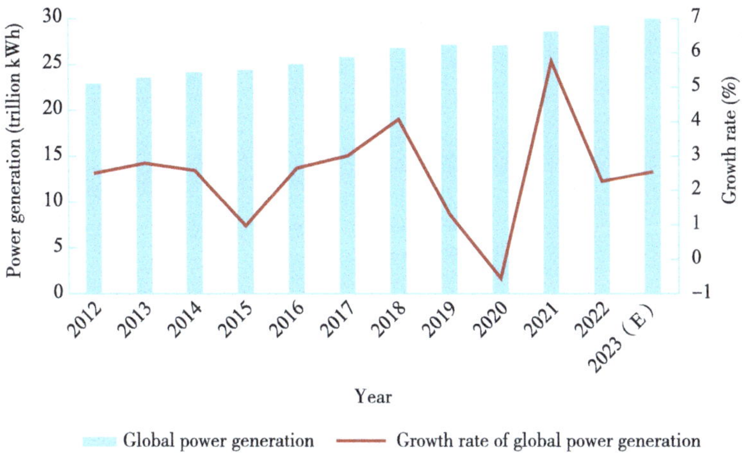

Fig. 5 Global electricity generation and growth rate from 2012 to 2023

points from the previous year, maintaining the average level of the past decade (Fig. 5).

Continuous Optimization of Global Electricity Structure

Fossil fuels remain the main source of global electricity, accounting for 60% of total electricity generation, but this proportion is gradually shrinking, down 0.6 percentage points from the previous year, marking a continuous decline for three years since 2020. Among them, the proportion of oil-fired power generation decreased by 0.3 percentage points, natural gas power generation remained unchanged, and coal-fired power generation decreased by 0.3 percentage points. The proportion of non-fossil energy increased by 0.6 percentage points, historically approaching 40%, with the proportion of nuclear power remaining unchanged, hydroelectric power decreasing by 0.9 percentage points, and the proportion of non-hydro renewable energy such as wind and solar power increasing by 1.4 percentage points (Fig. 6).

In 2023, the global fossil energy power generation was 180 trillion kilowatt-hours, with a growth rate of only 1.6%, significantly lower than that of non-fossil energy power generation. The global non-fossil energy power generation was 120 trillion kilowatt-hours, an increase of 4.1% year-on-year. Among them, hydropower generation experienced a historic decline due to droughts in some regions, down 3.1% year-on-year; nuclear power generation rebounded due to the end of maintenance periods in several countries, increasing by 2.2% year-on-year; non-hydro renewable energy power generation led by wind and solar power continued to grow rapidly, with a growth rate of 12.6%, and new non-hydro renewable energy power generation accounted for 71% of all new power generation (Fig. 7).

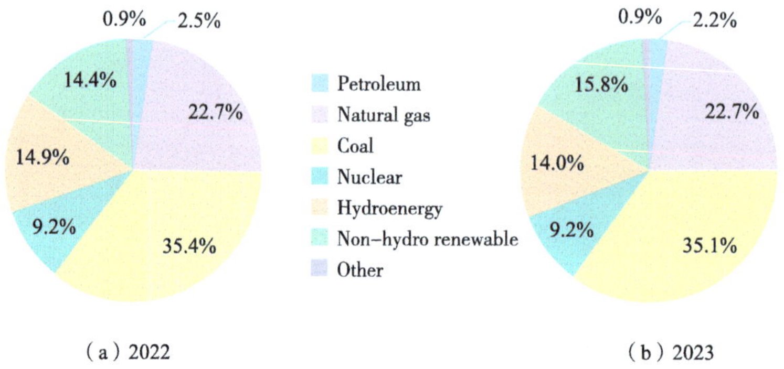

Fig. 6 Global electricity generation structure in 2022 and 2023

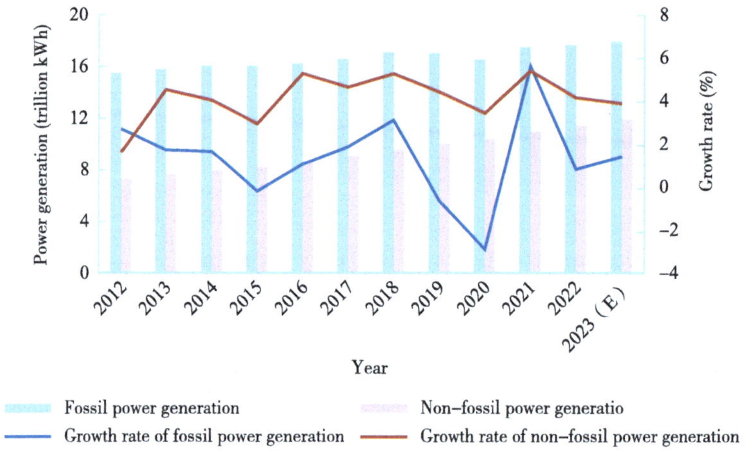

Fig. 7 Global fossil energy and non-fossil energy power generation and growth rate from 2012 to 2023

Asia–Pacific Region Leads Global Growth in Electricity Generation

The Asia–Pacific region remains the main driver of global electricity generation growth. Driven by the economic growth of China and India, the electricity consumption in the Asia–Pacific region reached 15.2 trillion kilowatt-hours, with a growth rate of 4.3%, increasing from 49.9% in 2022 to 50.7% in 2023. The demand for electricity in major developed economies slowed significantly due to weak economic growth, with North America's electricity consumption slightly increasing by 0.2%, and electricity consumption in Europe and the Eurasian continent decreasing by 0.3%. The Middle East and Central and South America saw faster growth in electricity consumption, with growth rates of 8.1% and 2.4% respectively; Africa's electricity consumption growth rate decreased by 2.2% (Fig. 8).

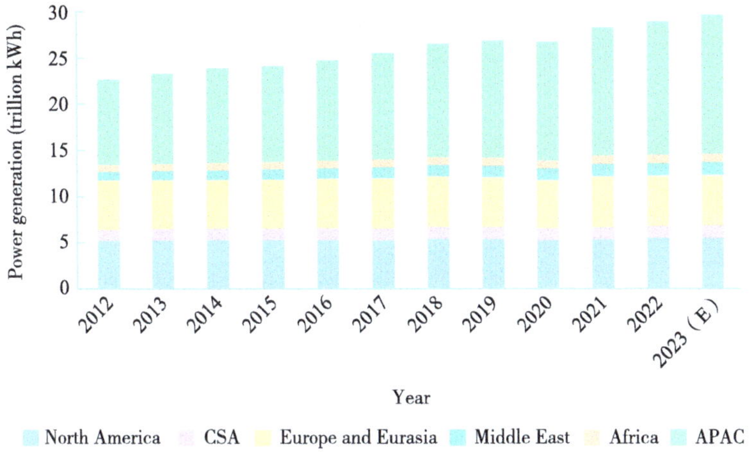

Fig. 8 Global electricity generation by region from 2012 to 2023

New Non-hydro Renewable Energy Installation Capacity Hits a New High

Under the trend of energy transition, countries are paying more attention to the development of non-fossil energy, based on their own resource endowments, and accelerating the construction of renewable energy projects. In 2023, the installed capacity of non-hydro renewable energy power generation in the world grew rapidly, estimated to increase by nearly 434 gigawatts, with a cumulative total exceeding 2560 gigawatts, a significant year-on-year increase of 20%, up 5 percentage points from the previous year. Solar power accounted for more than 70% of the new renewable energy installation capacity, followed by wind power at 25%. Global solar power installations increased by 65% year-on-year to 312 gigawatts, with a cumulative installation of 1374 gigawatts; wind power installations increased by 49% to 109 gigawatts, with a cumulative installation exceeding 1000 gigawatts. Offshore wind power installations added 73 gigawatts, accounting for 7% of the total new wind power installations worldwide, with more than half of the growth coming from China (Fig. 9).

Development Status of China's Energy Industry

In 2023, China's GDP exceeded 126 trillion yuan, with a growth rate of 5.2%, an increase of 2.2 percentage points from the previous year. China's energy security and stable supply capacity have steadily improved. The overall energy supply and demand in the country is stable and orderly, coal production continues to grow, and the ability to guarantee supply at the bottom continues to strengthen; the results of oil and gas reserve increases and production are significant, with annual crude

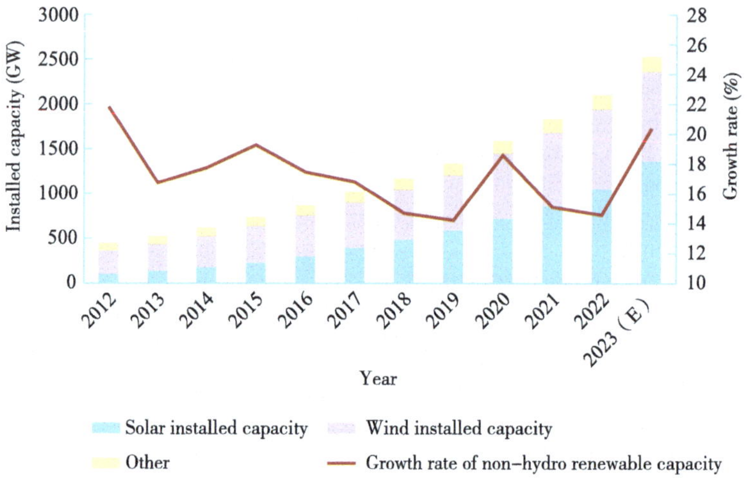

Fig. 9 Global renewable energy installation capacity and growth rate from 2012 to 2023. Author of this section: Zhou Xiwei Reviewer: Yuan Liuyan

oil production stabilizing at 200 million tons, and natural gas production exceeding 2300 billion cubic meters, achieving stable and increased crude oil production, and natural gas production increasing by more than 100 billion cubic meters for the seventh consecutive year. Domestic energy prices are generally stable. At the same time, China's energy structure transformation is accelerating, with renewable energy power generation installed capacity reaching 1.47 billion kilowatts, accounting for more than 50% of the total installed capacity of the country, historically surpassing thermal power installed capacity. The development of new energy such as wind power and solar energy is strong, and the scale of installations continues to grow rapidly.

Adhere to the Principle of "Establish First, Break Later", Diversified Supply Capacity Steadily Enhanced

The Core Supporting Role of Fossil Energy Continues to Strengthen

Currently, the energy structure in our country is being adjusted rapidly, but fossil fuels such as coal and petroleum still constitute the main body of energy supply, playing a core supporting role in ensuring energy security. By 2023, the production of major energy products in large-scale industries in our country will continue to grow, and the ability to ensure energy supply will steadily improve. Among them, coal production is 4.66 billion tons, an increase of 2.9% year-on-year; crude oil production is 209 million tons, an increase of 2% year-on-year; natural gas production is 235.3 billion cubic meters, an increase of 5.7%, and the increase has exceeded 10 billion cubic meters for seven consecutive years (Table 2 and Fig. 10).

In 2023, the coal price first suppressed and then rose, the fluctuation range narrowed, and the overall operation was within a reasonable range. The coal price (Qinhuangdao 5500 kcal thermal coal price) operated in the range of 807–1216 yuan/ton, among which the thermal coal price showed a V-shaped trend (Fig. 11). The average prices in the first, second, third, and fourth quarters were 1152 yuan/ton,

Table 2 China's energy production from 2013 to 2023

Year	Coal production (Billion tons)	Crude oil production (Billion tons)	Natural gas production (Billion cubic meters)
2013	42.44	2.10	1208.58
2014	41.36	2.11	1301.57
2015	39.98	2.15	1346.10
2016	38.88	2.00	1368.65
2017	39.14	1.92	1480.35
2018	39.75	1.89	1601.59
2019	40.19	1.91	1753.62
2020	40.49	1.95	1924.95
2021	41.30	1.99	2075.80
2022	45.60	2.05	2226.00
2023	46.60	2.09	2353.00

Data source National Bureau of Statistics, PetroChina Economic Research Institute

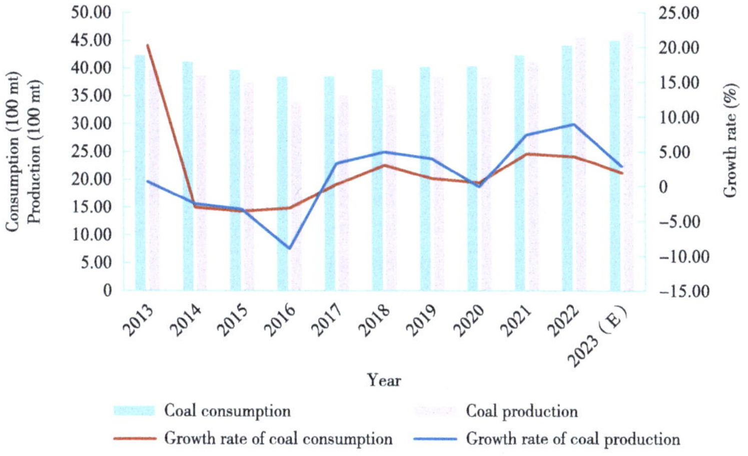

Fig. 10 China's coal consumption, production, and growth rate from 2013 to 2023. *Data source* National Bureau of Statistics, PetroChina Economic Research Institute

Fig. 11 Qinhuangdao 5500 kcal thermal coal price in 2023. *Data source* National Bureau of Statistics, Wind Consultation

935 yuan/ton, 874 yuan/ton, and 964 yuan/ton, respectively, a year-on-year decrease of 0.2%, 28.6%, 45%, and 50.1%.

In 2023, the coal inventory of national unified power plants remained above 200 million tons, the total amount of medium and long-term contracts for thermal coal signed was nearly 2.5 billion tons, and the available days of coal operated within 26 days. The inventory of key power plants nationwide reached 111 million tons, a year-on-year increase of 14.13%.

The Scale of Renewable Energy Continues to Grow Rapidly

In 2023, the overall power supply in the country was stable, fully leveraging the basic guarantee role of conventional power sources. The domestic total electricity consumption was 922.41 billion kilowatt-hours, an increase of 6.7% year-on-year, maintaining the characteristic of the growth rate of electricity consumption being higher than the growth rate of energy consumption over the past 10 years.

The scale of renewable energy installations achieved a new breakthrough, becoming a new force in ensuring power supply. In 2023, the national installed capacity of power generation was about 2.92 billion kilowatts, an increase of 13.9% year-on-year. Among them, the installed capacity of photovoltaic power generation was about 610 million kilowatts, an increase of 55.2% year-on-year; the installed capacity of wind power was about 440 million kilowatts, an increase of 20.7% year-on-year; the installed capacity of thermal power was about 1.39 billion kilowatts, an increase of 4.1% year-on-year; the installed capacity of hydropower was about 420 million kilowatts, an increase of 1.8% year-on-year; the installed capacity of nuclear power was about 60 million kilowatts, an increase of 2.4% year-on-year (Fig. 12).

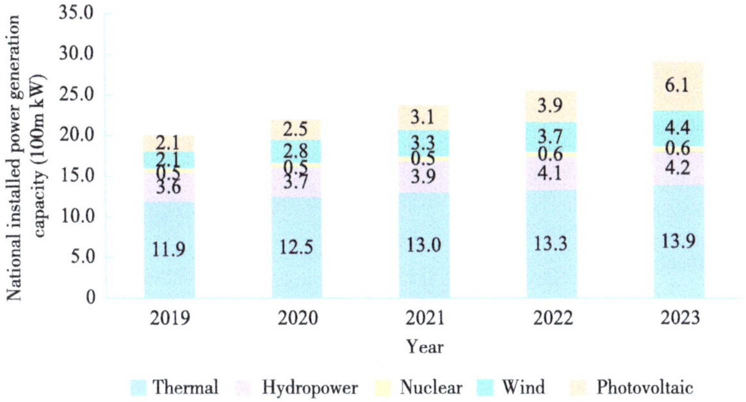

Fig. 12 National installed capacity of power generation from 2019 to 2023. *Data source* National Bureau of Statistics, PetroChina Economic Research Institute

In 2023, the utilization hours of power generation equipment in power plants of 6000 kilowatts and above nationwide was 3592 h, a decrease of 101 h from the same period of the previous year. Looking at the types, the average utilization hours of hydropower equipment nationwide was 3133 h, a decrease of 285 h from the same period of the previous year; the average utilization hours of thermal power equipment nationwide was 4466 h, an increase of 76 h from the same period of the previous year; the average utilization hours of nuclear power equipment nationwide was 7670 h, an increase of 54 h from the same period of the previous year; the average utilization hours of grid-connected wind power equipment nationwide was 2225 h, an increase of 7 h from the same period of the previous year. The average utilization hours of solar power generation equipment nationwide was 1286 h, a decrease of 54 h from the same period of the previous year (Fig. 13).

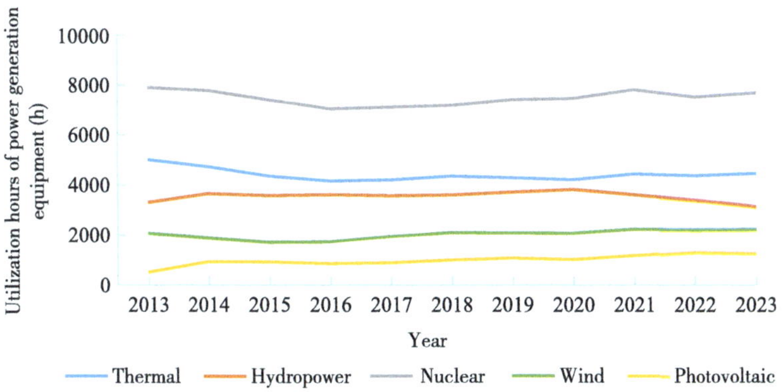

Fig. 13 Utilization hours of power generation equipment by type from 2013 to 2023. *Data source* National Bureau of Statistics, PetroChina Economic Research Institute

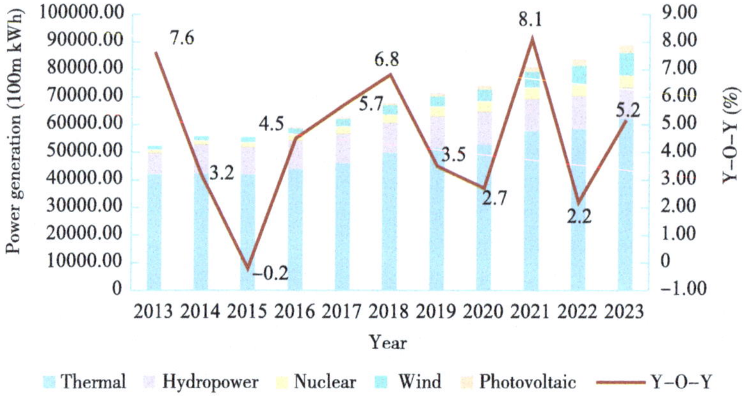

Fig. 14 China's power generation from 2013 to 2023. *Data source* National Bureau of Statistics, PetroChina Economic Research Institute

The power generation of renewable energy steadily increased. In 2023, the power generation of large-scale industries nationwide reached 89 trillion kilowatt-hours, an increase of 5.2% year-on-year. Among them, photovoltaic power generation was 3 trillion kilowatt-hours, an increase of 17.2% year-on-year; wind power generation was 8 trillion kilowatt-hours, an increase of 12.3% year-on-year; thermal power generation was 62 trillion kilowatt-hours, an increase of 6.1% year-on-year; hydropower generation was 11 trillion kilowatt-hours, a decrease of 5.6% year-on-year; nuclear power generation was 4 trillion kilowatt-hours, an increase of 3.7% year-on-year (Figs. 14 and 15). In 2023, the newly added installed capacity of power generation in the country broke through 300 million kilowatts for the first time, of which the newly added non-fossil energy power generation installed capacity exceeded 250 million kilowatts. The proportion of wind and photovoltaic power generation in the total electricity consumption of the whole society broke through 15%.

Scale of Wind and Solar Industry, Diversification of Zero-Carbon Technologies

In 2023, new energy represented by wind and solar, under the promotion of policies, markets, and technology, achieved significant results, with the scale of installations ranking first in the world, forming global leading advantages in scale, technology, cost, etc., showing the following characteristics.

First, the market scale of wind power and photovoltaics has doubled, and the growth rate continues to maintain a steady upward trend. In 2023, the newly added installed capacity of photovoltaics was about 216.88 million kilowatts, an increase of 148.1% year-on-year (Fig. 16). Looking at the regions, the top five in new installations are mainly concentrated in Hebei, Yunnan, Xinjiang, Shandong, Jiangsu

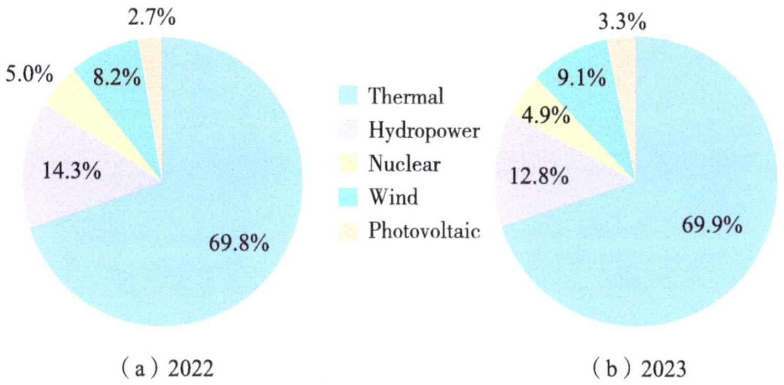

Fig. 15 China's power generation structure in 2022 and 2023. *Data source* National Bureau of Statistics, PetroChina Economic Research Institute

and other provinces. The newly added photovoltaic installed capacity in the top ten provinces is 130.12 million kilowatts, accounting for 60.2% of the total newly added photovoltaic installed capacity in the country. The newly added installed capacity of wind power was 75.9 million kilowatts, an increase of 101.7% year-on-year (Fig. 17). Looking at the regions, the newly added wind power installations are further concentrated in the "three north" and other provinces with better resource conditions. The top five provinces are Inner Mongolia, Xinjiang, Yunnan, Gansu, Hebei and other provinces. The newly added wind power installed capacity in the top ten provinces is 59.13 million kilowatts, accounting for 78% of the total newly added wind power installed capacity in the country. In terms of offshore wind power, as of the end of September 2023, China's offshore wind power installed capacity reached 31.89 million kilowatts, ranking first in the world. The localization rate of core key components of China's offshore wind power units has reached 100%, and China's offshore wind power projects will further develop towards deep and far sea in the future. In 2024, it is expected that the newly added wind power and photovoltaic installed capacity in the country will be about 200 million kilowatts.

Second, the cost of equipment has dropped significantly, and the economy of new energy projects has generally improved. Since 2023, the prices of the entire photovoltaic industry chain have dropped significantly. The annual price of photovoltaic modules is about 1.4 yuan/watt, and the average price of large-scale EPC is 3.76 yuan/watt, with annual decreases of 48% and 10% respectively; the price of onshore wind power units is stable and declining, with the annual average bid price of onshore wind power (including tower drum) about 2127 yuan/kilowatt, and the average bid price of offshore wind power (including tower drum) about 3653 yuan/kilowatt, with annual decreases of 5.2% and 7.1% respectively. In 2024, without major technological breakthroughs, the prices of wind power and photovoltaic equipment are expected to stabilize.

Third, the level of new energy consumption has steadily increased, and the rate of wind and light abandonment has decreased compared to the previous year. In

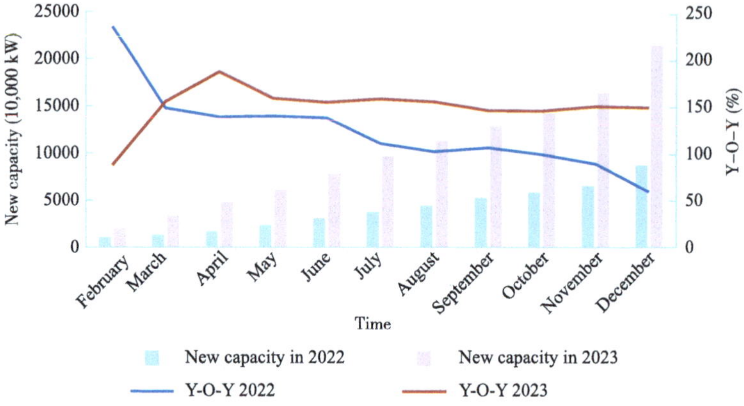

Fig. 16 Photovoltaic power generation installed capacity and growth rate in 2023. *Data source* National Bureau of Statistics, PetroChina Economic Research Institute

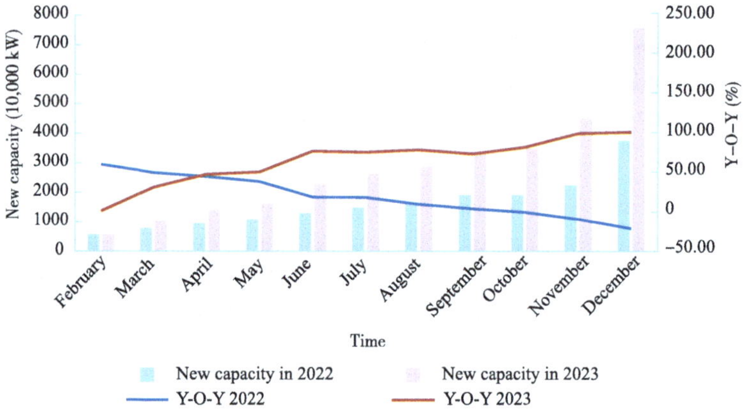

Fig. 17 Installed capacity and growth rate of wind power generation in 2023. *Data source* National Bureau of Statistics, PetroChina Economic Research Institute

2023, the utilization rates of wind power and photovoltaics nationwide were approximately 97.3% and 98% respectively, and the utilization rates of wind power and photovoltaics in Tianjin, Shanghai, Jiangsu, Zhejiang, Anhui, Fujian, Chongqing, Sichuan, and Guangxi reached 100%. The overall level of consumption utilization is relatively high, but at the same time, some provinces still have wind and light abandonment. Among them, the utilization rates of wind power in Hebei, Qinghai, and West Mongolia are all below 95%; the utilization rate of photovoltaics in Tibet is below 90% (Fig. 18).

Fourth, the green and low-carbon industry is accelerating its upgrade, and the iteration of energy technology is accelerating. In the field of wind power, the size of wind turbines has set a record. By the end of 2023, Sany Heavy Energy

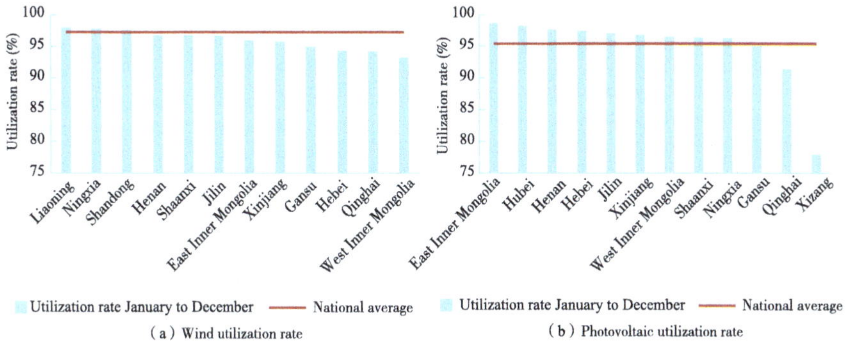

Fig. 18 Utilization rate of wind power and photovoltaics in 2023. *Data source* National New Energy Consumption Monitoring and Warning Center, PetroChina Economic Research Institute

released a 15-megawatt onshore wind turbine unit, and Mingyang Intelligent 20-megawatt offshore wind turbine went offline and announced the development of a 22-megawatt offshore wind power, refreshing the world's highest single-machine capacity record for onshore and offshore wind turbines. In just three years, the single-machine capacity of our country's wind turbines has been developed from 2.6 megawatts onshore and 4.8 megawatts offshore to the world's highest. In the field of photovoltaic power generation, the iteration of N-type cells to P-type cells is accelerating, and the domestic purchase scale of N-type modules in the first half of the year exceeded 24 million kilowatts, accounting for more than 20%. Several perovskite photovoltaic cell efficiencies have refreshed the world record, among which the conversion efficiency of Longi Green Energy's silicon-perovskite tandem cell reached 31.8%.

Fifth, new models and new formats are emerging one after another, and integrated development has become a new trend. Since the "Fourteenth Five-Year Plan", the exploration of multi-energy complementarity has continued to deepen, from the integration of water, wind, and light in river basins to the application of new models of renewable energy hydrogen production projects, the new formats formed by industry integration are showing more diverse characteristics. Taking the Kubuqi large base as an example, in September 2022, the state approved Huaneng Group to take the lead in undertaking the construction of the new energy base in the southern part of the Kubuqi Desert in Inner Mongolia, and the Three Gorges Group to take the lead in undertaking the construction of the new energy base in the central and northern parts of the Kubuqi Desert in Inner Mongolia, with the scale of wind and light power generation projects both being 12 million kilowatts. Relying on the construction of new power systems in the western and northern regions and the improvement of grid consumption capacity, innovative development methods and application models, through multi-energy complementarity and source-network-load-storage integration, wind-light-storage-number, wind-light-storage-park, wind-light-hydrogen production, a batch of wind power and photovoltaic projects that are consumed nearby are constructed.

Table 3 Total primary energy consumption and energy consumption structure in China from 2016 to 2023

Year	Total energy consumption (Billion tons of standard coal)	Energy structure (%)			
		Coal①	Oil	Natural gas	Non-fossil energy
2016	43.6	62.2	18.7	6.1	13
2017	44.9	60.6	18.9	6.9	13.6
2018	47.1	59	18.9	7.6	14.5
2019	48.7	57.7	18.9	8.1	15.3
2020	49.8	56.8	18.9	8.4	15.9
2021	52.4	56	18.5	8.9	16.6
2022	54.1	56.2	17.9	8.4	17.5
2023 (E)	57.2	55.3	18.3	8.7	17.7

Data source National Bureau of Statistics, PetroChina Economic Research Institute
①The change in the calorific value of coal is not considered

Energy Consumption Continues to Grow, and the Green and Low-Carbon Transition is Accelerating

Primary Energy Consumption Grows Rapidly

In 2023, it is estimated that the total primary energy consumption of the country will be 5.72 billion tons of standard coal, an increase of 5.7% year-on-year. Among them, non-fossil energy accounted for 17.7% of the total energy consumption, an increase of 0.2 percentage points from the previous year; the proportion of coal consumption steadily declined, accounting for 55.3%, a decrease of 0.9 percentage points; the proportions of oil and natural gas were 18.3% and 8.7%, respectively, an increase of 0.4 percentage points and 0.3 percentage points (Table 3).

Acceleration of Electrification Substitution in the Transportation Sector

In 2023, China's production and sales of new energy vehicles ranked first in the world for nine consecutive years, becoming an important force leading the transformation of the global automotive industry. The production and sales of new energy vehicles in China were 9.587 million and 9.495 million respectively, an increase of 35.8% and 37.9% year-on-year, and the market penetration rate reached 31.6%, which was 5.9 percentage points higher than the same period last year (Fig. 19).

China's new energy vehicle ownership reached 20.41 million, accounting for 6.07% of the total number of vehicles. In terms of driving form, in 2023, sales of pure electric vehicles were 6.685 million, a year-on-year increase of 24.6%; sales of plug-in hybrid vehicles were 2.804 million, a year-on-year increase of 84.7%; sales of fuel cell vehicles were 6000, a year-on-year increase of 72%.

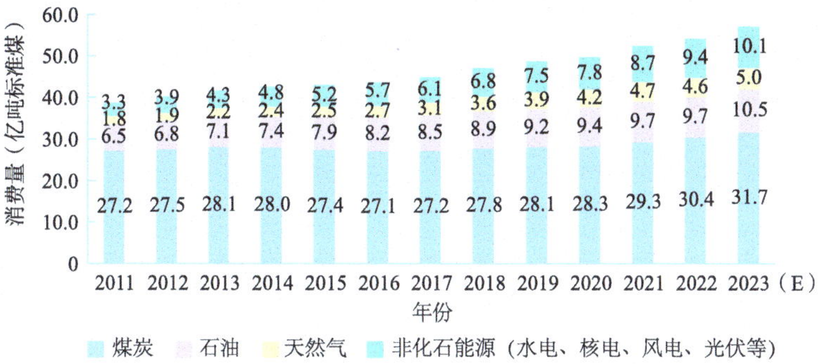

Fig. 19 2013–2023 China's new energy vehicle sales, growth rate and market penetration rate. *Data source* PetroChina Economic Research Institute, China Association of Automobile Manufacturers

China's oil companies are actively building "carbon–neutral gas stations". At present, they have been built in Jiangsu, Guangzhou, Sichuan, Shanghai, Tianjin, Fujian, Shandong, Shanxi, Henan, Zhejiang, Yunnan and other places. Taking the development of photovoltaic technology as an opportunity to promote the carbon–neutral "light-storage-charging" integrated operation of gas stations, and build a carbon–neutral park-type energy supply demonstration complex integrating refueling, gas filling, hydrogen filling, charging, battery swapping, photovoltaic power generation, auto service, and commerce.

Unit GDP Energy Consumption and Carbon Emissions Continue to Decline

In 2023, China's unit GDP energy consumption was 0.44 tons of standard coal/10,000 yuan, a decrease of 2.23% compared with 2022, maintaining a downward trend (Fig. 20). Since 2012, the national unit GDP carbon dioxide emissions have decreased by 34.4%, reversing the situation of rapid growth in total carbon dioxide emissions. As of 2022, the carbon emission intensity has decreased by more than 51% compared to 2005, and the forest coverage rate has reached 24.02%.

China has made positive contributions to global governance in response to climate change. In 2023, the annual trading volume of carbon emission quotas in the national carbon market was 212 million tons. While the scale of green electricity transactions is steadily expanding, the issuance of green certificates has entered a new stage of "full coverage". In 2023, about 176 million green certificates were issued throughout the year, and the cumulative green electricity transaction volume reached about 611 billion kilowatt-hours, which were 7.8 times and 10.5 times that of 2022, respectively.

(This section is written by: Duan Peiyi Reviewer: Yuan Liuyan)

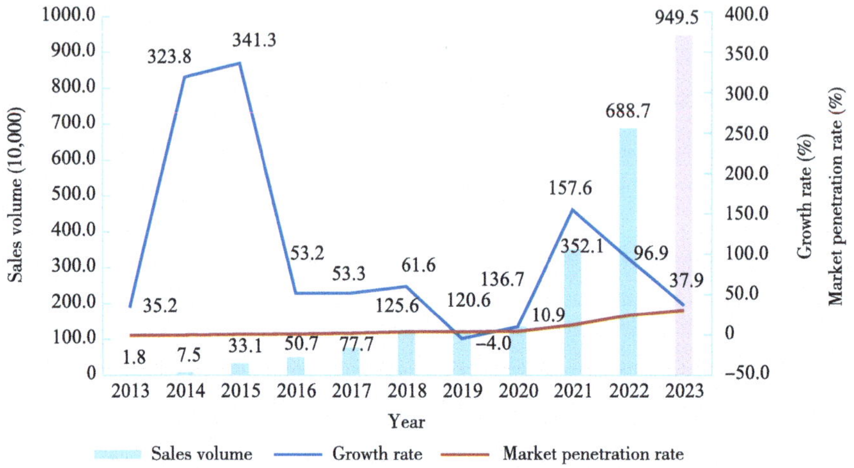

Fig. 20 China's GDP, total energy consumption and unit GDP energy consumption. *Data source* National Bureau of Statistics, PetroChina Economic Research Institute

Review of International Energy Policy

Since the outbreak of the Russo-Ukrainian conflict, the energy and climate policies of major countries have shown a four trends of accelerated transformation, universal subsidies, group confrontation and diversified technologies. In 2023, the energy transformation policy system of major countries continued to improve, with more evident government intervention in the energy industry and international cooperation. The "green barrier" is still increasing, and the challenge of achieving global temperature control targets is becoming increasingly difficult. On November 15, China and the United States issued "The Sunnylands Statement on Enhancing Cooperation to Address the Climate Crisis," representing the resumption of substantive dialogue between China and the United States on climate action. This provided a positive signal for global climate cooperation and consensus at COP28.

Accelerated Energy Transformation and More Ambitious New Energy Development Goals

The energy crisis has strengthened countries' determination to reduce their dependence on fossil fuels. Many developed countries have revised their medium and long-term energy development strategies to enhance the development of renewable energy and emission reduction targets.

The EU Has Basically Established a Low-Carbon Transformation Policy and Regulatory System

On July 25, 2023, the EU Council passed new rules, and member states will jointly ensure that final energy consumption in 2030 is at least 11.7% less than the 2030 energy consumption forecast set earlier (2020). On October 9, the EU Council passed the revised the Renewable Energy Directive, raising the target share of renewable energy in EU energy consumption in 2030 from the current 32 to 42.5%, and adding an additional 2.5% indicative supplement to achieve a 45% target. The industrial sector is required to achieve an annual growth rate of renewable energy usage of 1.6%, and by 2030, the proportion of renewable hydrogen in total industrial hydrogen consumption will reach 42%. In the construction sector, the new rules set an indicative target of at least 49% of renewable energy in buildings by 2030. Together with the simultaneously passed ReFuelEU Aviation Regulations, the last two pillars of the "Fit for 55" legislative proposal have been built. The EU now has set climate targets covering all key economic sectors legally.

The United States Accelerates Technological Innovation to Promote Net-Zero Transformation

On April 20, 2023, the White House Office of Science and Technology Policy, the Department of Energy, and the State Department jointly released the U.S. National Innovation Pathway, aimed at accelerating the innovation of key clean energy technologies. The report points out that the Biden administration is promoting the "innovation, demonstration, deployment" three-step method, expanding the deployment and research of technologies needed for the U.S. transformation. Priority development areas mainly include: advanced batteries, advanced nuclear energy, advanced solar energy, carbon capture, utilization and storage (CCUS), carbon removal, clean shipping fuels and ships, enhanced geothermal systems, nuclear fusion energy, hydrogen energy, long-duration energy storage, net-zero buildings, offshore wind power, sustainable aviation fuels. The goal is to achieve 50% zero-emission vehicle sales by 2030, zero carbon emissions in the power sector by 2035, and national net-zero emissions by 2050.

The UK Will Expand Domestic Clean Energy Scale and Accelerate Green Industry Development

On March 30, 2023, the UK government announced the "Green Energy Plan" to ensure the successful achievement of the 2050 climate target. The plan simplifies the planning permission process for building photovoltaic systems and offshore wind power, invests an additional £10 billion to increase capacity to promote exports, invests £160 million in emerging floating offshore wind power, contributes £240 million to establish a net-zero hydrogen energy fund, and invests £380 million to develop the UK's electric vehicle charging infrastructure. The plan declares the government's commitment to stop selling new petrol and diesel cars and vans by 2030, and from 2035, all new cars and vans must achieve zero emissions at the tailpipe. On October 26, the 2023 Energy Act was enacted, establishing energy production, energy security, energy regulatory systems, mainly involving carbon capture and storage (CCS), hydrogen energy, low-carbon heating, natural gas and electricity, nuclear fusion energy, aviation fuel, offshore wind power, oil and natural gas and other fields, aiming to fulfill the UK government's key long-term commitments in clean energy, climate change, and energy security and resilience.

Government Increased Interventions on the Market, Giving Industrial Subsidies

In order to seize the commanding heights of the new round of energy technology revolution and enhance energy autonomy, major powers have introduced high financial and tax subsidies, expanded climate financing, and promoted the localization and friendliness of the new energy industry chain. Some Western countries politicize economic, energy, and technological cooperation, increase market regulation and restrictions on multinational corporations for national security reasons, which will accelerate the restructuring of the global industrial chain, delay the cost reduction and deployment of clean energy technology, and make it more difficult to achieve global temperature control targets.

The United States Implemented a Record Subsidy Plan

Based on the $62 billion provided for energy sector investment by the Bipartisan Infrastructure Bill passed in 2022, the U.S. Inflation Reduction Act officially took effect on January 1, 2023, providing up to $369 billion in subsidies over 10 years to boost industries such as solar energy, wind energy, and electric vehicles. The bill includes up to 9 tax incentives, predicated on production and sales within the United States or North America, gradually achieving a goal of reducing carbon emissions by about 40% by 2030 compared to 2005, while also alleviating consumer pressure from rising traditional energy prices. According to a report by the Rocky Mountain Institute, the bill has brought $372 billion in new private investment and over 210,000 new jobs to the clean energy sector in its first year of implementation. The Department of Energy has over $100 billion in loan authorization to help companies deploy and expand innovative clean energy, advanced transportation, and tribal energy projects in the United States, and has established a new loan authorization of $250 billion to help companies readjust or reuse energy infrastructure to adapt to low-carbon economic development.

The EU Launched a Large Subsidy Plan to Build Domestic New Energy Industry Chain

On February 1, 2023, the European Commission officially proposed the EU Green Industrial Plan, deciding to allocate 250 billion euros from existing EU funds for industrial greening, to continue to maintain a leading position on the path to climate neutrality. This includes providing tax relief for companies investing in net-zero emission technologies, accelerating investment and financing for European clean technology production, and conducting global cooperation to make trade serve the green transformation. The plan includes four main pillars: establishing a predictable and simplified regulatory system; accelerating the speed of obtaining funds; enhancing skills applicable to the green transformation; and promoting open trade measures that contribute to supply chain resilience.

On March 16, 2023, the European Commission released three important proposals: the Net Zero Industrial Act, the Critical Raw Materials Act, and the Electricity Market Design Reform, to enhance the EU's competitiveness in global green industrial technology. The Net Zero Industrial Act aims to achieve at least 40% of the net-zero technology manufacturing capacity required for the EU to reach its energy and climate targets by 2030. The eight net-zero technologies supported by the Act include: solar photovoltaics and solar thermal energy, onshore wind energy and offshore renewable energy, batteries and storage, heat pumps and geothermal energy, electrolyzers and fuel cells, sustainable biogas and biomethane, carbon capture utilization and storage, and grid technology. Among them, the domestic manufacturing capacity for photovoltaics should reach at least 30 gigawatts, accounting for 45% of the EU's total demand; the domestic manufacturing capacity for wind energy and heat pumps should reach 36 gigawatts and 31 gigawatts respectively, accounting for 85% and 60% of demand respectively; battery manufacturing capacity should meet 90% of EU demand, reaching 550 gigawatts; the total installed capacity of hydrogen electrolyzers should reach at least 100 gigawatts, accounting for 100% of total demand.

The "Critical Raw Materials Act" proposes to establish a framework within the EU to ensure the secure and sustainable supply of critical raw materials. The EU includes the list of critical raw materials and strategic raw materials in EU law through the Act, and has set a capacity benchmark for strategic raw materials in the EU region by 2030: first, the EU should increase the utilization of its own geological resources of strategic raw materials, producing at least 10% of the materials consumed by the EU's strategic raw materials; second, in order to establish a complete value chain, enhance processing capacity, and be able to produce at least 40% of the strategic raw materials consumed annually; third, the recycling of strategic raw materials should reach at least 15% of annual consumption; fourth, efforts should be made to ensure that dependence on the supply of any strategic raw material (unprocessed and at any stage of processing) does not exceed 65% in a single third country.

On October 17, 2023, the EU Council reached a general plan for electricity market design reform. The main content includes: long-term electricity market contracts, capacity electricity price, user-side protection, and energy price limits during crises.

In terms of long-term electricity market contracts, it promotes power purchase agreements and difference contracts, and improves the liquidity of the long-term electricity market, with the aim of decoupling electricity market prices from fossil energy prices as much as possible. In terms of capacity electricity price, the carbon emission cap for generator sets has been modified to ensure that lower-carbon generator sets can receive compensation for capacity electricity price. In terms of user-side protection, it requires electricity suppliers to protect customers from the impact of wholesale market price fluctuations in their price hedging strategies.

Multiple Countries Launched Green Industry Subsidy Policies

On August 9, 2023, the German Federal Government approved the 2024 expenditure plan and the draft budget plan until 2027 for the "Climate and Transformation Fund". The expenditure of the Climate and Transformation Fund is expected to reach 57.6 billion euros in 2024, of which 47.4 billion euros come from the German Federal Ministry for Economic Affairs and Climate Action. From 2024 to 2027, the federal government plans to invest a total of 211.8 billion euros through this fund, making a key contribution to achieving Germany's energy and climate policy goals, developing future technologies, and transitioning to a sustainable climate-neutral economy.

On October 11, 2023, the French Parliament passed the "Green Industry Act". The aim is to accelerate the re-industrialization of France, promote ecological transformation, and counter the challenge of the American Inflation Reduction Act to European industry. The Act proposes to promote and accelerate the establishment of industrial bases domestically, mobilize public and private investment to finance green industries, plan to invest 1 billion euros from 2023 to 2027 for the development of industrial land, and through green industry tax relief measures, it is expected to bring 23 billion euros of investment and achieve 41 million tons of carbon reduction by 2030.

On December 13, 2023, the South Korean government announced the "Plan to Strengthen the Competitiveness of the Battery Recycling Industry", providing 380 trillion won (approximately 29 billion US dollars) of support within the next 5 years, giving comprehensive support to the battery industry from multiple perspectives such as standard regulations, financing, upstream minerals, overseas factories, battery recycling, carbon footprint, etc., to comprehensively enhance the competitiveness of the South Korean battery industry.

Energy Governance is Becoming Factionalized, Creating Green Trade Barriers

The intensification of great power competition has widened the divergence of interests between developed and developing countries in the field of international energy and climate governance. Some Western countries play the "climate cards", both Europe and America have introduced carbon tariff policies, shifting emission reduction responsibilities to developing countries. The United States is keen to rally allies and mineral resource countries to create a "democratic camp" supply chain alliance, turning global cooperation into regional and factional cooperation, and the geopolitical game around new energy and key minerals is becoming more intense. On the occasion of the Sino-US summit in November 2023, the two countries issued a joint statement on addressing the climate crisis, bringing positive signals to global climate governance.

Building a Multilateral Governance Mechanism for the Clean Energy Industry Chain

Following the "Energy Mineral Resources Governance Initiative" issued by the US State Department on June 11, 2019, in June 2022, the United States established a "Mineral Security Partnership" with countries and organizations such as Canada, Australia, Finland, France, Germany, Japan, South Korea, Sweden, the United Kingdom, and the European Union. In 2023, the United States further sought to establish a "Key Minerals Buyers Club" with the European Union, aiming to alleviate the rift in the US-EU alliance relationship in the Inflation Reduction Act and consolidate the Mineral Security Partnership, making the Western camp's key mineral security cooperation platform-based, normalized, and project-based. On April 16, 2023, the Group of Seven (G7) held a meeting on climate, energy, and environmental policy in Japan and issued a clean energy initiative, emphasizing diversification in the supply chains of key minerals and green products, setting technical standards for key minerals, weakening China's economic advantage in the battery field, and increasing investment in the new energy industry in Vietnam and India and other "de-risking" measures. The European Union also proposed in the Key Raw Materials Act to promote the establishment of a centralized procurement mechanism for key minerals such as bauxite, cobalt, lithium, and nickel, and welcomes the United States or other trade partners to join, to promote investment in a wider supply chain, thereby reducing China's market share.

Europe and America Built International Trade Carbon Barriers

On April 18, 2023, the European Parliament passed the EU Carbon Border Adjustment Mechanism (CBAM), the first batch of industries included steel, cement, aluminum, fertilizers, electricity, and hydrogen, from October 1, 2023, to December 31, 2025, as a transition period, and substantive implementation from January 1, 2026. The introduction of the EU carbon tariff has triggered a global chain reaction. On December 18, 2023, the British government announced that it will implement the UK CBAM from 2027, and the United States and others have also proposed related proposals.

On August 17, 2023, the EU New Battery Law officially came into effect, requiring that from July 2024, power batteries and industrial batteries must declare product carbon footprints, and from 2027, power batteries exported to Europe must hold a compliant "battery passport", recording the manufacturer of the battery, material composition, carbon footprint, supply chain, and other information. Power batteries entering the European market will face the challenge of "green trade barriers". Export companies must at least overcome the three obstacles of supplementing and improving carbon footprint declarations, meeting EU requirements for battery material recycling and regeneration, and dealing with the challenge of information disclosure in the battery passport.

Sino-US Climate Cooperation Returns to a Constructive Track

On November 15, 2023, China and the United States issued the Sunnylands Statement on Enhancing Cooperation to Address the Climate Crisis. According to the declaration, China and the United States reaffirmed their commitment to cooperate and work together with other countries to address the climate crisis, and launched the "21st Century 2020s Enhanced Climate Action Working Group"; China and the United States plan to fully accelerate the deployment of renewable energy in both countries from now until 2030, based on the 2020 level, to speed up the replacement of coal, oil and gas power generation; China and the United States will implement their respective national methane action plans, immediately launch technical working group cooperation, conduct policy dialogues, exchange technical solutions and build capacity, formulate their respective methane emission reduction action targets included in their 2035 national independent contributions, and support the progress of methane emission reduction control in both countries. They strive to advance at least five large-scale cooperation projects in areas such as industry and energy for carbon capture, utilization and storage by 2030. The declaration represents that China and the United States will carry out more substantive dialogues and cooperation in energy policy, strategy, technology and broader climate action, providing important momentum for solving the global climate problems at the United Nations Climate Change Conference in Dubai (COP28).

Subsequently, at COP28 from November 30 to December 13, various countries made positive progress in many aspects: on the first day, a decision was made to establish a loss and damage fund, representatives of the contracting parties conducted the first global stocktaking, important decisions concerning developing countries such as the global adaptation goal framework and the just transition path work plan were passed, but the issue of funding is still unresolved; the final agreement passed by the conference requires "to gradually abandon fossil fuels from the energy system in a fair, orderly and equal manner, accelerate action in this critical decade, so as to achieve net-zero emissions in line with science by 2050." "This is the first global climate agreement in history calling for the abandonment of fossil fuels", the executive secretary of the convention, Simon Stiell said in his closing remarks, and that "the COP28 agreement marks the beginning of the end of the fossil fuel era".

Diversification of Technological Routes, Multiple Energy Sources Are Having Opportunities

In 2023, countries paid more attention to adopting energy development paths suitable for their own national conditions according to local conditions. They adopted diversified strategies to avoid the risks brought by over-reliance on a single type of renewable energy. Major countries all attached importance to the development of green hydrogen energy, and supported the development of their own hydrogen energy industry from the policy and financial level. Most countries' nuclear energy policies have turned around, choosing to develop new generation nuclear power and nuclear fusion technology to achieve energy independence, and biofuels also receive policy support from European and American countries.

Release of New Hydrogen Energy Development Strategy

On June 5, 2023, the United States released the National Clean Hydrogen Energy Strategy and Roadmap. This is the first time the U.S. government officially releases a clean hydrogen energy national strategy and roadmap since the "Hydrogen Economy Strategy of the Beginning of this Century," covering the entire industry chain of clean hydrogen energy production-storage-transportation-use and the coordinated regulatory responsibilities of 11 departments, proposing $95 billion in financial support, clean hydrogen tax reductions and exemptions, and clean hydrogen energy standards and other powerful measures, to promote the achievement of clean hydrogen production capacity of 10 million tons/year, 20 million tons/year and 50 million tons/year in 2030, 2040 and 2050 respectively.

On July 26, 2023, Germany released a new version of the National Hydrogen Energy Strategy, updating the hydrogen energy development goals for 2030, proposing three-stage key measures, doubling the electrolyzer production capacity target for 2030 to 10 gigawatts, and emphasizing the development of hydrogen energy infrastructure. Through the funding of important projects of common interest in Europe, Germany will establish a hydrogen energy start-up network with more than 1800 km of hydrogen pipelines; promote the wide application of hydrogen energy in various industries to accelerate the formation of a globally leading hydrogen energy industry.

On June 6, 2023, the Japanese Ministry of Economy, Trade and Industry released a revised draft of the Basic Hydrogen Energy Strategy, proposing to invest 15 trillion yen in the next 15 years, planning to install 15 gigawatts of water electrolysis hydrogen production equipment by 2030, build 1000 hydrogen refueling stations, promote 800,000 hydrogen energy vehicles; by 2040, the hydrogen supply volume will be 12 million tons, while ensuring the achievement of carbon neutrality targets, enhance industrial competitiveness, and expand overseas markets.

In March 2023, the Indian government announced the draft of National Hydrogen Energy Mission, planning to deploy 45 gigawatts of renewable energy hydrogen production capacity by 2030, with a total investment of about 750 billion rupees (about 8.3 billion U.S. dollars), planning to deploy 5 gigawatts of renewable energy hydrogen production capacity in 2024–2025, the remaining 40 gigawatts capacity will be announced in subsequent plans.

Accelerate the Construction of Nuclear Power to Ensure Power Supply

The global energy crisis since 2022 has prompted many countries to change their nuclear power policies. Most European countries, as well as Japan and South Korea, are accelerating their nuclear power construction, with only Germany closing its last three nuclear reactors as scheduled on April 15, 2023, fulfilling its commitment to phase out nuclear power. On April 16, 2023, Unit 3 of the Olkiluoto Nuclear Power Plant in Finland was put into operation, the first new nuclear power unit to be put into operation in Finland in over 40 years, meeting about 14% of Finland's electricity demand. Unit 3 of the Mochovce Nuclear Power Plant in Slovakia was connected to the grid in February 2023, supporting 13% of the country's electricity consumption. Meanwhile, Sweden, France, the UK, and other European countries are also planning to develop a new generation of nuclear power units. On February 10, 2023, Japan passed the "Basic Policy for Achieving Green Transformation," clearly aiming to maximize the use of renewable energy and nuclear power, extending the operating time of existing reactors from 40 to 60 years, and developing the next generation of innovative reactors to replace retired ones. On April 14, 2023, the Japanese Cabinet approved the first National Nuclear Fusion Strategy, calling for private companies to participate more widely in fusion energy research and development, aiming to achieve nuclear fusion power generation around 2050. On July 10, 2023, the South Korean Ministry of Trade, Industry and Energy stated that in order to meet the growing demand for electricity and achieve carbon reduction targets, the government is assessing the feasibility of new nuclear power projects, and has set a goal of developing small reactors and exporting 10 nuclear power units by 2030. On October 26, 2023, the UK enacted the 2023 Energy Act, allowing the Minister for Energy Security and Net Zero Emissions to directly appoint the British Nuclear Energy Agency to oversee and promote government participation in nuclear energy projects, stipulating the development of a virtual prototype of a nuclear fusion power plant by 2040, making UK the first country to legislate for nuclear fusion energy regulation.

Improve the Layout of Alternative Fuel Technology Planning

In March 2023, the United States released the Clear Goals for U.S. Biotechnology and Bio-manufacturing, proposing specific paths to accelerate the widespread application of biotechnology and the explosive development of bio-manufacturing. The main points include: producing more than 20 commercially viable bio-products within 5 years, reducing lifecycle greenhouse gas emissions by more than 70% compared to now; meeting at least 30% of chemical demand through bio-manufacturing within

7 years, reducing agricultural methane emissions by 30%; expanding the scale of cell therapy through synthetic biology, bio-manufacturing, and engineered biology within 10 years, and reducing manufacturing costs by 10 times; replacing plastics and commercial polymers with bio-based substitutes within 20 years, replacing more than 90% of plastics; extending the tax incentive policy for biodiesel, renewable diesel and alternative fuels until 2024, and extending the income tax incentive policy for second-generation biofuels until 2025, setting the validity period for tax credits for sustainable aviation fuel until 2024. At the same time, the U.S. Department of Energy has increased funding for biomass energy technology innovation, announcing several projects in succession, investing more than $140 million to support biomass energy technology research and development, to promote the development of biomass refineries, and produce low-emission sustainable fuels.

On October 9, 2023, the EU Council passed the ReFuelEU Aviation Regulation, aiming to meet the EU's climate targets by increasing the demand and supply of sustainable aviation fuels. Sustainable Aviation Fuels (SAF) include certified biofuels, renewable fuels from non-biological sources (including renewable hydrogen), and aviation fuels that meet the sustainability and emission reduction standards of the Renewable Energy Directive.

Outlook for 2024

In 2024, the energy policies of developed countries need to focus on three aspects:

(a) **Continuous Support for New Energy Industries and Frontier Technologies**: Countries are likely to compete to increase subsidies and preferential policies to attract industry landing.
(b) **Continued Spread of Trade Protectionism**: Exclusive industry policies and carbon tariff bills like CBAM are expected to be implemented in more countries, which may trigger trade wars between some economies.
(c) **Methane Emission Control Becomes a Policy Hotspot**: The EU's methane emission control bill has entered the legislative stage, and countries participating in the "Global Methane Pledge" such as the United States may have further policy arrangements in fulfilling their emission control commitments.

(This section was written by Zhang Pengcheng and reviewed by Yuan Liuyan).

Review of China's Energy Policy

In 2023, relevant national departments focused on ensuring energy resource security, promoting high-quality economic development, and prudently advancing carbon peak and carbon neutrality. They adhered to problem-oriented and goal-oriented approaches, successively introduced or revised a series of industry policies and reform suggestions, and accelerated system reforms to adapt to market conditions, emphasized the role of energy investment in stimulating the economy, solidified the primary responsibility for energy conservation and carbon reduction, improved the future-oriented industry standard system, accelerated the international alignment of carbon markets and standards, and propelled the high-quality development of the energy industry.

Deepen the Reform of the Energy System

On July 11, the Opinions on Promoting the Gradual Transition from Dual Control of Energy Consumption to the Dual Control of Carbon Emissions, the Implementation Opinions on Further Deepening the Reform of the Oil and Gas Market System and Improving the National Oil and Gas Security Guarantee Capabilities, and the Guiding Opinions on Deepening the Reform of the Power System and Accelerating the Construction of a New Power System were approved at the Second Meeting of the Commission for Deepening Overall Reform after consideration. These three documents, revolving around the core issues of China's energy development, have milestone significance for further deepening the reform of the energy system.

Prudently Advancing the Gradual Shift from Dual Control of Energy Consumption to Dual Control of Carbon Emissions

The Opinions on Promoting the Gradual Transition from Dual Control of Energy Consumption to the Dual Control of Carbon Emissions focus on "carbon reduction" as a strategic direction and require the improvement of basic capabilities and the construction of supporting systems, to actively create conditions for the establishment and implementation of a dual control system for carbon emissions. The dual control of carbon emissions aims to effectively avoid the limitations of total energy control. While controlling the consumption of fossil fuels, it encourages the development of renewable energy, loosens the constraints of energy consumption on the economy, and steadily promotes green and low-carbon development under the premise of ensuring economic development and energy security. Considering the conditions for dual control of carbon emissions, it is required to maintain a steady work pace, coordinate the relationship between development and emission reduction, be pragmatic and act within our capabilities, appropriately adjust and optimize policy measures, guide and promote energy transition and green low-carbon development across society.

First, promote the full coverage of green power certificates for renewable energy. On August 3, the National Development and Reform Commission, the Ministry of Finance, and the National Energy Administration jointly issued the Notice on Completing the Full Coverage of Renewable Energy Green Power Certificate Work to Promote Renewable Energy Power Consumption, which further clarifies the authority, uniqueness, and universal applicability of green power certificates. Tradable green power certificates will be issued for renewable energy sources such as photovoltaics, wind power, newly operational hydropower, biomass power generation, geothermal power generation, and ocean energy power generation, excluding stock hydropower (produced before January 1, 2023), achieving full coverage of green power certificate for the power generated from renewable energy sources. The document regards green certificates as the sole proof of power users' green power consumption, establishes standards, systems, and identification frameworks for green power consumption certification based on green certificates, and achieves the unity of certificates and power, providing an accounting basis for not including the renewable energy in the dual control of energy consumption. Currently, it can bring a green benefit of RMB 0.03–0.05/kWh to wind power and photovoltaic projects, and can effectively expand the demand for green certificate consumption, stimulating the vitality of the green power consumption market. By the end of October 2023, China had cumulatively reached a green power trading volume of 87.8 billion kWh, and issued 148 million green certificates.

Second, improve the carbon trading market mechanism. After soliciting opinions in the first half of the year, the Ministry of Ecology and Environment and the State Administration for Market Regulation jointly issued the Administration Measures for Voluntary Greenhouse Gas Emission Reduction Trading (Trial Implementation) on October 19. This is a fundamental system ensuring the orderly operation of the national voluntary greenhouse gas emission reduction trading market and is an important indicator of the launch of the voluntary reduction market. On October 24, a notice on four methodologies including Methodology for Voluntary Greenhouse Gas Reduction Projects—Afforestation Carbon Sink was issued on the website of the Ministry of Ecology and Environment, which clarifies the principal basis for the design, implementation, examination and approval, accounting and verification of emission reductions for voluntary greenhouse gas reduction projects in four specific areas: afforestation carbon sink, grid-connected solar power generation, grid-connected offshore wind power generation, and mangrove cultivation. On October 25, the Notice on the Arrangements for the Work of National Greenhouse Gas Voluntary Emission Reduction Trading Market was released on the website of the Ministry of Ecology and Environment, which reveals the possibility of restarting CCER within the year. Since CCER can be used to offset a maximum of 5% of the carbon emission quotas, the national greenhouse gas voluntary emission reduction trading market will complement the national carbon emission rights trading market and together constitute China's comprehensive carbon trading system.

Third, explore the construction of national carbon peak pilots. On November 6, the National Development and Reform Commission issued the National Carbon Peak Pilot Construction Plan, proposing the selection of 100 representative cities and industrial parks nationwide as carbon peak pilots, with the first batch selected from 15 provinces and regions. The document requires that pilot cities accelerate the establishment and improvement of financial, fiscal, investment, pricing policies, and standard systems that favor green development; and innovate carbon emission accounting, assessment, and management mechanisms. Pilot zones should expedite the establishment of carbon emission control-oriented management mechanism. This move can effectively stimulate the initiative and creativity of cities and parks in low-carbon development, providing the whole country with practical and effective experiences and approaches.

Fourth, issue the first top-level design document for methane emissions control. On November 7, the Ministry of Ecology and Environment along with 11 other departments released the Methane Emissions Control Action Plan, proposing that by 2030, efforts should be intensified to build a monitoring, accounting, reporting, and verification system for methane emissions, to advance the control of methane emissions in the energy sector, to promote methane emission control in the agricultural sector, to strengthen methane emission control in waste and sewage treatment, to enhance the coordinated control of pollutants and methane, to boost technological innovation and the supervision of methane emission control, to accelerate the development of a regulatory, standards, and policy framework, and to reinforce global methane management and collaborative initiatives.

Advancing the Market-Oriented Reform of the Oil and Gas Industry in an Orderly Manner

The Implementation Opinions on Further Deepening the Reform of the Oil and Gas Market System and Improving the National Oil and Gas Security Guarantee Capabilities is another important document that systematically deploys the oil and gas system reform, following the Opinions on Deepening the Reform of the Oil and Gas System released in 2017. In response to potential risks in the future, it sets oil and gas security as the primary goal of the policy, prudently advances system and mechanism reform, strengthens effective supervision, improves the production, supply, storage, and sales system, focuses on deepening the reform of the oil and gas reserve system, and gives full play to the emergency response and adjustment capabilities of the reserves.

First, optimize mineral rights management and unleash the vitality of the upstream market. The Ministry of Natural Resources has issued several important documents to deepen the reforms to "streamline the government, delegate power, and improve government services" in respect of mineral resources and accelerated a new round of mineral exploration breakthrough strategic actions. The Notice on Deepening the Reform of Consolidated Examination and Certificates for Land-use Planning released by the Ministry of Natural Resources on May 4 is conducive to streamlining the approval procedures for land use in energy company projects, improving approval efficiency and regulatory service levels, and thereby accelerating energy project construction. On April 15, the Ministry of Natural Resources, in conjunction with the Ministry of Finance and the State Administration of Taxation, issued the Measures for the Collection of Proceeds from the Transfer of Mineral Rights, which clearly stipulates that the proceeds from the transfer of mineral rights should be divided between the central and local governments at a ratio of 4:6 and are included in the general public budget management. The notice also includes regulations on determining the places from which the proceeds should be collected in complex situations, such as those concerning cross-provincial, cross-city and cross-county mineral rights, as well as oil and gas mineral rights. The payment of proceeds from mining rights transfer should be made "by an amount" and "by a specific rate year by year", which changes the past practice of one-time collection of mineral rights transfer proceeds, slows down the collection rhythm, alleviates the financial pressure of some geological exploration enterprises, and encourages enterprises to accelerate the transition to mining production. The Notice on Further Improving the Registration and Management of Prospecting and Mining of Mineral Resources released by the Ministry of Natural Resources on May 6 clearly states that mining right holders can carry out exploration work in the deep and upper parts of the mining area without the need to register for new exploration rights. The transfer of exploration rights obtained through market competition methods such as bidding, auction, and listing is not subject to the restriction of holding exploration rights for 2 years. The restriction period for the transfer of mineral rights obtained by agreement has been shortened from 10 to 5 years. The above measures will further optimize the business environment in the field of mineral resources, simplify the procedures for mining rights application and transfer, and invigorate the secondary market for mining rights. On July 26, the Opinions on Several Matters of Deepening the Reform of Management of Mineral Resources was released by the Ministry of Natural Resources. It extends the reservation period for exploration rights from 2 to 5 years. The deduction base is changed from the area initially specified in the permit to the area specified in the exploration permit at the time of renewal. The deduction ratio is adjusted from 25 to 20%. Moreover, the scope of proven geological reserves of oil and gas that have been submitted is not included in the deduction base. The area of deep or upper exploration for which mining rights have already been granted is not deducted, effectively relieving the pressure of the existing exploration rights being withdrawn and reduced too quickly.

Second, continue to rectify the domestic oil market. On July 3, the Ministry of Finance and the State Administration of Taxation issued the Announcement on the Execution Standards of Some Refined Oil Consumption Tax Policies, incorporating 17 types of products including iso-octane, stabilized light hydrocarbons, and light coal tar into the taxable range of refined oil consumption tax. This measure strengthens tax supervision and further purifies the market environment.

Third, clarify and optimize the order of natural gas utilization. On September 28, the Comprehensive Department of the National Energy Administration issued a notice soliciting public comments on the Natural Gas Utilization Policy (Draft for Comment). The document defines the natural gas usage priority for preferred, allowed, restricted and prohibited projects, proposes that for preferred projects, local governments at all levels are to be supported through policies addressing planning, land use, financing, and fiscal matters, encourages independence in relevant technology and equipment used in natural gas projects, encourages and supports the construction of natural gas refueling facilities and equipment for vehicles and ships. It also specifies restrictions like limiting the construction of base-load gas-fired power projects in large coal bases such as Shaanxi, Inner Mongolia, Shanxi, and Xinjiang, among 14 regions. The document reflects the orientation of determining the demand based on the supply and utilizing natural gas economically and efficiently, and focuses on protecting livelihoods, key areas, and development, to promote a greater role for natural gas in the construction of a new energy system.

Fourth, standardize the pricing mechanism for natural gas. On June 3, the National Development and Reform Commission issued the Guiding Opinions on Establishing and Improving the Price Linkage Mechanism Between Upstream and Downstream Natural Gas Markets, which for the first time proposes to accelerate the establishment of a price linkage mechanism between upstream and downstream natural gas markets, to improve regulatory rules, price adjustment announcements, and information disclosure systems, effectively alleviating cost pressures along the industry chain. Since the issuance of the Opinions, regions such as Hebei, as well as several cities including Nanjing and Qingdao, have adjusted residential gas prices in coordination, with increases ranging from RMB 0.13 to RMB 0.40 per cubic meter. The Opinions help ensure that end-user gas prices sensitively reflect changes in the market situation, play a regulatory role of prices on supply and demand, and enhance the enthusiasm for ensuring supply by gas providers. On December 5, the National Development and Reform Commission issued the Notice on Ratifying the Prices of Cross-provincial Pipeline Transportation of Natural Gas. For the first time, the price of trans-provincial natural gas pipeline transportation operated by the China Oil and Gas Pipeline Corporation is approved by regions. The numbers transportation rates was significantly reduced from 20 to 4, simplifying the business costs of transactions between users and pipeline operators for transportation services. This is conducive to the interconnection and fair opening of pipeline facilities, promoting the free flow of natural gas resources and market competition.

Deepening the Reform of the Power System and Accelerating the Construction of a New Type of Power System

The Guiding Opinions on Deepening the Reform of the Power System and Accelerating the Construction of a New Power System for the first time elucidate the "five characteristics" of the new type of power system—clean and low-carbon, safe and abundant, economically efficient, coordinated in terms of supply and demand, flexible and intelligent. This marks another milestone in China's power sector reform following the State Council's issuance of the Several Opinions on Further Deepening the Power System Reform, aiming to eliminate the bottlenecks of traditional policy mechanisms, and, based on secure and reliable substitution of new energy, improve the institutional mechanisms that adapt to the new type of power system, promote the combination of effective markets and well-functioning government, encourage technological innovation, market mechanism innovation, and business mode innovation, and ensure the power supply can meet the basic needs of the public.

First, advance the construction of the electricity spot market. On September 18, the National Development and Reform Commission and the National Energy Administration jointly issued an important document for the construction of a unified national electricity market system—Basic Rules for the Electricity Spot Market

(Trial). The aim is to establish a market pricing mechanism under which the price can rise and fall, utilize the time-sharing price signals of the electricity spot market, enhance operational flexibility of thermal power units, promote the priority consumption of new energy with lower variable costs, and encourage peak cut, promote the coordinated interaction of energy, grid, load and storage, fully release the overall adjustment capacity of the system, enhance the power supply capacity, and support the construction of a new type of power system. On November 1, the General Office of the National Development and Reform Commission and the Comprehensive Department of the National Energy Administration issued the Notice on Accelerating the Construction of Electricity Spot Market, which proposes to accelerate the construction of regional electricity markets and orderly achieve full coverage of the electricity spot market, while ensuring it facilitates the secure and stable supply of electricity, and to expedite the opening up of various types of power sources to participate in the electricity spot market. Since the launch of the pilot construction of the electricity spot market in 2017, the spot market has been underperforming in promoting the optimal allocation of interprovincial electricity resources, exerting price signals, and facilitating the consumption of renewable energy. The aforementioned document will fully leverage the decisive role of the market in the allocation of electricity resources, better play the role of the government, further stimulate the vitality of business entities in all links, and assist in the construction of a new energy system.

Second, strengthen the capacity to ensure the safe operation of power grids. On September 27, the National Development and Reform Commission, along with five other departments, released the Measures for the Administration of Electricity Demand Side (2023 Edition), and the National Development and Reform Commission and the National Energy Administration issued the Measures for the Administration of Power Load (2023 Edition), clarifying the connotation of power load management, the rights and responsibilities of all parties, and the implementation process, clearly stating that the main methods involve demand response and orderly electricity usage to play a dual role in power load management: ensuring the secure and stable operation of the power grid and maintains a steady order of power supply and consumption and promoting the consumption of renewable energy and enhances energy use efficiency. The two documents will help further deepen the reform of market-oriented electricity, establish more incentivizing electricity energy markets, ancillary service markets, and capacity markets, and fully leverage the synergistic effect of coordination among the government, power grid, and users to effectively safeguard the security of the electricity system and the legitimate rights and interests of electricity consumers.

Third, the State establishes a coal power capacity pricing mechanism. On November 10, the National Development and Reform Commission and the National Energy Administration jointly issued the Notice on Establishing a Coal Power Capacity Pricing Mechanism, deciding to establish a coal power capacity pricing mechanism from January 1, 2024. Coal power capacity prices will be subject to a two-part tariff system and determined based on a proportion of fixed costs of coal power units recovered. The fixed costs are subject to a national unified standard, set at RMB 330 per kilowatt per year. From 2024 to 2025, the proportion of fixed costs recovered through capacity tariffs in most regions will be about 30%; starting from 2026, the proportion of fixed costs recovered through capacity tariffs in various regions will increase to no less than 50%. The establishment of the coal power capacity pricing mechanism and the implementation of a two-part tariff policy for coal power are necessary not only for the rapid development of new energy in recent years in China but also for further acceleration of the development of new energy and the transformation towards green and low-carbon energy in the next step.

Enhance Oil and Gas Supply Security Capacity

Various ministries and commissions thoroughly implement the guiding principles of the report to the 20th National Congress of the Communist Party of China regarding the security of food, energy resources, and key industrial and supply chains, and work to enhance the resilience of the oil and gas industry's supply chain.

Intensify Exploration and Development Efforts to Enhance Oil and Gas Supply Capacity

In January, the State-owned Assets Supervision and Administration Commission of the State Council issued the Notice on Investment Management of Central Enterprises in 2023 and Further Expansion of Effective Investment. It requires increased investment in food, energy, and strategic mineral resources, strengthened efforts on domestic exploration, development and increase in reserves and production of important energy and mineral resources, promotion of the construction of overseas energy resource supply bases, and investment in capacity reserves in key links of the industrial chain in areas where conditions permit.

On April 12, the National Energy Administration issued the 2023 Energy Work Guidance, which requires to put energy supply and price stability in the first place and strengthen efforts on domestic exploration, development and increase in reserves and production of important energy and mineral resources, and further consolidate the guarantee foundation of fossil energy, to ensure a steady increase in the self-sufficiency rate of energy. On February 27, the National Energy Administration issued the Action Plan for Accelerating the Integrated Development of Oil and Gas Exploration and Development with New Energy (2023–2025), aiming to reduce oil and gas consumption in the production process by using clean alternatives. It requires by 2025 an accumulated increase in the supply of natural gas commodities by about 4.5 billion cubic meters, an accumulated increase in natural gas production by about 3 billion cubic meters, and an accumulated increase in crude oil production of more than 2 million tons. The fundamental goal is to enhance the clean substitution of self-use energy and enhance the supply capacity of oil and gas.

Promote Investment in the Refining and Chemical Industry and Its Transformation and Upgrade

In September 2023, during his inspection at Sinopec Jiujiang Branch, General Secretary Xi Jinping pointed out, "The petrochemical industry is an important pillar of the national economy. We must adhere to green and intelligent development, be down-to-earth and enterprising, and make new contributions to ensure national energy security and promote high-quality development of the petrochemical industry." This provides a fundamental guideline for the transformation and upgrading of the refining and chemical industry.

On August 25, Seven ministries including the Ministry of Industry and Information Technology released the Work Program for Stabilizing Growth in the Petrochemical and Chemical Industries, pointing out that from 2023 to 2024, the annual growth rate of the industrial added value of the petrochemical and chemical industries should be around 5%. Focusing on such five aspects as expanding effective investment, enriching high-quality supply, stabilizing foreign investment and trade, strengthening factor supply, and stimulating enterprise vitality, it proposes 11 key tasks, including promoting the commencement of a batch of projects that "recude oil consumption and increase chemicals use", encouraging petrochemical enterprises to implement comprehensive technical renovations of old facilities, guiding export enterprises to enhance overseas warehousing capabilities and delivery levels, supporting leading enterprises to become "chain masters", creating sources and centers of innovation and technological research and development, and accelerating to cultivate a batch of world-class comprehensive chemical enterprises and integrated energy enterprises.

On October 12, the National Development and Reform Commission and three other departments jointly issued the Guidelines on Promoting the Green, Innovative, and High-quality Development of the Refining Industry, which stipulates that by 2025, the domestic crude oil primary processing capacity should be controlled to within one billion tons, with the proportion of refining capacity at the ten-million-ton level being around 55%. By 2030, the capacity structure and productivity layout should be further optimized, significantly enhancing the security capabilities for chemical raw materials and special products. Technologies such as green hydrogen refining and CCUS are expected to complete industrial and large-scale demonstration validation. The Guidelines have also deployed 17 measures in four areas: promoting industrial upgrading, advancing the efficient use of energy and resources, accelerating green and low-carbon development, and strengthening the leadership of scientific and technological innovation. In 2022, the annual revenue of China's refining industry accounted for approximately 31% of the petroleum and chemical industry and 4% of the national industrial enterprises above the designated size. The Guidelines will guide the transformation and upgrading, optimization and reorganization, energy saving and emission reduction, and technological innovation of refining enterprises, and promote the refining industry towards green innovation and high-quality development.

On December 28, the Ministry of Industry and Information Technology and seven other departments issued the Guideline on Accelerating the Transformation and

Upgrade of Traditional Manufacturing Industries, calling for the improvement of industrial energy-saving management systems and the advancement of energy-saving and carbon reduction technological transformation. Improve the cross-regional capacity replacement mechanism, provide policy support for capacity replacement ratios to technology renovation projects with high energy efficiency and low carbon emissions; carry out synergistic innovation initiatives for pollution reduction and carbon reduction, and pilot demonstration projects for carbon capture, sequestration, and comprehensive utilization; orderly advance the substitution of coal with reduced consumption in key industries, guide the reasonable growth in industrial gas use, enhance the level of electrification in industrial end-use energy; promote the coupled development of petrochemical, steel, non-ferrous metals, building materials, and power industries, popularize models such as combined steel production, integrated refining and chemical processing, and collaborative resource utilization, and drive inter-industry connectivity, mutual supply and demand, and interlinking of production facilities to achieve cascading utilization of energy resources and cyclic industry linkage. In the future, the oil and gas industry needs to strengthen cross-sector collaboration with upstream and downstream enterprises in the industry chain, establish alliances for the chemical and CCUS industrial chains, and promote the circular utilization of resources and low-carbon development.

Strengthen Energy Conservation and Consumption Reduction at the Source

The 2023 Government Work Report requires "promoting the green transformation of development methods, accelerating the construction of a new energy system, increasing the proportion of renewable energy, and focusing on controlling the consumption of fossil energy". National ministries and commissions have issued a series of supporting policies that distribute the pressure of energy conservation and emission reduction to specific enterprises, projects, and equipment.

Clarify the Energy-Saving and Carbon-Reduction Standards for Enterprises

On March 17, the National Development and Reform Commission and the State Administration for Market Regulation issued the Notice on Further Strengthening the Updating, Upgrading, Application and Implementation of Energy Conservation Standards. The notice requires energy-consuming enterprises to take primary responsibility, and strictly implement energy-saving standards in aspects such as project energy-saving review, energy efficiency labeling management, energy-saving and low-carbon product certification, and government procurement. It also requires carrying out effective supervision and inspection of the implementation of energy-saving standards through measures such as energy-saving supervision, product quality spot checks, and credit management, promoting energy-saving and carbon-reducing updates and transformations in key industries and product equipment, phasing out approximately 20% of outdated products and production capacities based on energy-saving transformations, improving supporting measures for governmental procurement, taxation, finance and otherwise, and further strengthening positive incentives for enterprises and projects with leading energy efficiency. On July 4, five departments including the National Development and Reform Commission issued the Energy Efficiency Benchmark and Baseline Levels for Industry Key Areas 2023, which mandates that by the end of 2025, 25 key sectors including refineries, ethylene, para-xylene, synthetic ammonia upgrade their energy efficiency to above the baseline level, provides a basis for formulating and implementing policy measures such as the elimination of outdated production capacities, differential electricity pricing and punitive electricity prices, energy saving review for fixed asset investment projects, and fiscal and tax incentives for energy-saving products, and also sets clear targets for enterprises to carry out energy conservation and efficiency enhancement.

Strengthen the Energy-Saving Preview Before Investment

On April 5, the National Development and Reform Commission announced the Measures for the Energy Conservation Preview of Fixed Asset Investment Projects, requiring that construction units of projects invested by enterprises must obtain energy-saving review opinions issued by the energy-saving review authorities before starting construction. The measures effectively link with the improvement of total energy consumption and intensity control policies, highlighting the guidance to control the consumption of fossil energy. They require the provision and application of data on fossil energy and renewable energy consumption, energy use by raw materials, and carbon emissions in the preparation and review of energy-saving reports. These measures play a significant role in improving the energy efficiency level of new projects, curbing the blind development of high-energy, high-emission and low-level projects, and promoting the transformation and upgrade of industrial structure.

Incorporate the Requirements for Energy Conservation and Carbon Reduction into the Feasibility Study Report of Projects

On April 6, the National Development and Reform Commission issued the Outline and Explanation for the Compilation of Feasibility Study Reports on Investment Projects. It stipulates that feasibility reports should include an "analysis of utilization effects of resources and energy" and an "analysis of carbon peak and carbon neutrality", propose comprehensive resource utilization plans and resource conservation measures, reflect the policy orientation of controlling energy consumption from the source and reducing carbon from the design stage. Moreover, the emphasis is on considering the project's full life cycle to coordinate project financing and implementation plans, integrating new requirements such as expanding domestic demand, achieving carbon peak and neutrality, fostering independent innovation, and digitalization into the feasibility study system and norms organically. When enterprises plan new projects and compile feasibility study reports, they must adhere to relevant requirements, calculate the total amount and intensity of resource consumption, account for the total annual carbon emissions and the carbon emission intensity of main products throughout the project's lifecycle, and propose emission control schemes and emission reduction pathways.

Improve the Standard System in the Energy Field

Since the proposal of the dual carbon goals, China has issued multiple documents to advance the establishment of standards in the energy industry. By 2023, a national energy standard establishment policy system has been built, with clear objectives, a comprehensive framework, and consideration of both long-term and short-term goals.

Comprehensively Deploy the Establishment Work for the Carbon Peak and Carbon Neutrality Standard System

Since 2023, China has successively issued the Outline for Building a Powerful Country with Quality, 2023 National Standard Project Guidelines, the Guidelines for the Establishment of a Standard System for Carbon Peak and Carbon Neutrality, and 2023 Energy Industry Standard Project Guidelines. These documents have progressively clarified the project plans for near and medium-term dual carbon standards and built a national dual carbon standard establishment policy system with clear objectives, a comprehensive framework, and consideration of both long-term and short-term goals. Among them, the Guidelines for the Establishment of a Standard System for Carbon Peak and Carbon Neutrality issued in April 2023 by 11 ministries and commissions including the National Development and Reform Commission, the National Energy Administration, the Standardization Administration, and the Ministry of Industry and Information Technology, has established the framework of China's dual carbon standard system, including such four primary sub-systems as basic and general sub-system, carbon emission reduction sub-system, carbon removal sub-system, and market mechanism, which are further subdivided into 15 secondary sub-systems and 63 tertiary sub-systems, drawn the "construction blueprint" for the revision work of China's dual carbon standards over the next three years, aiming to address key questions of carbon emission measurement such as "how to make calculation," "how to ensure accuracy" "how to make reduction" "how to achieve neutralization" and "how to trade" and provide comprehensive standard support for carbon peak and carbon neutrality in critical industries and sectors.

On November 13, such five departments as the National Development and Reform Commission, the Ministry of Industry and Information Technology, the State Administration for Market Regulation, the Ministry of Housing and Urban–Rural Development, and the Ministry of Transport jointly issued the Opinions on Accelerating the Establishment of a Product Carbon Footprint Management System, offering a systematic deployment of key tasks for product carbon footprint management. The Opinions also involve organizing relevant industry associations, leading enterprises, and research institutes to formulate accounting rules and standards for the carbon footprint of key products, and encouraging enterprises to voluntarily carry out product carbon footprint certification based on market principles. It is planned to introduce around 200 key product carbon footprint accounting rules and standards at the national level by 2030.

Accelerate the Advancement of Industrial Standardization

On August 8, to implement the national strategy for developing the hydrogen energy industry and fully leverage standards in regulating and guiding the development of this sector, six departments, including the Standardization Administration and the National Development and Reform Commission, jointly issued the Guidelines for the Construction of the Hydrogen Energy Industry Standard System (2023 Edition). The Guidelines are the first comprehensive national-level guidelines for constructing a hydrogen energy industry chain standard system. They specify the key tasks in domestic and international hydrogen energy standardization work for the next three years, arrange actions for the development of core standards and the enhancement of international standardization, which will be beneficial in raising the safety levels and synergistic effects in all sectors, as well as strengthening the core competitiveness of China's hydrogen energy industry. On October 12, the National Energy Administration issued the Notice on the Issuance of the 2023 Revision Plan for Industry Standards in the Energy Sector and the Foreign Language Translation Plan, proposing 665 items for the standard formulation plan, 455 items for the standard revision plan, and 64 items for the foreign language translation plan in the energy sector for the year 2023. Among them, there are 218 items for the coal industry, 32 items for the photovoltaic and solar-thermal industries, 14 items related to the hydrogen energy industry, and 8 items pertaining to carbon emission reduction. The planned item number increased by 170 compared to 2022, reflecting the high priority China places on the standardization of the energy industry and its acceleration towards alignment with international norms, thereby gaining influence.

Plan in Advance for the Establishment of Future Industrial Standard Systems

On August 22, 4 ministries including the Ministry of Industry and Information Technology jointly issued the Implementation Plan for New Industry Standardization Pilot Project (2023–2035), determining the "8 + 9" key areas of new industry standardization. Among them, the emerging industries focus on eight major areas: new-generation information technology, new energy, new materials, high-end equipment, new energy vehicles, green development and environmental protection, civil aviation, and ship and marine engineering equipment; the future industries focus on nine major areas including bio-manufacturing and new types of energy storage. The goal is, by 2025, to establish more than 2000 new national and industrial standards, cultivate more than 300 advanced group standards, participate in the formulation of more than 300 international standards, build a new enterprise-led, government-guided, open and integrated industrial standardization system, and accelerate the improvement of the standardization and internationalization levels of new industries. Energy companies need to focus on improving mandatory national standards for energy security, strategic mineral resource security, and energy consumption limits, accelerate the research and development of national standards for carbon monitoring, accounting, verification, and key basic materials, and plan ahead for new industry standards while strengthening the alignment with international standards and quickly filling in the relevant gaps.

Outlook for 2024

In 2024, relevant government departments will continue to implement the new strategy for energy security, deeply embody the spirit of the 20th National Congress of the Communist Party of China and the Central Economic Work Conference, promote stability through progress, focus on economic construction and high-quality development, coordinate development and emissions reduction, aiming to establish a new energy system. First, we will leverage economic stimulus through energy investment, expand effective investment and moderately preemptive investment, promote the transformation and upgrading of traditional high-energy-consuming industries towards intelligence, green development, and high-end sectors to achieve the goals of the "14th Five-Year Plan steadily"; second, we will perfect the construction of the three major markets for carbon trading, green certificates, and electricity, as well as their collaborative mechanisms, to facilitate the market-based realization of ecological value and accelerate the alignment of carbon-related standards with international norms; third, we will enhance the emergency response capability for oil and gas and speed up the construction of major energy bases to ensure national energy security.

1. This consumption includes net gas extraction from gas storage facilities and LNG tanks, corresponding to an apparent consumption of 3949 billion cubic meters, an increase of 259 billion cubic meters, an increase of 6.8%.
2. The corresponding apparent consumption is 4189 billion cubic meters.

International Oil Market Analysis

Review and Outlook of the International Oil Market

In 2023, the global oil demand gradually recovered, "OPEC+" continued to actively control supply, the market supply and demand maintained a tight balance, oil inventories remained low, the impact of geopolitical premiums gradually faded, international oil prices fell year-on-year, but still remained high, with the annual average price of Brent crude oil futures at 82.17 USD/barrel, a year-on-year decrease of 17%.

In 2024, it is expected that the growth rate of global oil demand will slow down, supply growth will be faster than demand growth, the surplus of the oil market fundamentals will increase slightly, and low inventories will still support international oil prices. The baseline scenario predicts that the annual average price of Brent crude oil futures in 2024 will be 75–80 USD/barrel.

International Oil Price Trend in 2023

International oil prices fell year-on-year, showing a trend of low before high, and high before falling (Fig. 1). In 2023, geopolitical premiums gradually faded, and international oil prices fell significantly year-on-year. In the first half of the year, under the background of the Federal Reserve's continuous interest rate hikes and the turmoil in the European and American banking industry, concerns about the macroeconomic situation dominated market sentiment, and international oil prices fluctuated downwards; in the third quarter, US economic data remained resilient, the Federal Reserve's interest rate hike cycle was nearing its end, and macro pressures eased; at the same time, Saudi Arabia and Russia continued to voluntarily cut production until the end of the year, causing a significant tightening of market supply, oil inventories fell to a low level, supporting international oil prices to continue to rise. In the fourth quarter, oil demand entered the off-peak season, the speed of oil

X. Qian et al., *Global Oil and Gas Industry Development Report: 2023*,
https://doi.org/10.1007/978-981-95-2375-7_3

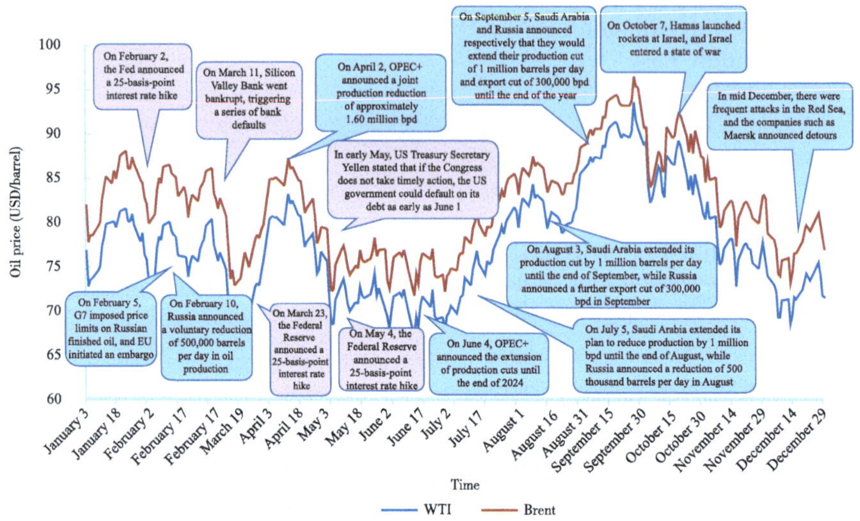

Fig. 1 2023 international oil price trend and important influencing factors

inventory reduction slowed down, and international oil prices fell. Throughout the year, the average prices of WTI and Brent crude oil futures were 77.60 USD/barrel and 82.17 USD/barrel, respectively, a year-on-year decrease of 16.73 USD/barrel (17.74%) and 16.88 USD/barrel (17.04%).

Brent-WTI price spread narrows. In 2023, the Brent-WTI crude oil futures price spread was 4.57 USD/barrel, a year-on-year narrowing of 0.14 USD/barrel (Fig. 2). On the one hand, after the Ukraine crisis, the United States became the biggest beneficiary of the international oil market after the Ukraine crisis and the "OPEC+" production cut, continuously increasing its crude oil export efforts to Europe, Asia and other places. Compared with 2022, the US's seaborne crude oil exports increased by 650,000 barrels/day to 4.02 million barrels/day (19%) in 2023. On the other hand, as WTI crude oil enters the European Brent pricing system, the influence of US-produced oil on the international oil market pricing system is becoming more and more significant.

Russian crude oil and Dubai crude oil price trends are stronger than European and American benchmark oil prices. In 2023, trade flows changed, Russian crude oil prices trended stronger than benchmark oil prices, and exceeded the Western limit price of 60 USD/barrel for Russian crude oil. Especially supported by the procurement demand of countries such as India, the price of Urals crude oil significantly strengthened, the Urals-Brent crude spot price spread averaged − 18.34 USD/barrel, a year-on-year narrowing of 7.88 USD/barrel (Fig. 3); ESPO-Dubai crude spot price spread averaged − 8.27 USD/barrel, a year-on-year narrowing of 6.64 USD/barrel (Fig. 4). The "production cut alliance" led by Saudi Arabia continued its policy of limiting production to protect prices, which supported the price of Middle Eastern

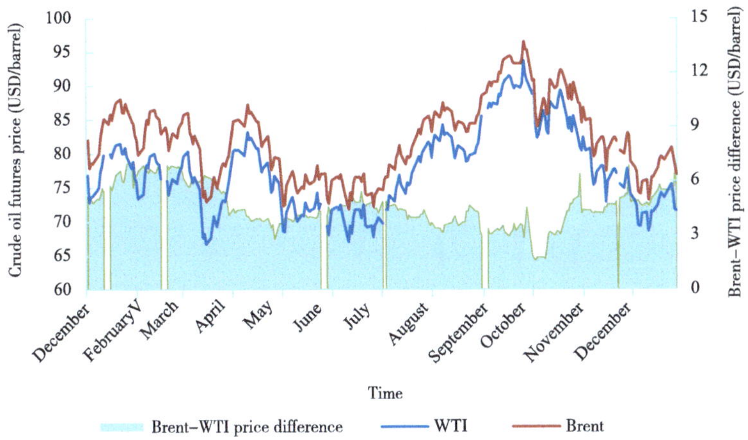

Fig. 2 2023 Brent and WTI crude oil futures price spread

crude oil, and the price trend of Dubai crude oil was also stronger than that of benchmark oil. The Brent-Dubai crude spot price spread averaged 0.51 USD/barrel, a year-on-year narrowing of 3.84 USD/barrel (Fig. 5).

The spread between Shanghai crude oil and international benchmark oil prices generally widened. In 2023, the average price of Shanghai crude oil futures was 582.39 yuan/barrel (82.58 USD/barrel), a year-on-year decrease of 70.87 yuan/barrel (14.65 USD/barrel) (Fig. 6). The average price spread between Shanghai and Brent crude oil futures was 0.41 USD/barrel, compared to − 1.81 USD/barrel in 2022. The trend of domestic prices being stronger than foreign prices was mainly driven by

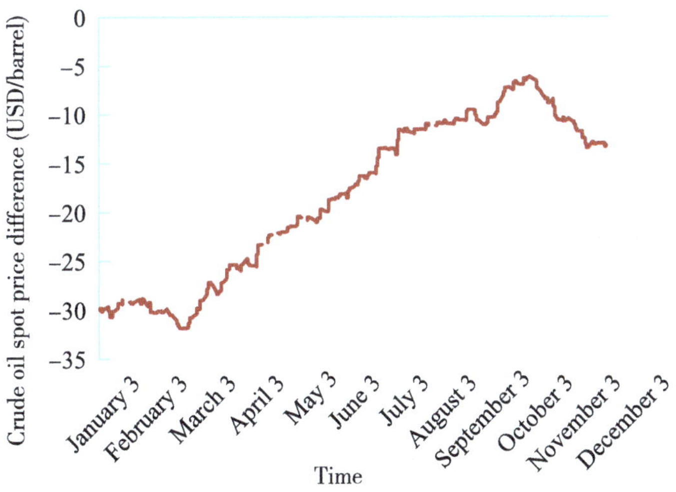

Fig. 3 2023 Urals-Brent crude spot price spread

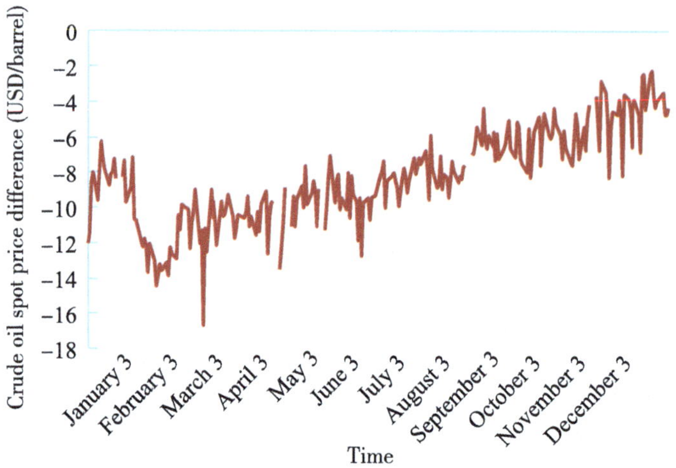

Fig. 4 2023 ESPO-Dubai crude spot price spread

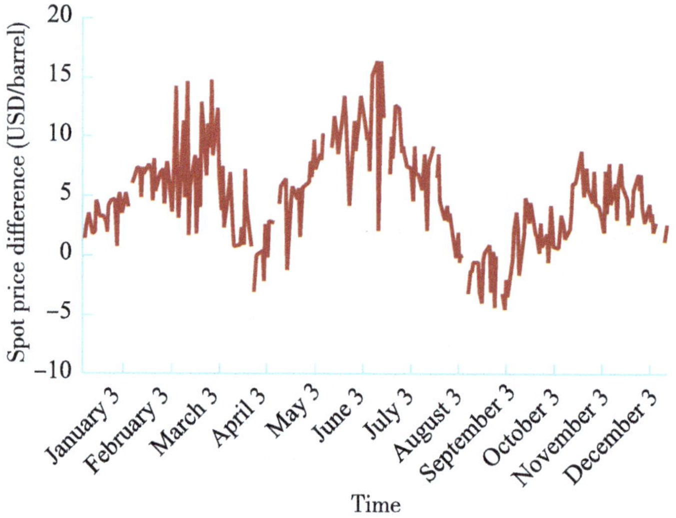

Fig. 5 2023 Brent-Dubai crude spot price spread

factors such as the depreciation of the RMB exchange rate and the boost in domestic demand.

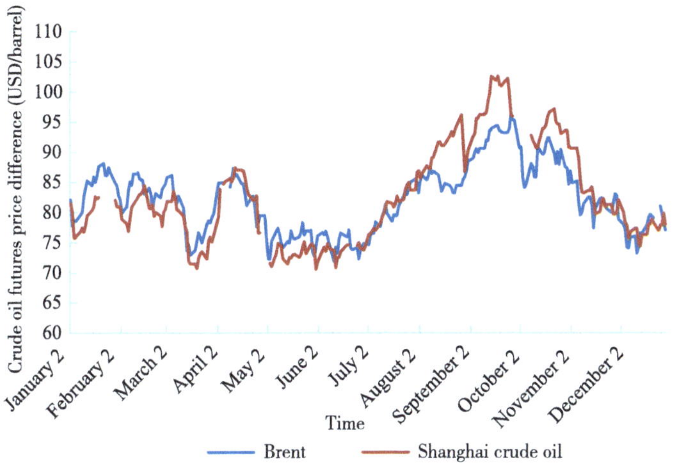

Fig. 6 2023 Brent and Shanghai crude oil futures prices

Analysis of the Characteristics of the International Oil Market in 2023

The Federal Reserve's Interest Rate Hike Cycle Ends, Macro Concerns Gradually Ease

The Federal Reserve began its current round of interest rate hikes in March 2022, raising rates 11 times for a total increase of 525 basis points. In the first half of 2023, the Fed raised rates by 25 basis points in January, March, and May. The stability of the global financial system was tested in a high-interest-rate environment, with Silicon Valley Bank and Credit Suisse among those declaring bankruptcy in April. In the second half of the year, the inflation situation in the United States improved, and the Fed only raised rates by 25 basis points in July, then paused rate hikes three times in a row. The macro-side pressure on international oil prices significantly eased (Fig. 7).

China's Demand Steadily Recovers, Global Oil Demand Exceeds Pre-COVID-19 Levels

In 2023, developing countries led the growth in global oil demand. China's economy continued to recover, and since the Politburo meeting of the Communist Party of China in July, policies to stabilize growth have been introduced one after another. After the end of the COVID-19 pandemic, the demand for compensatory travel by residents increased, and oil demand steadily recovered, with an annual increase of 650,000 barrels/day. High oil prices suppressed U.S. oil demand, gasoline demand

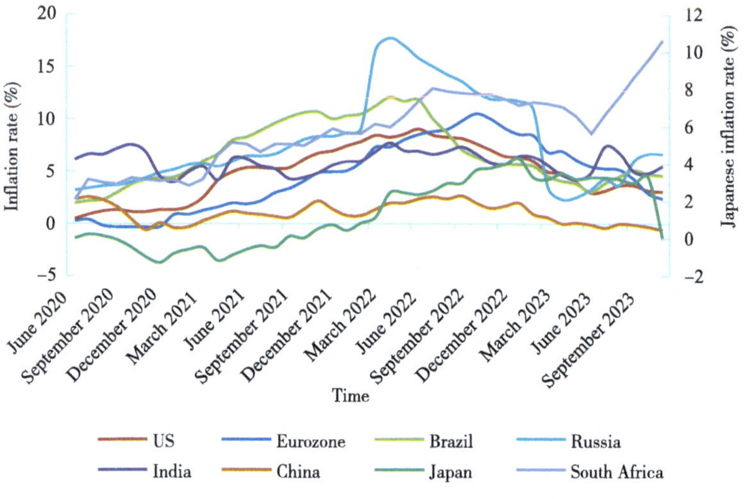

Fig. 7 Inflation situation in major economies

during the driving season was below expectations, and winter temperatures were higher than average, resulting in demand below previous years. U.S. oil demand in 2023 only slightly increased by 100,000 barrels/day compared to the previous year. Global oil demand for the year was 101.7 million barrels/day, recovering to 101% of pre-COVID-19 levels (Fig. 8), an increase of 2.1 million barrels/day year-on-year, with China's demand increase accounting for 30%.

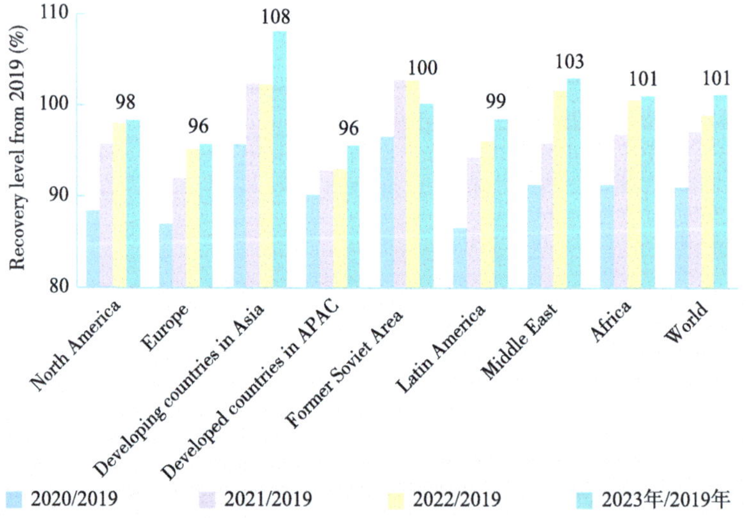

Fig. 8 Recovery of global oil demand by region from 2020 to 2023

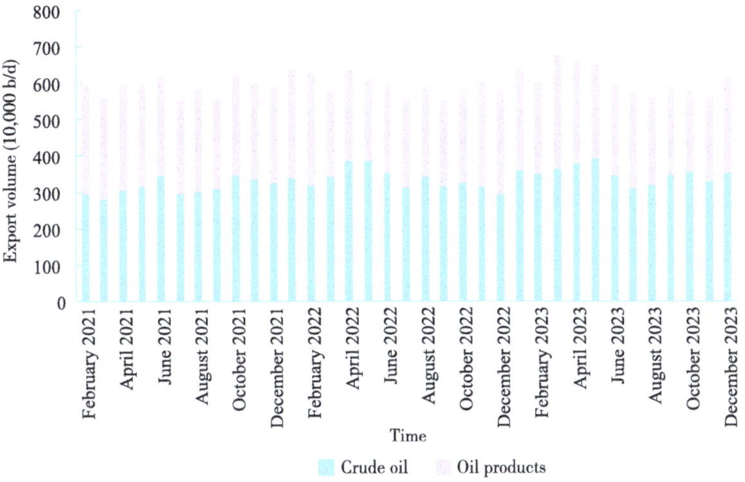

Fig. 9 Russian seaborne crude oil and oil product exports

Russia's Export Direction Changes, Export Volume Increases Instead of Decreasing

Western sanctions did not have a serious impact on Russian oil exports, and Russian oil exports basically achieved a shift from Europe to Asia–Pacific. In 2023, Russia's seaborne crude oil exports increased by 150,000 barrels/day year-on-year to 3.49 million barrels/day, an increase of 150,000 barrels/day year-on-year, 320,000 barrels/day higher than before the Ukraine crisis in February of the previous year (Fig. 9). Among them, exports to China and India were 1.24 million barrels/day and 1.75 million barrels/day, respectively, an increase of 240,000 barrels/day and 920,000 barrels/day year-on-year. Starting in the second quarter, Russian oil exports showed a downward trend, mainly due to the country's voluntary production cuts, restrictions on refined oil exports, and refinery maintenance.

"OPEC+" Continues to Cut Production to Protect Prices, Saudi Arabia and Russia Strengthen Consistency of Actions

"OPEC+" actively implemented multiple rounds of production cuts throughout the year, especially Saudi Arabia and Russia, which started voluntary production cuts or export reductions in the second half of the year, providing significant support for oil prices. Specifically, on April 2, nine members of "OPEC+" announced an unexpected voluntary collective production cut of 1.66 million barrels/day, effective from May until the end of 2023; on June 4, Saudi Arabia announced an additional voluntary production cut of 1 million barrels/day after the 35th ministerial meeting of "OPEC+", which was extended several times and announced in September to be extended until the end of the year, while other member countries extended their existing production cut plans until the end of 2024; Russia voluntarily cut production by 500,000 barrels/day until the end of the year in March, and cut exports by 500,000 barrels/day in August, changed to a cut of 300,000 barrels/day in September and extended until the end of 2023; on November 30, after the 36th ministerial meeting of "OPEC+", some member countries announced a voluntary production cut decision of up to 2.193 million barrels/day.

Non-"OPEC+" Production Cut Countries "Hitch a Ride", Iran, U.S. and Others Significantly Increase Production

With "OPEC+" continuously cutting production, some oil-producing countries increased production to grab market share. Iran's crude oil production rose to a five-year high, reaching about 3.15 million barrels/day at the end of the year, an increase of about 600,000 barrels/day from the end of 2022, and 1 million barrels/day higher than the lowest level since 2018, with remaining production capacity only 600,000 barrels/day (Fig. 10). Despite a continuous decline in the number of drilling rigs in the United States, the efficiency of single wells improved and the number of DUC wells decreased. The annual crude oil production still achieved growth, with a year-on-year increase of more than 1 million barrels/day, to 12.92 million barrels/day (Fig. 11).

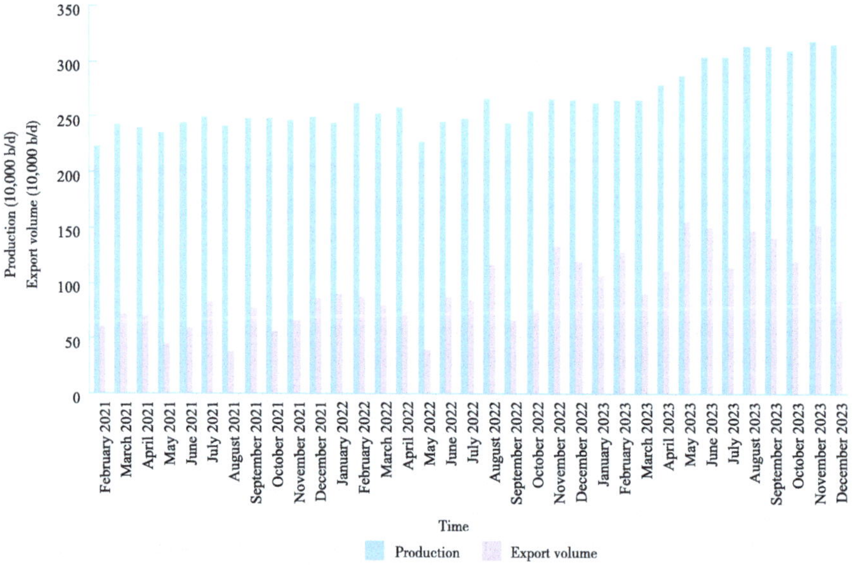

Fig. 10 Iran's crude oil production and export volume from 2021 to 2023

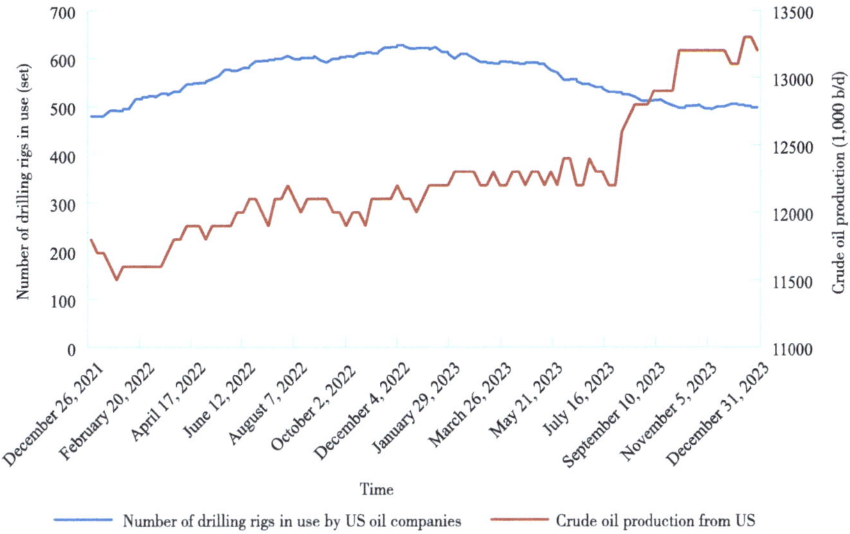

Fig. 11 U.S. crude oil production and number of active drilling rigs

The Oil Market Fundamentals Maintain a Tight Balance, with Inventories Rapidly Declining in the Second Half of the year

With "OPEC+" deepening and extending production cuts, global oil inventories remained low in 2023, and the market fundamentals maintained a tight balance. In the first and second quarters, the global oil market produced more than it needed by 1.75 million barrels/day and 700,000 barrels/day, respectively, while in the third and fourth quarters, it produced less than it needed by 950,000 barrels/day and 100,000 barrels/day, respectively, with an oversupply of 350,000 barrels/day for the year (Fig. 12). Especially in the second half of the year, the oil market entered a rapid inventory reduction phase, with oil inventories at a low level of the five-year average.

Outlook for the International Oil Market in 2024

Economic Growth Still Faces Challenges, but Inflation Issues Are Expected to Ease

In 2024, under the influence of positive factors such as a rebound in investment and increased consumer confidence, the world economy will continue the slow recovery trend of 2023. At the same time, technological innovation and the digital economy will provide new momentum for economic growth. However, the endogenous growth power of the world economy is still insufficient, the pressure of demand contraction still exists, and geopolitical conflicts, adjustments in relations between major powers,

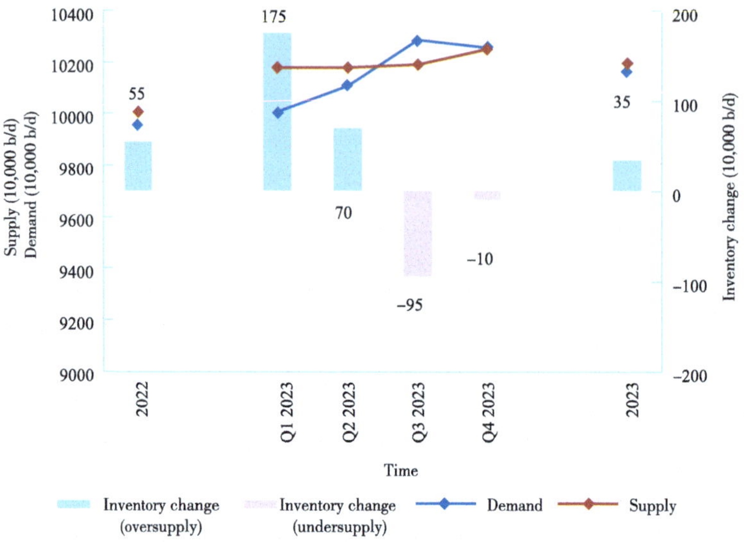

Fig. 12 Balance of supply and demand and inventory changes in the global oil market in 2023

high interest rates, and debt risks continue to restrict economic growth. It is expected that the global economic growth rate in 2024 will be 3.0%, slightly lower than that in 2023 (Table 1).

Table 1 Latest economic growth forecasts by three major institutions

Country or region	OECD		United Nations		World Bank	
	2023 (E)	2024 (E)	2023 (E)	2024 (E)	2023 (E)	2024 (E)
World	2.9	2.7	2.3	2.5	2.1	2.4
China	5.2	4.7	5.3	4.5	5.6	4.6
United States	2.4	1.5	1.1	1.0	1.1	0.8
Euro Area	0.6	0.9	0.9	1.5	0.4	1.3
Japan	1.7	1.0	1.2	1.0	0.8	0.7
India	6.3	6.1	5.8	6.7	6.3	6.4
Developed countries	–	–	1.0	1.2	0.7	1.2
Developing countries	–	–	4.1	4.2	4.0	3.9

The Federal Reserve May Start Cutting Interest Rates in the Second Half of the year, Gradually Reducing the Pressure on Oil Prices from the Financial Side

The Federal Reserve predicts that core inflation will fall below 3% year-on-year by 2024, with the Federal Open Market Committee's forward-looking rate for 2024 at 4.25–4.5%. The Fed is expected to start cutting interest rates in the second half of 2024. Historically, there is no significant pattern in the impact of interest rate cuts on oil prices, but the end of the rate-hiking process means that the pressure on oil prices will ease. After the process of inflation is completed, the Fed's policy is likely to return to a rule-following decision-making mode, and the correlation between monetary policy and oil prices will decrease. The main driver of oil prices will still be changes in demand brought about by economic prospects.

China's Demand Growth Has Significantly Slowed, and the Rapid Recovery Period of Global Demand Has Ended

The period of rapid recovery of global oil demand in the post-pandemic era will end, and the rapid development of electric vehicles will increase the impact on oil demand. It is estimated that global oil demand in 2024 will be 102.7 million barrels/day, an increase of 950,000 barrels/day year-on-year, a decrease of 55% compared to the previous year (Fig. 13). Among them, China's demand growth accounts for nearly 35% of the total global increase. Regionally, demand growth will still mainly come from developing Asian countries, accounting for nearly 50% of the total increase. OECD countries' oil demand has reached a turning point and has fallen by 400,000 barrels/day to 46.7 million barrels/day, still below pre-COVID-19 levels.

"OPEC+" Cautiously Increases Production, Non-OPEC Oil-Producing Countries' Production May Significantly Increase

At the end of 2023, "OPEC+" reached a non-binding voluntary production reduction agreement with difficulty, and Angola announced that it will withdraw from OPEC. The unity and cohesion of this organization will face certain tests in 2024. Saudi Arabia and "OPEC+" are expected to gradually exit production cuts to maximize oil revenue; non-OPEC countries led by the United States will increase oil production by 1.2 million barrels/day, of which U.S. crude oil production will increase by 200,000 barrels/day, and oil production will increase by more than 400,000 barrels/day (Fig. 14). Canada, Brazil, and Guyana's oil production will all increase by about 200,000 barrels/day, and global oil supply will increase by 1.3 million barrels/day.

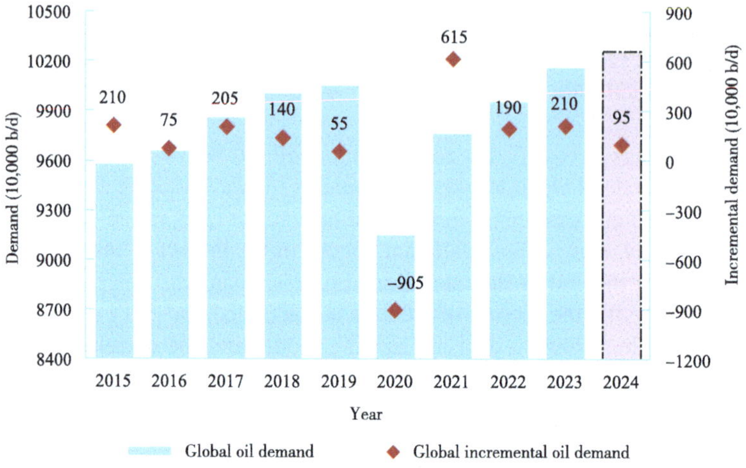

Fig. 13 Increase in global oil demand from 2015 to 2024

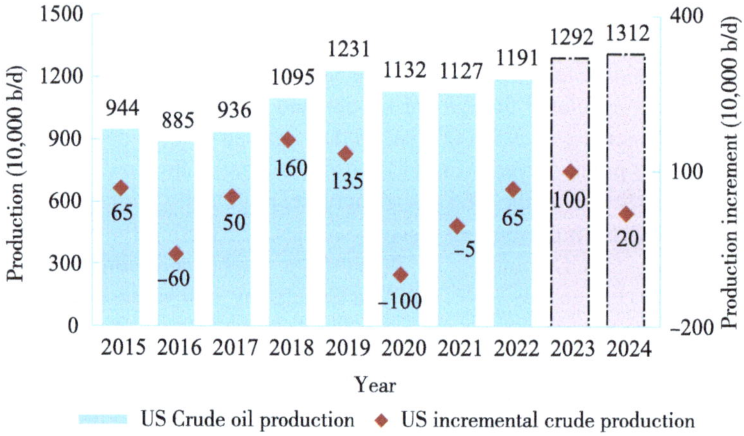

Fig. 14 U.S. crude oil production and forecast

The Oversupply of the World Oil Market Expands, and Oil Inventories Remain at a Low Level

Under the premise of "OPEC+" cautiously increasing production, it is expected that the global oil market will maintain an oversupply situation in 2024, and the degree of oversupply will intensify. The four quarters are oversupplied by 200,000 barrels/day, 800,000 barrels/day, 200,000 barrels/day, and 500,000 barrels/day, respectively, with an average annual oversupply of 500,000 barrels/day (Fig. 15). Although the level of oil inventory will rise, it will generally remain at a low level of the 5-year

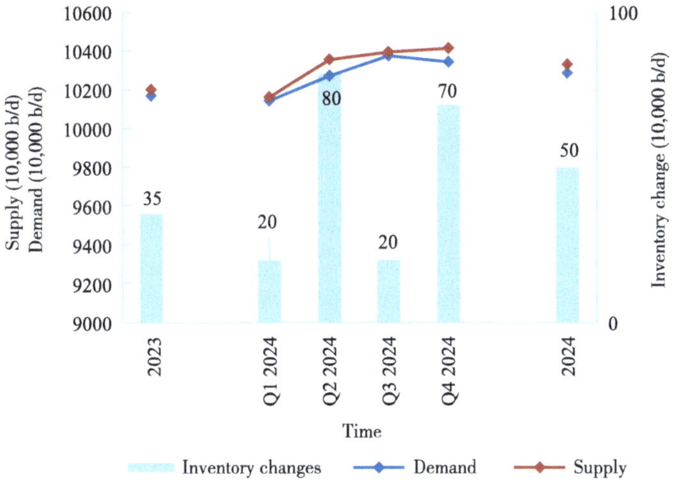

Fig. 15 Balance of supply and demand and inventory changes in the global oil market in 2024

average, and the market's inventory accumulation speed in the second half of the year is slower than in the first half.

The Operating Center of International Oil Prices Moves Down, with Brent's Annual Average Price at 75–80 USD/barrel

Under the baseline scenario, although the inflation problem in Europe and the United States is expected to further ease, the global economy still faces challenges in a high-interest-rate environment; the rapid recovery period of global oil demand after the COVID-19 pandemic ends, and oil demand growth slows down; several major oil-producing countries in the Americas significantly increase production, and oil supply growth will be faster than demand growth; the market fundamentals are more relaxed than in 2023, oil inventory levels rise overall, but still remain at a relatively low level in the past 5 years, providing some support for oil prices; Europe and the United States will start to cut interest rates, and the pressure on oil prices from the financial side is expected to ease, with Brent's annual average price at 75–80 USD/barrel, and the price center has moved down compared to 2023 (Fig. 16). The optimistic scenario predicts that the global economy will recover steadily, relations between major powers will ease, the situation in Russia-Ukraine and Palestine-Israel will further deteriorate, causing a significant reduction in oil supply, and international oil prices still have the possibility of rising again to above 100 USD/barrel. The pessimistic scenario predicts that a major systemic risk event will occur in the financial market, the global economy will slow down significantly, the situation in Russia-Ukraine and the Middle East will gradually ease, and international oil prices still face the possibility of falling below 65 USD/barrel. In addition, the United States will hold elections in 2024, which may become an important uncertainty factor affecting the oil market. Apart from the rise in geopolitical uncertainty, election pressure will usually make the Biden administration strive to control oil prices within the year, putting some pressure on oil prices.

Key focus

The impact of WTI crude oil entering the European Brent pricing system.

West Texas Intermediate (WTI) Midland crude oil produced in the United States became part of the Brent crude oil price benchmark in June 2023. Brent crude oil is one of the important benchmark crude oils in the world. According to "Energy Information" (Energy According to data provided by Intelligence, approximately 78% of global physical crude oil trades are directly or indirectly priced based on Brent crude oil prices. The pricing system for Brent crude oil involves five types of light, low-sulfur crude oil, including Brent and Forties crude oil produced in the UK, Oseberg and Ekofisk crude oil produced in Norway, and Troll crude oil also produced in Norway, commonly referred to as BFOET.

The core reason for the appraisal company to include WTI in the Brent pricing system is that the immediate production of Brent crude oil is gradually declining, while the United States is the world's largest oil producer. WTI is continuously exported to Europe, and including WTI can more comprehensively reflect the global crude oil supply and demand situation. After WTI crude oil joined the Brent price basket and became its main source of supply, the existing crude oil pricing system was impacted, and the United States is expected to dominate the international oil market pricing system. The release of data on U.S. oil supply and demand, imports and exports, will play a leading role in the international oil market, directly and indirectly

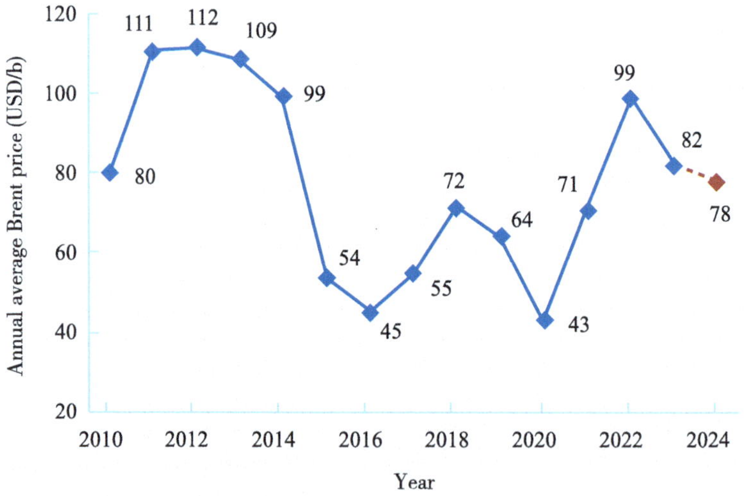

Fig. 16 Brent's annual average price trend from 2010 to 2023 and forecast for 2024

determining the trend of all three major price benchmarks in the international oil market. Not only will Brent crude oil prices be affected by U.S. oil production and exports, but the Dubai/Oman crude oil price benchmark, although located in the Middle East, mainly follows Brent crude oil and will also be affected.

(This section was written by Shi Hongyu and reviewed by Shan Weiguo)

Review and Outlook of the International Oil Market

In 2023, gasoline, diesel, and naphtha supply and demand will rise, regional supply imbalances and "OPEC+" production cuts will significantly reduce the market share of heavy oil, diesel supply will tighten, and inventories will be at a low level, supporting high refining margins. The pattern of refined oil trade flows continues to evolve after the Ukraine crisis. Wholesale and retail prices of refined oil in three places have significantly fallen.

It is expected that in 2024, the global oil product supply and demand will further rebound, the oil product demand of OECD countries will reach a turning point of decline, fuel oil will be the only major oil product with declining demand, and the rise of refining capacity in the Middle East will ease the market supply tension. The market will seek new trade rebalancing under the background of geopolitical conflicts.

Global Oil Product Supply and Demand Situation

Gasoline, Diesel, and Naphtha Supply and Demand Rise, Fuel Oil Demand Declines

In 2023, the global economy further recovers from the COVID-19 pandemic, especially the recovery growth of the Chinese economy, driving the supply and demand of major oil products to rise. The total demand for the year is 87.46 million barrels/day, an increase of 3.2% year-on-year; the total supply is 90.84 million barrels/day, an increase of 3.4% year-on-year, with an oversupply of 3.38 million barrels/day. All major oil products (including gasoline, diesel, jet fuel, naphtha, and fuel oil) are oversupplied. Looking at it by region, North America, Russia-Central Asia, and the Middle East are generally oversupplied, while Asia–Pacific, Europe, Latin America, and Africa have a certain supply gap (Fig. 17).

Gasoline supply and demand both slightly increased. In 2023, the global demand for gasoline (including ethanol gasoline) is 26.22 million barrels/day, a year-on-year increase of 1.2%; the supply is 26.92 million barrels/day, a year-on-year increase of 1.5%, with an oversupply of 700,000 barrels/day. The development of electric vehicles and alternative energy sources puts certain pressure on the demand side of

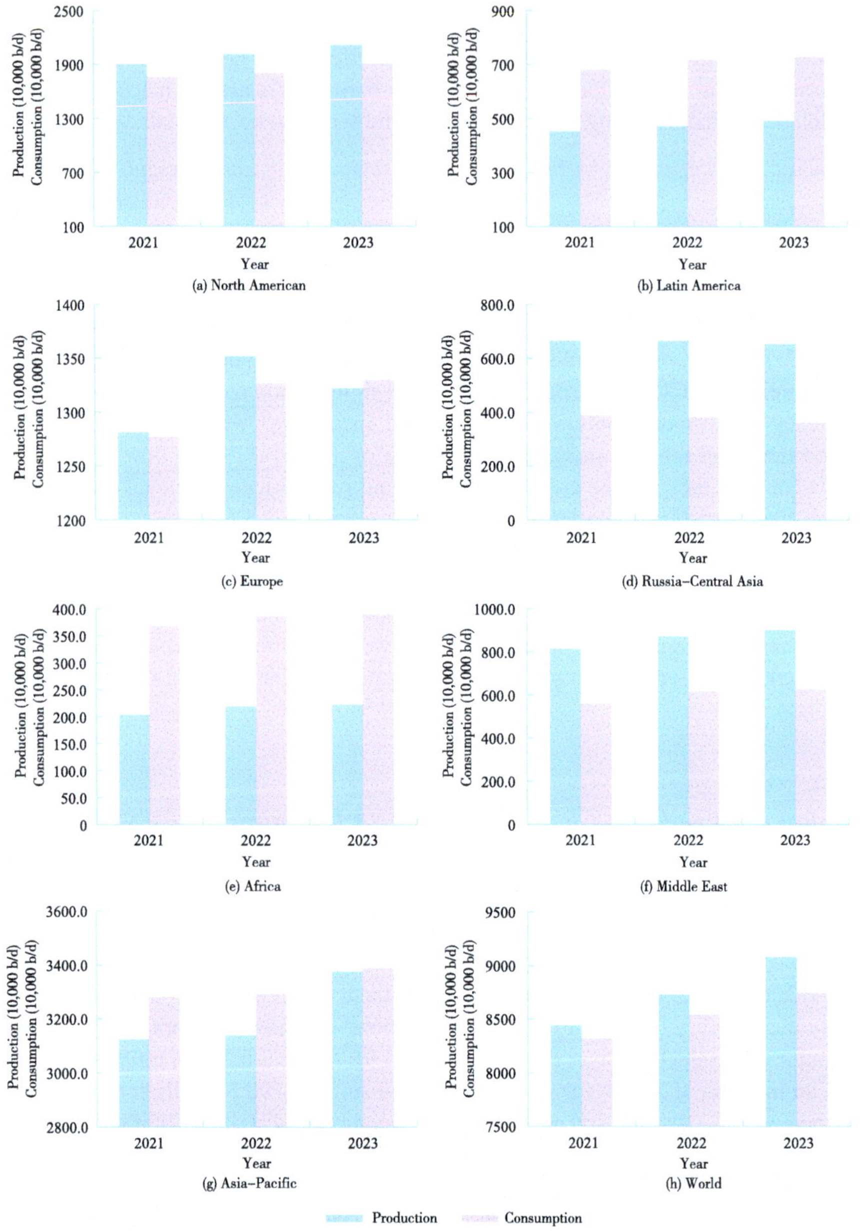

Fig. 17 Balance of oil product supply and demand by region, 2021–2023. *Data source* PetroChina Economic Research Institute, S&P Global

gasoline. U.S. gasoline demand is lower than expected, and the demand for gasoline fell significantly after the end of the driving season, while the refinery operating rate dropped significantly during the same period, resulting in a slight decline in the annual refinery operating rate.

Diesel supply is regionally imbalanced. In 2023, the global demand for diesel is 29.10 million barrels/day, a year-on-year increase of 3.2%; the supply is 30.34 million barrels/day, a year-on-year increase of 3.8%, with an oversupply of 1.24 million barrels/day. After the COVID-19 pandemic, the recovery speed of diesel demand is faster than that of gasoline and jet fuel, and it has risen to 101% of the pre-pandemic level. Since the Ukraine crisis, the trade flow of diesel has been rebalanced, with Russia's exports of diesel to Europe decreasing and the United States' exports of diesel to Europe increasing, resulting in a tight supply of diesel in Europe. European middle distillate oil inventories and U.S. distillate fuel oil inventories remain low.

Jet fuel demand continues to recover. In 2023, the global demand for aviation fuel (including some kerosene and lamp oil) was 7.23 million barrels/day, an increase of 17.8% year-on-year; the supply was 7.33 million barrels/day, an increase of 12.1% year-on-year, with a surplus of 100,000 barrels/day. The annual demand for aviation fuel only recovered to about 92.7% of the level before the COVID-19 pandemic, the United States has already recovered to pre-pandemic levels, and Europe has recovered to more than 90% of pre-pandemic levels. After the Asian market "reopened", the global demand for aviation fuel accelerated its recovery, but the recovery of routes associated with the Asian market was generally lower.

The demand for naphtha has slightly increased. In 2023, the global consumption of naphtha was 7.55 million barrels/day, an increase of 2.3% year-on-year; the supply was 7.94 million barrels/day, an increase of 3.3% year-on-year, with a surplus of 390,000 barrels/day. The global increase in naphtha supply and demand is mainly due to the major trend of the refining industry transitioning from "fuel type" to "chemical products and organic materials type". With the continuous development of the chemical product chain in the refining industry and the increase in naphtha demand, refineries have adjusted their structures to meet market demand, and the yield of naphtha has been improved.

The demand for fuel oil has slightly declined. In 2023, the global consumption of residual fuel oil was 6.37 million barrels/day, a decrease of 1.2% year-on-year; the supply was 6.74 million barrels/day, basically flat year-on-year, with a surplus of 370,000 barrels/day. Shipping, power generation, and refining are the three major consumption areas of fuel oil, but the shipping industry is performing poorly, and the Baltic Dry Index has consistently been below the average level of recent years. The increase in alternative fuels such as methanol and LNG has also led to a decrease in the demand for fuel oil. The Middle East is the region with the most oil-fired power generation in the world (mainly high-sulfur fuel oil), and the seasonal decline in power generation demand also puts pressure on the basic supply and demand of fuel oil (Table 2).

Table 2 Balance of supply and demand for major oil products in the world in 2023

Region	Gasoline		Diesel		Kerosene		Naphtha		Residual fuel oil	
	Supply	Consumption	Supply	Consumption	Supply	Consumption	Supply	Consumption	Supply	Consumption
North America	1005	982	622	446	196	191	46	23	35	35
Europe	378	212	597	701	91	124	95	91	87	88
Asia Pacific	811	807	1030	970	262	242	445	570	199	276
Russia-Central Asia	145	125	213	127	27	36	63	14	90	20
Africa	45	114	96	192	21	23	23	6	36	32
Latin America	139	216	189	285	26	43	20	31	73	42
Middle East	169	166	287	189	110	64	102	20	154	144
Global total	2692	2622	3034	2910	733	723	794	755	674	637

Unit 10,000 barrels/day
Data Source PetroChina Economic Research Institute, S&P Global

There Are Structural Contradictions in the Supply and Demand of Oil Products in the Asia–Pacific Region. Gasoline, Diesel, and Coal Maintain a Surplus, While There Are Supply Gaps for Naphtha and Fuel Oil

In 2023, the total demand for oil products in the Asia–Pacific region will be 34.85 million barrels/day, an increase of 5.8% year-on-year; the total supply of oil products will be 33.76 million barrels/day, an increase of 7.6% year-on-year, with a supply–demand gap of 1.09 million barrels/day. Gasoline, diesel, and aviation coal continue to maintain a surplus, while naphtha and fuel oil still have supply gaps. In the first half of the year, China's demand quickly recovered, and the regional market was in a tight supply situation. In the second half of the year, the weakening macro expectations put pressure on the demand side, which to some extent alleviated the tight supply situation (Figs. 18 and 19).

Gasoline: In 2023, the demand for gasoline in the Asia–Pacific region will be 8.07 million barrels/day, an increase of 8.6% year-on-year; the supply will be 8.12 million barrels/day, an increase of 8.6% year-on-year, with a surplus of 50,000 barrels/day. In the first half of the year, China's demand recovered, supporting the gasoline market. In the second half of the year, China entered the off-season, and the post-epidemic era ended the period of rapid recovery of oil demand, and the overall gasoline market was weak.

Diesel: In 2023, the demand for diesel in the Asia–Pacific region will be 9.68 million barrels/day, an increase of 0.8% year-on-year; the supply will be 10.30 million

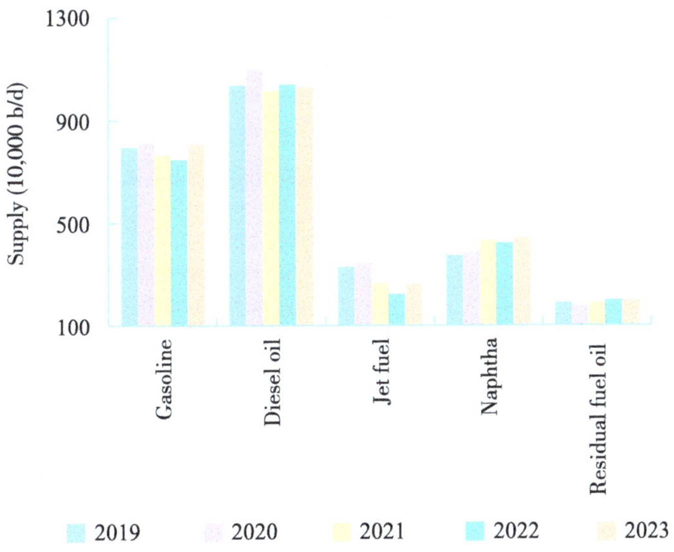

Fig. 18 Supply of major oil products in the Asia–Pacific region from 2019 to 2023. *Data Source* PetroChina Economic Research Institute, S&P Global

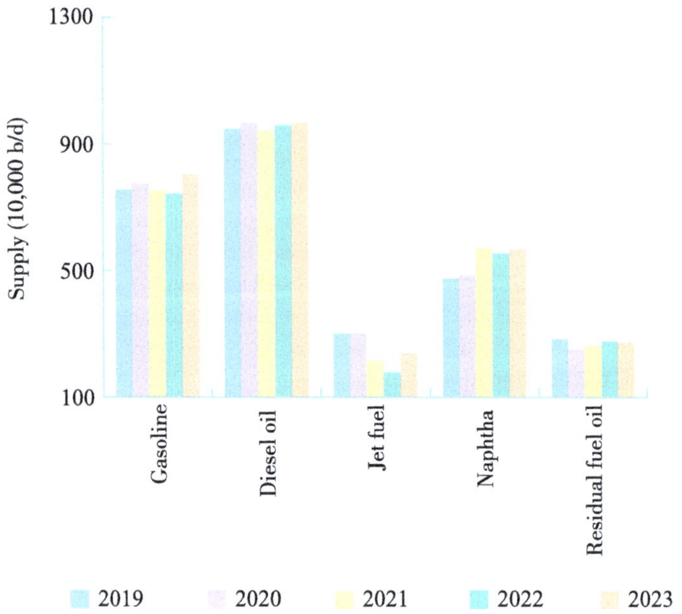

Fig. 19 Demand for major oil products in the Asia–Pacific region from 2019 to 2023. *Data Source* PetroChina Economic Research Institute, S&P Global

barrels/day, a decrease of 0.8% year-on-year, with a surplus of 620,000 barrels/day. Affected by the Western sanctions against Russia, Russia once banned oil product exports, more diesel resources in the region flowed to the Western market, and at the same time, Middle Eastern diesel flowed into Europe and reduced inflow to the region, which to some extent alleviated the situation of diesel supply exceeding demand in Europe.

Aviation Coal: In 2023, the demand for aviation coal in the Asia–Pacific region will be 2.42 million barrels/day, an increase of 34.4% year-on-year; the supply will be 2.63 million barrels/day, an increase of 18.0% year-on-year, with a surplus of 210,000 barrels/day. Driven by the rapid recovery of the aviation industry, aviation coal is the fastest growing oil product. Also affected by the high profit of diesel production, more refineries produce diesel and control the production of aviation coal, causing the supply of aviation coal in the region to tighten.

Naphtha: In 2023, the demand for naphtha in the Asia–Pacific region is 5.68 million barrels/day, an increase of 2.3% year-on-year; the supply is 4.48 million barrels/day, an increase of 6.7% year-on-year, with a gap of 1.2 million barrels/day. Under the pressure of carbon neutrality, the refining industry chain continues to transform from "fuel type" to "chemical products and organic materials type", especially the "oil reduction and chemical increase" in China is significant. Russia's naphtha exports to the Asia–Pacific region have declined, and South Korea, the world's largest naphtha

Table 3 Balance of oil product supply and demand in major countries and regions in the Asia–Pacific region in 2023 (supply–demand)

Country and region	Gasoline	Diesel	Jet fuel	Naphtha	Fuel oil	Total
China	29	26	18	– 20	– 32	21
Taiwan, China	7	16	2	– 16	0	9
Japan	1	6	– 5	– 31	5	– 24
South Korea	23	53	30	– 41	– 1	64
Singapore	18	13	2	– 15	– 55	– 37
India	31	58	16	11	– 6	110
Thailand	2	5	1	– 3	10	15
Indonesia	– 37	– 3	– 2	– 2	8	– 36
Malaysia	– 17	– 6	0	– 2	5	– 8
Pakistan	– 6	– 14	– 4	– 1	– 2	– 27
Philippines	– 12	– 5	0	1	3	– 13
Vietnam	– 4	– 10	– 3	– 1	– 1	– 19
Australia	– 20	– 48	– 14	0	– 1	– 83
Other Asia–Pacific regions	– 10	– 41	– 20	0	– 10	– 81
Total surplus/shortage	5	62	21	– 120	– 77	– 109

Unit 10,000 barrels/day
Data source PetroChina Economic Research Institute, Platts

importer, has significantly reduced its imports from Russia, resulting in a gap in regional naphtha demand.

Fuel Oil: In 2023, the demand for residual fuel oil in the Asia–Pacific region is 2.76 million barrels/day, a decrease of 0.4% year-on-year; the supply is 1.99 million barrels/day, basically the same year-on-year, with a gap of 770,000 barrels/day. Affected by the influx of resources from Russia and the Middle East, the supply and demand of high-sulfur fuel oil is loose; affected by the overall downturn in the shipping industry, the demand for fuel oil has also declined; affected by factors such as the further standardization of imported crude oil quotas, the imposition of consumption taxes on mixed aromatics, light cycle oil and diluted bitumen, and stricter regulatory policies, the demand for straight-run fuel oil imports from local refineries remains stable (Table 3).

Oil Prices in Major Regions of the World

In 2023, international oil prices fell overall, and oil prices in major regions of the world fell in sync. However, due to regional supply imbalances and the closure of some refineries, the cracking spread remained high. Among them, the decline in oil

Table 4 Average oil prices in the three major markets in 2023

Variety	New York Harbor, USA		Rotterdam, Europe		Singapore, Asia	
	Price (US dollars/barrel)	Year-on-year (%)	Price (US dollars/barrel)	Year-on-year (%)	Price (US dollars/barrel)	Year-on-year (%)
Gasoline	124.75	− 15.77	102.79	− 14.56	94.04	− 15.26
Diesel	101.89	− 21.28	111.07	− 21.42	106.41	− 21.47
Aviation fuel	123.02	− 23.53	111.55	− 19.59	104.57	− 17.59

Unit US dollars/barrel
Data source Reuters

prices in the US, Europe, and Asia was similar (Table 4), and retail prices of gasoline and diesel in major regions of the world varied.

The Spot Prices of Gasoline and Diesel in the Three Major Markets Are Basically Consistent with the Trend of International Oil Prices, and They Have Generally Fallen

United States: In 2023, the average annual prices of gasoline and diesel in New York Harbor were 124.75 USD/barrel and 101.89 USD/barrel, respectively, a year-on-year decrease of 15.77% and 21.28%. Since the United States announced a ban on the import of Russian energy, the weekly gasoline inventory in the United States has been below the 5-year average level. During the summer travel peak, record high temperatures in Texas led to a sharp drop in the output of refined oil products by American refiners for several weeks, down 2%, and gasoline prices soared to a new high for the year. After the end of the summer travel peak, due to the continuous decline in international oil prices, gasoline prices in the United States continued to fall, and the average gasoline prices in 11 states, including Alabama, Arkansas, Georgia, Texas, etc., fell below 3 USD/gallon.

Europe: In 2023, the average annual prices of gasoline and diesel in Rotterdam were 102.79 USD/barrel and 111.07 USD/barrel, respectively, a year-on-year decrease of 14.56 and 21.42%. The EU's ban on Russian oil products took effect on February 5, greatly changing the global flow of diesel. Diesel from Asia and the Middle East flowed to Europe, and Middle Eastern refineries became the main suppliers to replace Russian diesel. However, the diesel trade between Russia and Europe did not substantially break away in the short term. Against the backdrop of deepening sanctions, Russian diesel still flowed into Europe through ship-to-ship transfers and transits through third countries.

Asia: In 2023, the average annual prices of gasoline and diesel in Singapore were 94.04 USD/barrel and 106.41 USD/barrel, respectively, a year-on-year decrease of

15.26 and 21.47%. The Ukraine crisis has changed the global flow of refined oil trade. The United States has increased exports of refined oil to Europe, and Asia has been relatively less affected, with refined oil prices changing with international oil prices. Some diesel, due to arbitrage opportunities, flowed to Europe, and the inventory of medium distillates was at a low level.

Retail Prices of Gasoline and Diesel in Some Countries and Regions in the World

In 2023, retail prices of gasoline and diesel in most countries and regions around the world fell. The average retail prices of gasoline and diesel, including taxes, remained basically stable compared to the previous year, with gasoline falling by 1% and diesel by 2% (Table 5).

Different levels of tax burden lead to significant differences in retail prices in different countries and regions. In 2023, the tax share in gasoline prices in mainland China is about 32%, lower than the UK (56%), South Korea (55%), Japan (40%), and higher than Hong Kong, China (30%), Singapore (22%), the United States (15%); the tax share in diesel prices is about 28%, lower than the UK (53%), South Korea (43%), Japan (29%), and higher than the United States (14%), Singapore (7%), Hong Kong, China (6%). The countries with the cheapest retail gasoline and diesel prices in the world are Venezuela, Libya, and Iran, and the most expensive are Hong Kong,

Table 5 Average retail prices of gasoline and diesel in some countries and cities in 2023

Country and city	Product specification		Excluding tax		Including tax	
			Gasoline (Yuan/liter)	Diesel (Yuan/liter)	Gasoline (Yuan/liter)	Diesel (Yuan/liter)
China	92# (National V)	0# (National V)	5.26	5.39	7.75	7.53
Hong Kong, China	Gasoline	Diesel	12.43	15.68	17.87	16.67
South Korea	92	Ultra diesel	4.02	4.78	8.86	8.40
Japan	91	Diesel	5.19	5.47	8.67	7.66
Singapore	95	Ultra diesel	10.99	12.01	14.06	12.85
United States	Regular	Diesel	5.71	6.74	6.72	7.80
United Kingdom	95	Diesel	5.69	6.47	12.92	13.86
Average price	Gasoline	Diesel	7.04	8.08	10.98	10.68

Data source Official websites of retail prices of refined oil in various countries and regions

China, Central Africa, and Iceland. The retail price of refined oil in China is generally at a medium level in the world.

Trade Situation of Refined Oil in Major Countries and Regions in the World

In 2023, the scale of global refined oil trade expanded, and arbitrage activities between regions significantly increased. Looking at the oil products, gasoline mainly flows into Africa, Latin America, and the Middle East; diesel mainly flows from North America, Asia–Pacific, and Russia-Central Asia to Europe, Latin America, etc.; aviation coal mainly flows from Asia–Pacific and the Middle East to Europe, Africa, and South America.

North America Exports to Europe, with an Increase in Export Scale

In 2023, the export volume of refined oil in the United States increased to 6.1 million barrels/day, a year-on-year increase of 3.3%, of which the total volume of sea exports was 2.8 million barrels/day, a year-on-year increase of 12%. In the context of Russia's temporary ban on the export of refined oil, American refiners have increased their exports to Europe. The change in trade flow has led to a significant increase in the demand for long-haul shipping, which supports refining margins. The tight diesel market has also squeezed the production space for gasoline. Due to the United States' cessation of imports of distillate oil from Russia, the domestic refined oil market has further tightened.

European Diesel Supply Eases, Trade Flow Changes Significantly

Europe mainly imports diesel from Russia, the Middle East, and the United States, and excess gasoline needs to be digested through exports. After the Ukraine crisis, the source of diesel supply in Europe has undergone major changes. It has increased the import of diesel from the United States, the Middle East, and Asia–Pacific at a long distance, and increased intra-Atlantic shipping trade, leading to a significant increase in the transportation distance of diesel, which supports the cracking spread. Gasoline is partly flowing to the Asia–Pacific market.

Middle East Oil Product Exports to Asia and Africa Decrease, Export Scale Increases

The Middle East is an important source of supplementing global diesel supplies. In 2023, the export momentum of diesel in countries such as Saudi Arabia, Kuwait, and the UAE has increased, increasing exports to the EU, the Far East, and South Asia, and production capacity has gradually increased. The Middle East is becoming a major supplier to replace Russian diesel.

Intensified Competition in the Asia–Pacific Market, Reduced Imports from the Middle East

In the second half of 2023, with the weakening of macroeconomic expectations and the decline in gasoline demand, China's refined oil export quota is significantly higher than the previous year, and the capacity of Asia–Pacific refineries is increasing rapidly. The global refining capacity center is accelerating its shift to Asia, intensifying competition in the Asia–Pacific market. Due to the existence of regional arbitrage space, some resources flow to Europe, and imports from the Middle East decrease (Tables 6, 7 and 8).

Table 6 Main countries and regions of net gasoline imports and exports in the Asia–Pacific region in 2023

Ranking	Net gasoline exports (10,000 barrels/day)		Net gasoline imports (10,000 barrels/day)	
1	India	31	Indonesia	37
2	China	29	Australia	20
3	South Korea	23	Malaysia	17
4	Singapore	18	Philippines	12
5	Taiwan, China	7	Pakistan	6

Table 7 Main countries and regions of net diesel imports and exports in the Asia–Pacific region in 2023

Ranking	Net diesel exports (10,000 barrels/day)		Net diesel imports (10,000 barrels/day)	
1	India	58	Australia	48
2	South Korea	53	Pakistan	14
3	China	26	Vietnam	10
4	Taiwan, China	16	Philippines	5
5	Singapore	13	Indonesia	3

Table 8 Gasoline and diesel import and export situation of major countries in the Asia–Pacific region in 2023

Main exporting countries	China	On September 1, the third batch of China's refined oil export quotas was officially issued, totaling 12 million tons, a month-on-month increase of 33%, a year-on-year increase of 140%, and 3 million tons of low-sulfur marine fuel oil. In 2023, three batches of export quotas were issued domestically, with a total of 39.99 million tons, an increase of 2.74 million tons from the previous year's total. From January to September, the total export volume of gasoline, diesel, and coal was 31.89 million tons, accounting for nearly 80% of the total quota, and the export capacity at the end of the year was limited
	India	In 2022, the Indian government increased the export tax on petroleum fuels such as gasoline, diesel, and jet fuel. At the same time, the Indian government requires gasoline and diesel exporters to fulfill local sales commitments. Gasoline exporters need to promise that 50% of the bill of lading quantity has been or will be supplied to the domestic market in this fiscal year. For diesel, the requirement is to supply 30% of the quantity to the local market. In 2023, the Indian government announced that it would maintain export restrictions on diesel and gasoline
Main importing countries	Vietnam	The average import tariff on gasoline products has been reduced to 5.62%, and the average import tariffs on diesel and Mazut oil have been reduced to 0.58% and 1.38% respectively
	Australia	In 2023, Australia's imports of gasoline and diesel are estimated to increase by 2%, reaching a record level for the second consecutive year

Outlook for the World Oil Market

The global oil market will maintain a situation of both supply and demand rising. By 2024, the demand for major oil products will exceed the level before the COVID-19 pandemic (Fig. 20). Aviation kerosene will be the main source of demand growth, with a year-on-year increase of 690,000 barrels/day. Demand for gasoline and diesel will increase slightly, with year-on-year increases of 380,000 barrels/day and 180,000 barrels/day respectively; the demand for oil in OECD countries will reach a turning point and decline, with a year-on-year decrease of 400,000 barrels/day, to 46.7 million barrels/day. On the supply side, Asia–Pacific and the Middle East will continue to lead the global refining capacity growth, with the total supply expected to reach 92.27 million barrels/day, a year-on-year increase of 1.6%. Developed countries are actively promoting the transformation of traditional refineries into biodiesel refineries, the market will gradually seek a new balance in oil trade under the shortage of Russian oil, and more refined oil resources in the Middle East will flow out.

Trade flows will reach a rebalance. By 2024, the competition in the European diesel market will attract traditional export markets such as the United States, the Middle

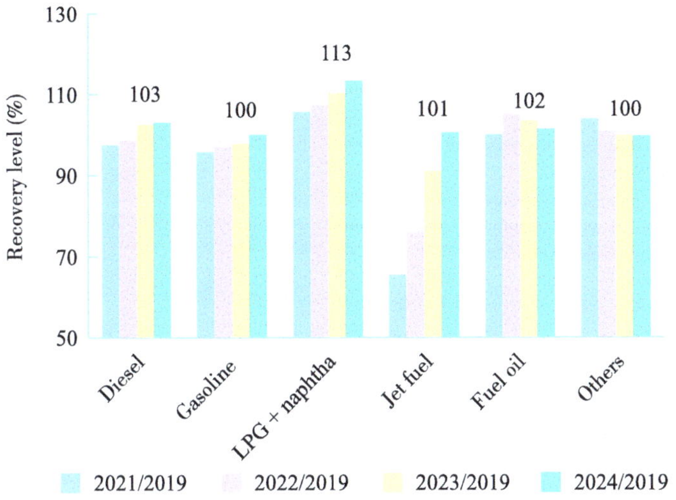

Fig. 20 Global major oil demand and forecast

East, and Asia–Pacific. It is expected that China's refined oil exports will remain stable year-on-year, India's exports to Europe will rise, and Asia–Pacific's refined oil exports are expected to increase. The United States will continue to increase exports to Europe. Africa's imports from India and the Middle East will decrease, and imports from Russia will increase. Refined oil trade flows may achieve a rebalance, and the trend of regionalization and camp formation is emerging.

In the longer term, the global refining center will further shift to the east. New refineries in Asia–Pacific and the Middle East, with their advanced and environmentally friendly technologies, are more competitive than European refineries. The withdrawal of capacity in Europe and America will increase their dependence on imports. After Europe stops importing Russian refined oil, the global refined oil trade distance will further increase. The cracking spread will remain high to support long-distance transportation. Due to its own demand rise, the Middle East will not increase its exports too quickly. The United States remains a long-term important source of supply for Latin America.

(This section is written by: Liu Kailei Reviewer: Shan Weiguo)

Analysis of China's Oil Market

Review of China's Crude Oil and Petroleum Market

The Role of Increasing Reserves and Production Continues to Play, and Crude Oil Production is Stable at More Than 200 Million Tons

Crude oil production has been stable at more than 200 million tons for two consecutive years. In 2023, under the guidance of the clear requirement of the report of the 20th National Congress of the Communist Party of China to "increase the intensity of oil and gas resource exploration and development and increase reserves and production", the whole industry has unswervingly increased the intensity of oil and gas exploration and development and increased reserves and production, accelerated the green and low-carbon transformation, and ensured sustainable development with a high level of storage and extraction balance. The domestic crude oil production has maintained positive growth for five consecutive years, with an annual crude oil production of 209 million tons, a year-on-year increase of 2.1% (Fig. 1). Crude oil imports reached 564 million tons, a year-on-year increase of 11% (Fig. 2).

The Recovery of Production and Living Consumption, Apparent Consumption of Petroleum Sets a Historical Record

China's oil consumption accelerates recovery, and the degree of external dependence rises. The year 2023 is the first year of the normalization of economic development after the transition of China's COVID-19 epidemic prevention and control. The annual GDP achieved a growth of 5.2%. Nationwide, residents' travel and the start of construction of industrial and mining infrastructure have all recovered, driving the

X. Qian et al., *Global Oil and Gas Industry Development Report: 2023*,
https://doi.org/10.1007/978-981-95-2375-7_4

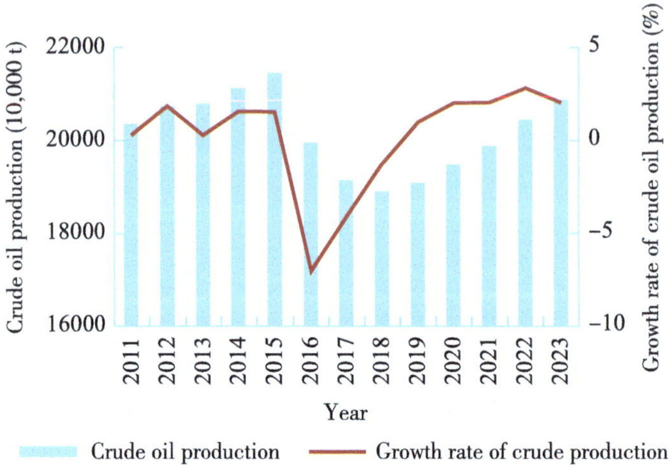

Fig. 1 China's crude oil production and growth rate

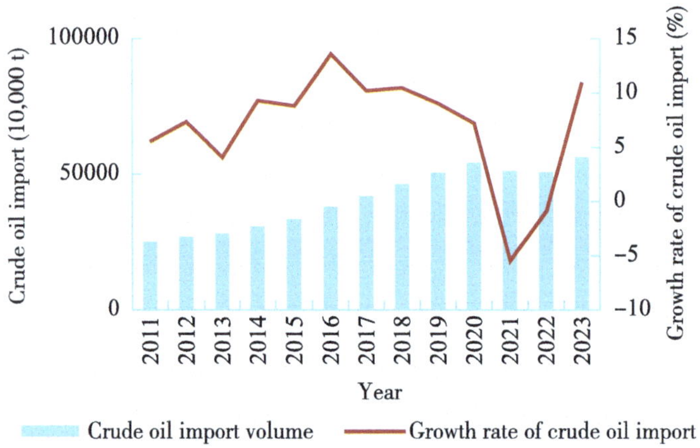

Fig. 2 China's crude oil import volume and growth rate

domestic oil market to warm up quickly. Thanks to the implementation of economic stimulus policies and the rapid recovery of the market after the end of the epidemic, the oil consumption in 2023 reached 756 million tons, an increase of 11.5% year-on-year, setting a record high. All four quarters achieved year-on-year positive growth, with year-on-year growth rates of 4.7%, 18.6%, 11.2%, and 11.9% respectively (Fig. 3). The annual foreign dependence rose to 72.4% (Fig. 4).

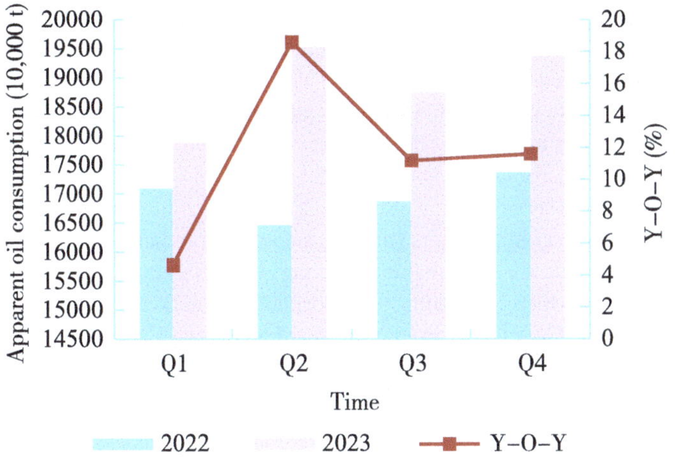

Fig. 3 Apparent consumption and growth rate quarterly oil in 2022 and 2023

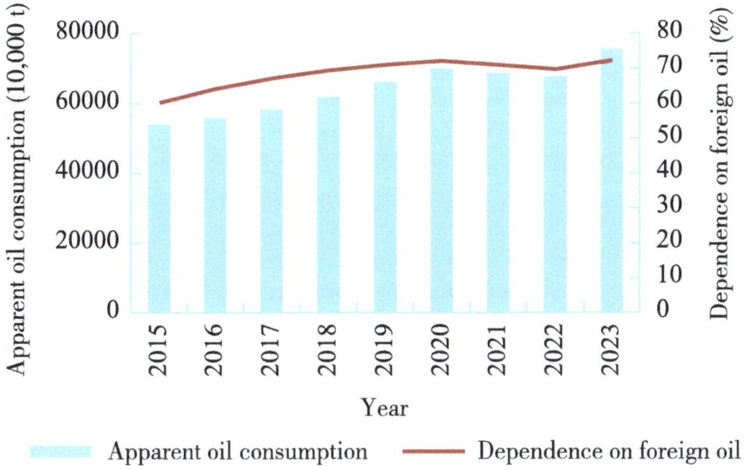

Fig. 4 China's apparent oil consumption and foreign dependence

The Supply and Demand of Refined Oil Recover, and The Market Adjustment Elasticity Increases

The development of alternative energy sources is clearly differentiated. In terms of new energy vehicles, the high-speed growth trend continues. According to data from the China Automobile Industry Association, the sales of new energy vehicles in 2023 reached 9.5 million units, setting a new historical high, with a market penetration rate of 31.6%, a holding volume of 20.41 million units, accounting for 6.1% of the total number of vehicles, replacing about 17 million tons of refined oil. In terms of natural gas, it returns to growth under the economic recovery. With the sharp drop in natural gas prices, natural gas has returned to the advantage range in terms of economy compared to oil products, and the sales of natural gas vehicles have rebounded sharply. It is estimated that the annual holding volume of natural gas vehicles is 6 million units, and the volume of gas used for transportation is 364 billion cubic meters, replacing about 29.65 million tons of refined oil (Table 1). In terms of biodiesel/aviation coal, Sinopec Zhenhai Refining and Chemical's full range of bio-aviation coal products (bio-aviation coal, biodiesel, bio-naphtha) have passed the Roundtable on Sustainable Biomaterials (RSB) three series certification for the first time, further expanding the international market. In terms of ethanol/methanol, after the COVID-19 epidemic, residents' travel showed compensatory growth, gasoline consumption quickly warmed up, and ethanol/methanol, as part of gasoline components, showed a certain increase in consumption.

Domestic refined oil consumption quickly recovered after the COVID-19 pandemic, reaching levels close to those of 2019 throughout the year. In 2023, the consumption of refined oil was 399 million tons, an increase of 9.5% year-on-year, recovering to 98.6% of 2019 (Fig. 5). Gasoline consumption warmed up, but the development of new energy vehicles will reduce gasoline consumption by 4 percentage points. The annual gasoline consumption was 163 million tons, an increase of 9.7% year-on-year, 0.6% lower than 2019 (Fig. 6). Diesel consumption has made steady progress. The oil use in the industrial, logistics, and construction industries showed a "one increase, one flat, one decrease". The annual diesel consumption was 202 million tons, an increase of 3.3% year-on-year, recovering to 98.7% of 2019 (Fig. 7). Kerosene consumption rebounded from the bottom, domestic business travel demand was released, the annual kerosene consumption was 34.75 million tons, an increase of 65.7% year-on-year, recovering to 94.6% of 2019 (Fig. 8).

Crude oil processing volume set a historical record, with production growth exceeding demand. In 2023, China's refining capacity increased to 936 million tons/year, ranking first in the world. The annual crude oil processing volume was 738 million tons, an increase of 8.6% year-on-year, setting a new historical record (Fig. 9). Looking at it monthly, with the peak of the epidemic and the recovery of oil product demand, from March onwards, the crude oil processing volume of domestic refineries rebounded sharply, breaking through 60 million tons, and maintained a high level until the end of the year. In 2023, the annual refined oil production was 444 million tons, an increase of 11.1% year-on-year, slightly lower than the 2019 production. The annual gasoline, diesel, and kerosene production were 175 million tons, 218 million tons, and 50.64 million tons, respectively, an increase of 8.5%, 4.9%, and 68.5% year-on-year, respectively, 3.0%, 2.0%, and 0.5% lower than 2019 (Fig. 10).

Refined oil exports rebounded for the first time after three consecutive years of decline, playing a role in adjustment. In 2023, a total of three batches of refined oil export quotas totaling 40 million tons were issued, with a relatively sufficient quota. The annual export volume of refined oil was 41.9 million tons, an increase of 21.8% year-on-year, ending the trend of continuous decline for three years (Fig. 11).

Table 1 Vehicle fuel substitution situation

Year	Natural gas		Fuel methanol (10,000 tons)		Biodiesel/Aviation coal (10,000 tons)		Coal-to-liquid (10,000 tons)	
	Physical volume (Billion cubic meters)	Substitution volume (10,000 tons)	Physical volume	Substitution volume	Physical volume	Substitution volume	Physical volume	Substitution volume
2020	331	2700	900	405	50	50	520	520
2021	341	2780	950	425	50	50	680	680
2022	327	2665	850	380	50	50	700	700
2023	364	2965	900	405	50	50	700	700

Year	Fuel ethanol (10,000 tons)		New energy vehicles (Electric, hydrogen fuel)		Shared travel (10,000 tons)		Total (10,000 tons)
	Physical volume	Substitution volume	Physical volume (10,000 vehicles)	Substitution volume (10,000 tons)	Physical volume	Substitution volume	
2020	270	162	492	403	0	60	4300
2021	290	174	784	640	0	60	4809
2022	300	180	1310	1070	0	60	5105
2023	300	180	2041	1670	0	60	6030

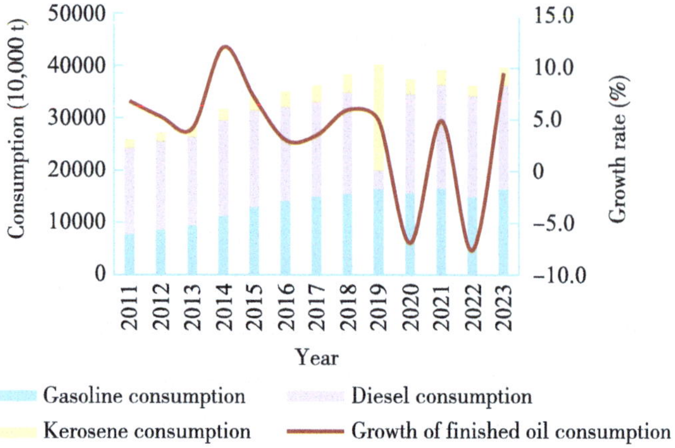

Fig. 5 Refined oil consumption and growth rate from 2011 to 2023

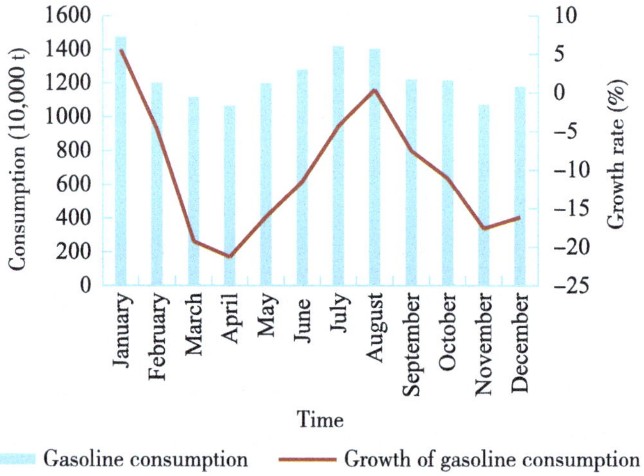

Fig. 6 Monthly gasoline consumption and growth rate in 2023

This effectively ensured that refineries maintained a high load, promoted the enhancement of the elasticity of domestic market regulation, and effectively played the positive role of "exporting instead of storing". From the perspective of supply and demand balance, the annual refined oil slightly increased in storage, and the market supply and demand showed a loose pattern.

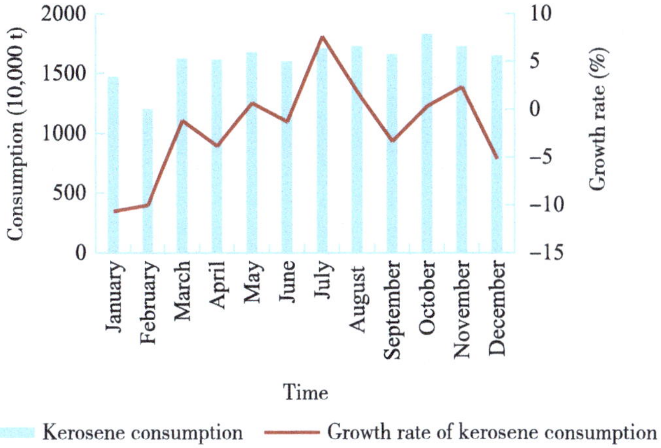

Fig. 7 Monthly diesel consumption and growth rate in 2023

Fig. 8 Monthly kerosene consumption and growth rate in 2023

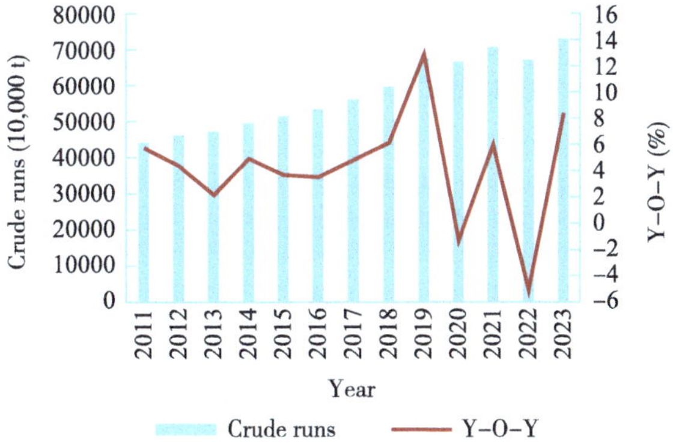

Fig. 9 Changes in crude oil processing volume and year-on-year growth rate

Fig. 10 Refined oil production and year-on-year growth rate

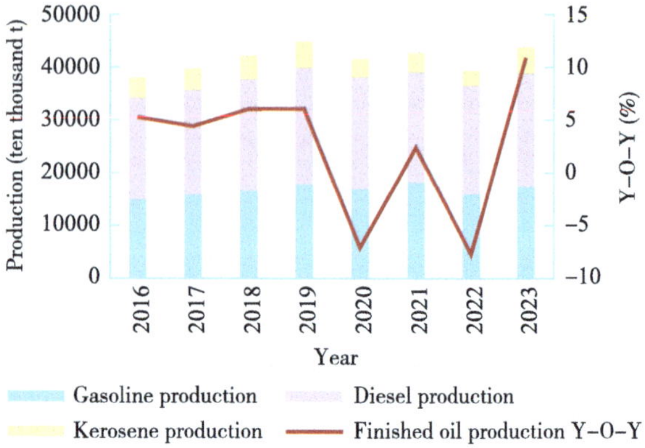

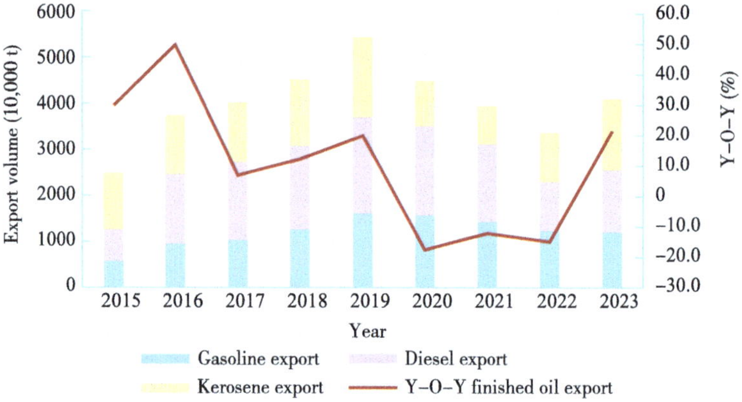

Fig. 11 Refined oil export volume and year-on-year growth rate from 2015 to 2023

Industry Regulation Continues to Strengthen, and The Market Environment is Purified

The scope of oil product taxation has been expanded, and the rectification continues with great intensity. Since 2021, the state has issued a series of industry regulatory policies with increasing intensity, and the standardization and concentration of the domestic refined oil market have significantly improved. In 2023, industry regulatory measures were further implemented, and the intensity of rectification continued to increase. Special rectification plans were implemented in many places to regulate the order of the refined oil market, including expanding the scope of oil consumption tax collection, strengthening the management of refined oil circulation, and comprehensive management of fuel dispensers. The continuous release of policies (Table 2) has led to a gradual decrease in the resources of blended gasoline and diesel, which has played an important role in the regulation and stable development of market order. According to statistics from market research institutions, the proportion of ticketless gasoline and diesel resources in circulation in the market has dropped from 60% (2020) before the market rectification to 40%, and the contradiction between supply and demand of domestic resources has been alleviated to a certain extent.

Off-balance-sheet resources continue to shrink, and resources return to the balance sheet. In 2023, regulatory policies were further strengthened, and off-balance-sheet resources further returned to the balance sheet. In 2023, the off-balance-sheet gasoline and diesel resources of local refineries were 13.75 million tons and 940,000 tons, respectively, a year-on-year decrease of 2.35 million tons (− 14.6%) and 15.83 million tons (− 94.4%). The resources of blended gasoline and diesel using imports as raw materials in 2023 were both 140,000 tons, a year-on-year decrease of 150,000 tons and 390,000 tons, respectively.

The domestic price drop is smaller than the international oil price, the gasoline price arrival rate hits a new high since 2016, and although the diesel price arrival rate has slightly decreased year-on-year, it is still the second highest level since 2015. In 2023, after three years of market rectification, the market has developed towards orderly, standardized, and fair development, and the price level of gasoline and diesel has returned to a reasonable range. According to the monitoring of PetroChina's Economic Research Institute (China National Petroleum Corporation Oil and Gas Market Simulation and Price Forecast Key Laboratory), the average wholesale price arrival rate of 92# gasoline of the main operating unit in 2023 was 90.1% (Fig. 12), an increase of 2.7 percentage points from the previous year (87.4%), 4.9 percentage points higher than the average of the same period in the past five years, the highest since 2016. The average wholesale price arrival rate of 0# diesel was 93.7% (Fig. 13), a decrease of 1.4 percentage points from the previous year (95.1%), 1.6 percentage points higher than the average of the same period in the past five years, the second highest level since 2015.

Table 2 Release of domestic oil market policies in 2023

Release time	Department/Meeting	Policy name
June 12	Ministry of Commerce	"Regulations on the Management of Refined Oil Circulation (Draft for Soliciting Opinions)"
June 30	Ministry of Finance and State Taxation Administration	"Announcement on the Implementation of Some Refined Oil Consumption Tax Policies"
July 11	Second Meeting of the Central Committee for Deepening Overall Reform	Clear opinions on oil and gas market supervision
August 1	State Administration for Market Regulation, Ministry of Public Security, Ministry of Commerce, State Taxation Administration	"Implementation Plan for Comprehensive Management of Fuel Dispenser Fraud"

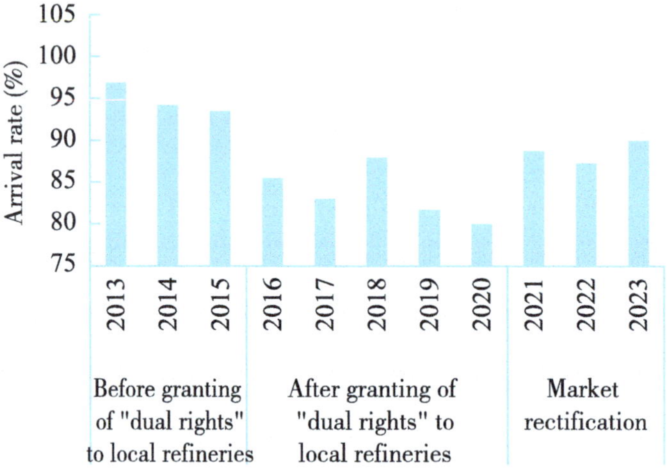

Fig. 12 92# gasoline wholesale price arrival rate

Fig. 13 0# diesel wholesale price arrival rate

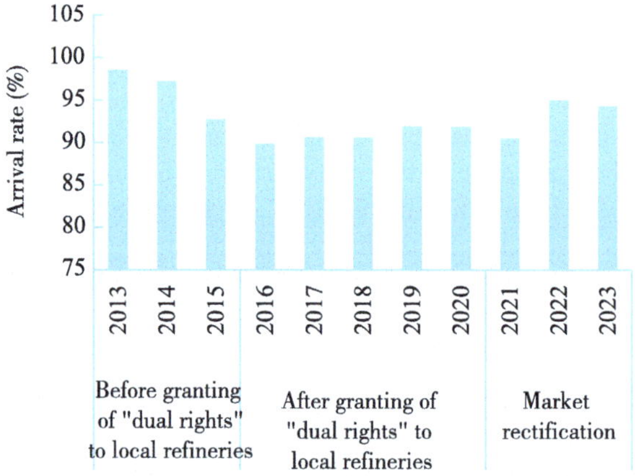

From the perspective of the annual trend, the wholesale price arrival rate of gasoline and diesel shows a trend of first high and then low (Figs. 14 and 15)**.** In terms of gasoline, by 2023, as the comprehensive control of the COVID-19 pandemic is lifted, the frequency of private car travel increases among the public, and driven by holidays such as the Spring Festival, "May Day", "National Day", etc., the radius of vehicle travel increases, driving up gasoline demand. In the first three quarters, the gasoline price realization rate remained above 89%, especially from April to July, the gasoline wholesale price realization rate was higher than the upper limit of the same period in the past five years; in the fourth quarter, as the weather turned cold, gasoline demand weakened and then stabilized. As for diesel, from January to April, the diesel price realization rate reached 96.5%, a historical high for the same period; in May, as the economic recovery slowed down, limiting the growth rate of diesel demand, the price realization rate began to decline; in October, the lack of support from industries such as real estate, logistics transportation, and infrastructure construction dragged down the recovery of diesel demand, and the diesel realization rate showed a downward trend.

(Authors of this section: Li Ran, Hu Yue, Fei Huawei Reviewer: Wang Lining)

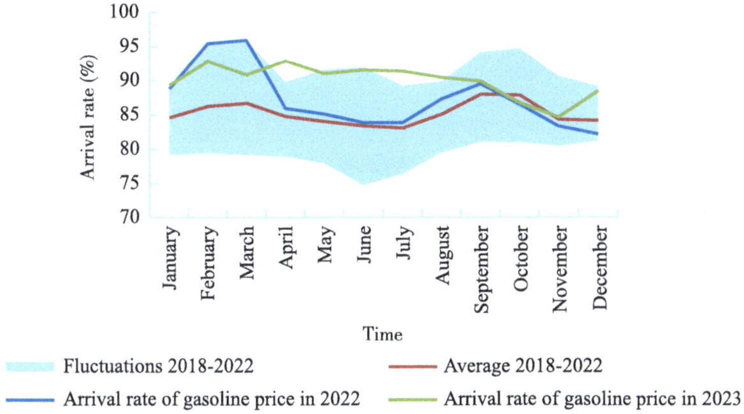

Fig. 14 Fluctuation of 92# gasoline wholesale price realization rate

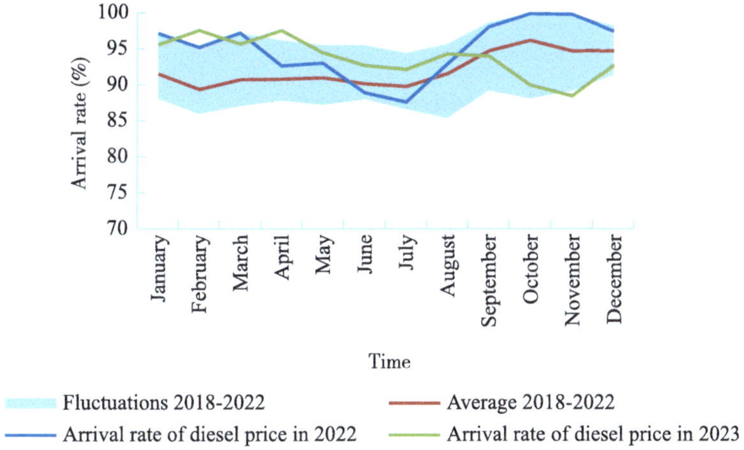

Fig. 15 0# diesel wholesale price realization rate fluctuation

Outlook for China's Petroleum and Refined Oil Market

The Central Economic Work Conference emphasized that "economic construction is the central task, and high-quality development is the primary task". In 2024, domestic macroeconomic policies will solidify the foundation for economic recovery, new productive forces will drive the improvement of total factor productivity and enhance new momentum for economic growth. It is expected that the year-on-year GDP growth rate will reach about 5%, and the consumption of petroleum and oil products will continue to grow.

Outlook for China's Petroleum Market

Crude oil production remains stable, and oil consumption returns to slow growth. In 2024, with the continuous improvement of China's oil production, supply, storage, and sales system, and the implementation of oil exploration, development, and reserve increase strategies, it is expected that crude oil production will still remain at the level of 200 million tons. In 2024, after the rapid recovery period of the COVID-19 pandemic, coupled with the accelerated substitution of oil for transportation by electric vehicles, it is expected that China's oil demand will enter a slow growth stage, and the annual apparent consumption of oil will reach 764 million tons, a year-on-year increase of 1.0%.

Outlook for China's Refined Oil Market

The growth rate of refined oil consumption will significantly fall back, and the total amount is close to peaking (Fig. 16)**.** In 2024, it is expected that domestic refined oil demand will reach 400 million tons, a year-on-year increase of 0.3%. **Gasoline** consumption continues to recover (Fig. 17), but the amount of new energy substitution will further increase. It is expected that the sales volume of new energy vehicles in 2024 will reach 12 million, a year-on-year increase of nearly 30%; the substitution volume for refined oil will rise to more than 20 million tons. It is comprehensively expected that the annual gasoline consumption in 2024 will be 165 million tons, a year-on-year increase of 1.3%; **Diesel** consumption will still be dragged down by the real estate industry, coupled with the continuous transformation and upgrading of the economic structure, and LNG still has a large-scale substitution for diesel consumption. It is expected that the annual consumption in 2024 will be 196 million tons, a year-on-year decrease of 2.8% (Fig. 18); **Kerosene** consumption benefits from the recovery of the tourism industry. It is expected that the annual consumption in 2024 will be 39.3 million tons, a year-on-year increase of 13.1% (Fig. 19).

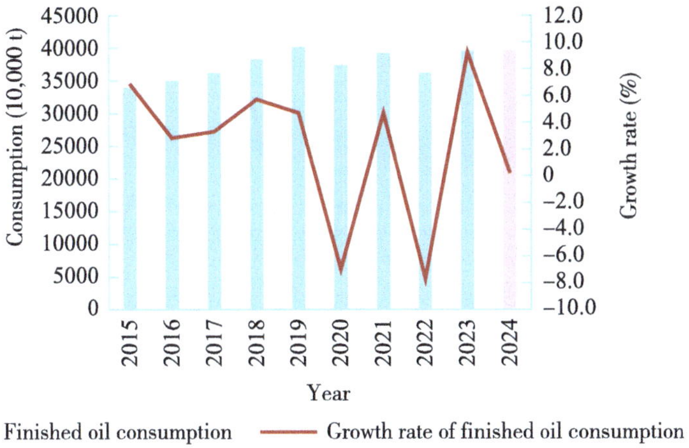

Fig. 16 Forecast of refined oil consumption in 2024

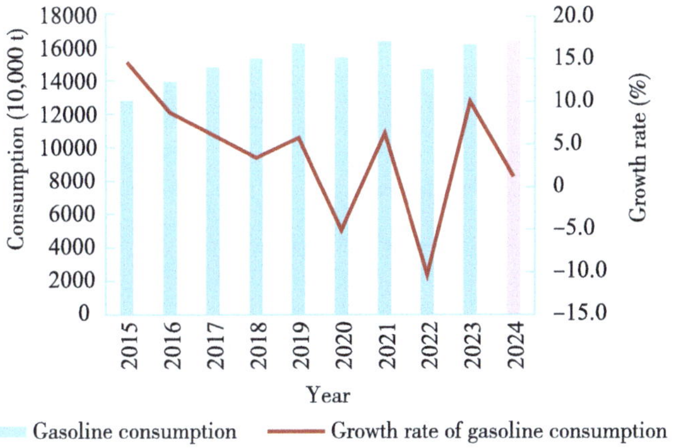

Fig. 17 Forecast of gasoline consumption in 2024

The growth of refined oil supply is faster than demand, and the refined oil market is in a loose pattern. In 2024, Zhenhai Refining (11 million tons/year) and Yulong Island Petrochemical (20 million tons/year) will be put into operation, and it is expected that refining capacity will increase to 961 million tons/year, crude oil processing volume 752 million tons, a year-on-year increase of 1.8%; refined oil production will reach 451 million tons, a year-on-year increase of 1.6%. The growth rate of domestic refined oil production is still faster than demand, the domestic production-demand gap is about 50 million tons, an increase of more than 10% compared to 2023, and the market is in a loose state.

(Authors of this section: Li Ran, Zhang Haoxu, Qu Ruiling Reviewer: Wang Lining)

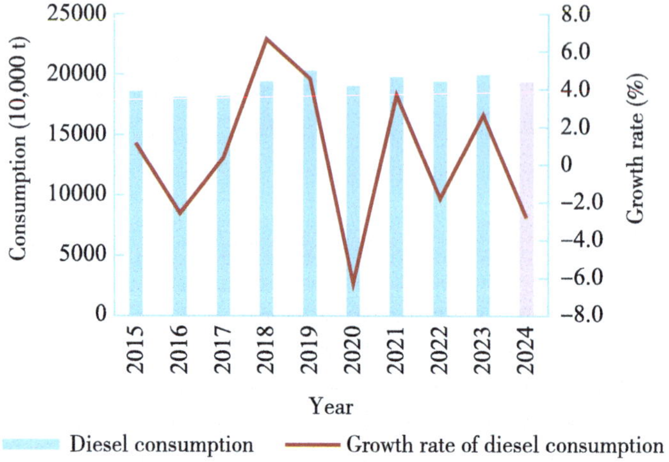

Fig. 18 Forecast of diesel consumption in 2024

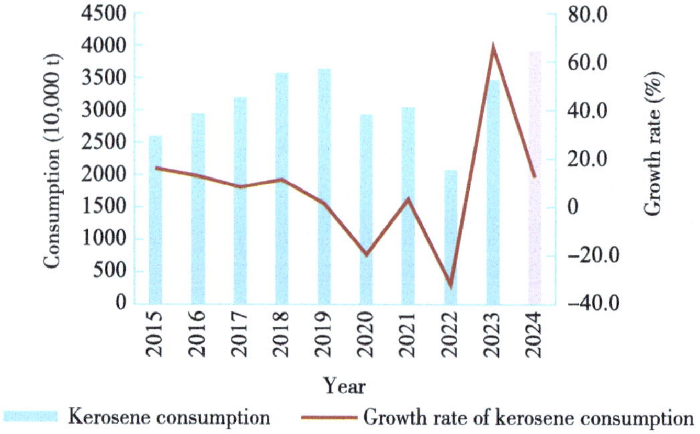

Fig. 19 Forecast of kerosene consumption in 2024

Analysis of the International Natural Gas Market

In 2023, the international natural gas pattern adjusted deeply, the geopolitical premium gradually faded, international gas prices fell from high levels, the natural gas market demand slowly recovered growth, supply and demand tended to be loose, and the main market inventory operated at high levels. The international natural gas trade pattern continues to adjust deeply, more Russian pipeline gas exit the European market, and Asia drove the growth of LNG trade. In 2024, the global natural gas market continues its fragile balance, uncertainty factors will increase, and international gas prices fluctuate more frequently.

Review of the International Natural Gas Market

Analysis of Global Natural Gas Demand

Global Natural Gas Consumption Slowly Recovers

In 2023, the global natural gas market slowly recovered, with an estimated annual consumption of 3.96 trillion cubic meters (Fig. 1), representing a 0.5% year-on-year increase, compared to the previous year's growth rate of − 3.1%. The main reasons are: firstly, the fall in international gas prices boosted demand in markets such as Asia; secondly, the global economy lacks growth momentum, leading to weak demand for industrial gas; thirdly, the increased use of alternative energy sources such as nuclear power and wind light suppresses the demand for gas in power generation; fourthly, the winter temperature is warmer than usual, limiting the increase in demand for heating gas.

X. Qian et al., *Global Oil and Gas Industry Development Report: 2023*,
https://doi.org/10.1007/978-981-95-2375-7_5

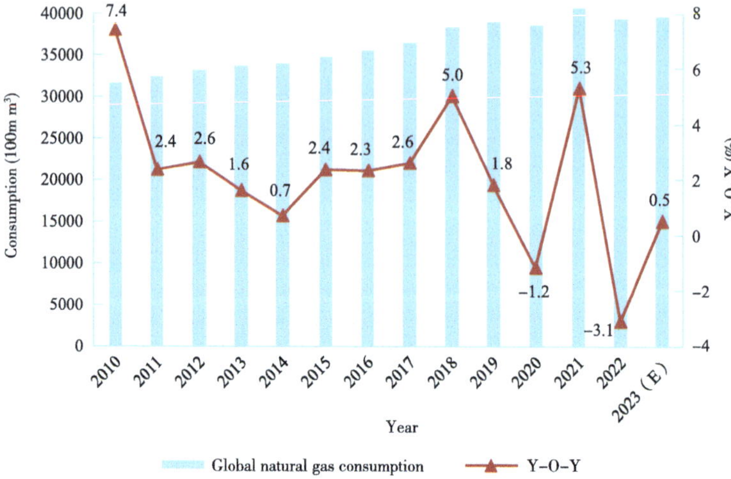

Fig. 1 Natural gas consumption and year-on-year growth rate in the world from 2010 to 2023. *Data source* Data from 2010 to 2022 comes from the "Energy Institute World Energy Statistics Yearbook 2023", and data for 2023 comes from the PetroChina Economic Research Institute

Europe Experienced Negative Growth, North America's Growth Rate Slowed Down, and Asia–Pacific Steadily Recovered

In 2023, the main market demand remained relatively weak, with Asia playing a significant role. Among them, Europe continued the negative growth trend observed in the previous year. The consumption growth rate in North America significantly slowed down, and the demand in the Asia–Pacific region significantly increased (Fig. 2).

In North America, the growth rate of natural gas consumption has significantly slowed down. In 2023, North America's natural gas consumption reached 1.1 trillion cubic meters, with the growth rate dropping from 4.7% in the previous year to 1.2%. Specifically, Canada's consumption was 123 billion cubic meters, an increase of 1.2% year-on-year; Mexico's consumption was 98.2 billion cubic meters, an increase of 1.7% year-on-year; the United States' consumption was 891 billion cubic meters, an increase of 1.1% year-on-year. This represents a decrease of 4.3 percentage points from the previous year, mainly due to factors such as a warm winter, slow industrial recovery, and increased use of renewable energy, which have suppressed gas demand.

Europe's consumption continued to decline. In 2023, Europe's natural gas consumption was 467.9 billion cubic meters, a year-on-year decrease of 6.2%, with the previous year's growth rate being − 13.0%. On the one hand, the economic growth of major European economies was stagnant, the manufacturing industry continued to be in the contraction zone, suppressing the demand for industrial gas. On the other hand, the rapid growth of wind and light installations, the return of hydropower to normal levels, France's revitalization of nuclear power, coupled with

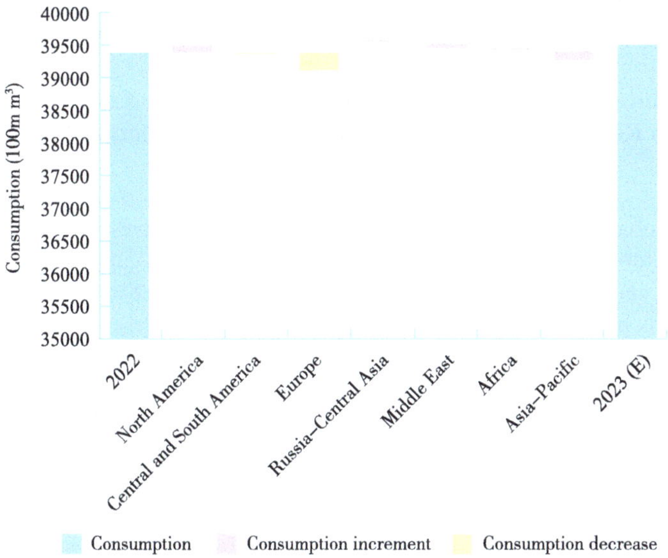

Fig. 2 Changes in natural gas consumption in major regions of the world in 2023. *Data source* Data for 2022 comes from the "Energy Institute World Energy Statistics Yearbook 2023", and data for 2023 comes from the PetroChina Economic Research Institute

the overall decline in electricity consumption, suppressed the demand for gas for power generation.

Asia–Pacific consumption steadily recovered. In 2023, Asia–Pacific's natural gas consumption was 923 billion cubic meters, an increase of 1.8% year-on-year, with the previous year's growth rate being − 3.2%. China and India contributed the main increase. Among them, China's natural gas consumption was 391.7 billion cubic meters, an increase of 6.6% year-on-year; Japan's natural gas consumption was 91.5 billion cubic meters, a decrease of 9.0% year-on-year; South Korea's natural gas consumption was 58 billion cubic meters, a decrease of 6.3% year-on-year; India's natural gas consumption was 62.5 billion cubic meters, an increase of 7.4% year-on-year; emerging countries' natural gas consumption was 226.3 billion cubic meters, an increase of 1.8% year-on-year.

Russia's consumption returned to positive growth, Central and South America's consumption decline narrows, Africa's consumption slowly recovered. In 2023, Russia's natural gas consumption was 424.5 billion cubic meters, an increase of 4.0% year-on-year, with the previous year's growth rate being − 14.0%. Central and South America's natural gas consumption was 160.7 billion cubic meters, a decrease of 0.6% year-on-year, with the previous year's growth rate being − 3.7%. Africa's natural gas consumption was 163.2 billion cubic meters, an increase of 0.4% year-on-year, with the previous year's growth rate being − 3.6%.

Analysis of Global Natural Gas Supply

Global Natural Gas Production increased Slightly, North America's Production Reached a New High, Russia's Production Continued to Decline

In 2023, global oil and gas exploration and development investment continued to grow, and global natural gas production reached 4.28 trillion cubic meters, with the growth rate increasing from 0.4% in the previous year to 0.8%. By region, North America's natural gas production was 1.36 trillion cubic meters, an increase of 3.4% year-on-year, with the United States' production being 1.15 trillion cubic meters, an increase of 3.3% year-on-year. The Middle East's production was 738.8 billion cubic meters, an increase of 0.8% year-on-year, with the previous year's growth rate being 3.7%, and Qatar's production being 182 billion cubic meters, a decrease of 0.7% year-on-year. Central Asia's production was 217.3 billion cubic meters, an increase of 1.0% year-on-year, with the previous year's growth rate being 0.3%. Russia's production is 642 billion cubic meters, a decrease of 4.5% year-on-year, with the previous year's growth rate being − 12.7%. Central and South America's natural gas production was 150.6 billion cubic meters, an increase of 0.5% year-on-year, with the previous year's growth rate being 1.6%. Africa's natural gas production was 263.8 billion cubic meters, an increase of 3.0% year-on-year, with the previous year's growth rate being − 4.3% (Fig. 3).

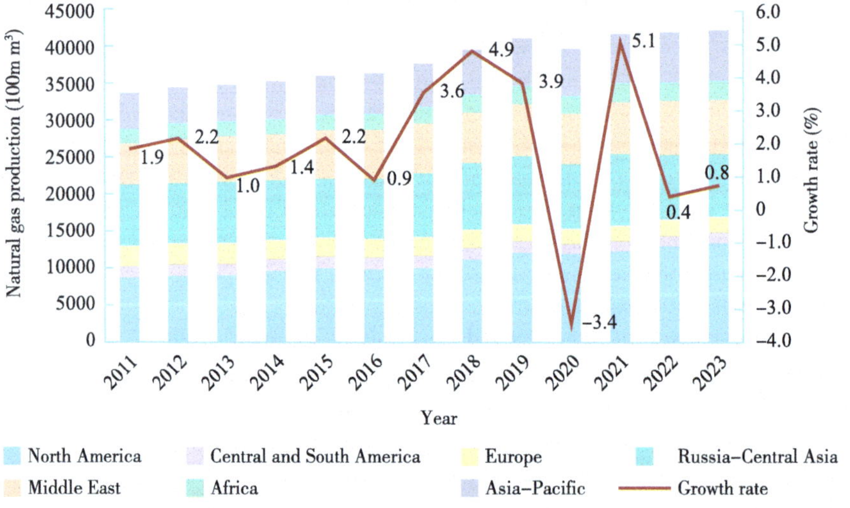

Fig. 3 Natural gas production and growth rate by region in the world from 2011 to 2023. *Data source* PetroChina Economic Research Institute

Global LNG Liquefaction Capacity Growth Rate Slowed Down

In 2023, the global LNG liquefaction capacity was 457 million tons/year, with the growth rate dropping from 2.9% in the previous year to 0.9% (Fig. 4), the lowest growth rate in nearly 10 years. The annual new capacity was 4 million tons/year, a decrease of 9.27 million tons/year from the previous year, mainly from Mozambique's Coral. The South FLNG (3.4 million tons/year) and Oman LNG (600,000 tons/year) projects have three production lines. Looking at it by country, the United States was the world's largest LNG producer, with a liquefaction capacity of 90.42 million tons/year. Australia, Qatar, Malaysia, and Russia ranked second to fifth, with capacity of 87.8 million tons/year, 77.4 million tons/year, 32 million tons/year, and 30.4 million tons/year, respectively.

Global Natural Gas Trade Analysis

In 2023, the global natural gas trade volume was about 12.5 trillion cubic meters, a year-on-year decrease of 1.0% (Table 1), among which pipeline gas trade volume decreases, and LNG trade continued to grow. The global natural gas trade pattern continued to adjust, with Russia's supply of pipeline gas to Europe continuing to decline significantly; European LNG imports fell from high levels, and Asian LNG imports resumed growth.

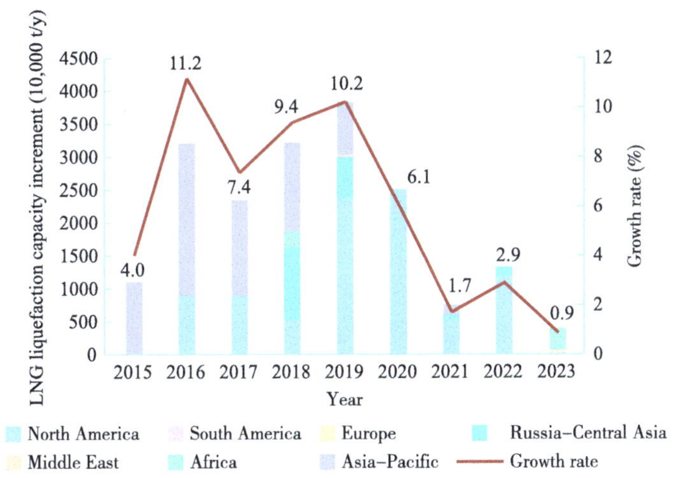

Fig. 4 Global regional additions to LNG liquefaction capacity and growth rate from 2015 to 2023. *Data source* S&P Global

Table 1 World natural gas trade volume and proportion from 2019 to 2023

Item	2019		2020		2021		2022		2023 (E)	
	Trade volume (Billion cubic meters)	Proportion (%)	Trade volume (Billion cubic meters)	Proportion (%)	Trade volume (Billion cubic meters)	Proportion (%)	Trade volume (Billion cubic meters)	Proportion (%)	Trade volume (Billion cubic meters)	Proportion (%)
LNG	4842	34.9	4901	37.7	5157	42.3	5424	43.0	5532	44.3
Pipeline gas	8015	65.1	7558	62.3	7044	57.7	7184	57.0	6944	55.7
Total	12,857	100	12,459	100	12,206	100	12,608	100	12,476	100
Growth rate (%)	4.0		– 3.1		– 2.1		3.3		– 1.0	

Data source Data from 2019 to 2022 is from the "Energy Institute World Energy Statistics Yearbook 2023", and data for 2023 is from the PetroChina Economic Research Institute

Decline in Pipeline Gas Trade

In 2023, the global pipeline gas trade volume was 694.4 billion cubic meters, a year-on-year decrease of 3.3%; it accounted for 55.7% of the global natural gas trade, a decrease of 1.3 percentage points from the previous year. Among them, Russia exported 49 billion cubic meters of pipeline gas to Europe, a year-on-year decrease of 42.6%; China's imports of pipeline gas increased by 6.4% year-on-year.

The Growth Rate of LNG Trade Volume Slowed Down

In 2023, the global LNG trade volume was 411 million tons, a year-on-year increase of 2.0%, and the growth rate slowed down by 3.0 percentage points from the previous year. Among them, Europe's LNG import volume was 126 million tons, a year-on-year decrease of 1.8%; Asia's import volume was 264 million tons, a year-on-year increase of 3.5%; South America's import volume was 6.22 million tons, a year-on-year decrease of 2.0% (Fig. 5). The global LNG spot trade volume was 131 million tons, a year-on-year increase of 5.4%, accounting for 32.1% of the total LNG trade volume.

European Imports Decrease, Asian Imports Stopped Falling and Rebound

In 2023, due to the weak economy, sluggish manufacturing industry, and high inventory, Europe's LNG imports turned from a sharp increase in the previous year to a decrease, with an annual import volume of 126 million tons, a year-on-year decrease of 1.8%. Germany put into operation 3 floating receiving stations, with an LNG import volume of 5.06 million tons. Looking at the main sources of imports, Europe

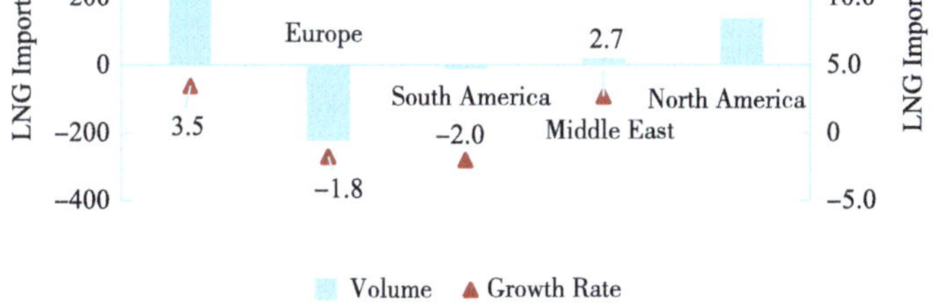

Fig. 5 Increase and growth rate of LNG imports by region in 2023. *Data source* S&P Global

imported 57.81 million tons from the United States, a year-on-year increase of 6.8%, and the growth rate slowed down significantly from the previous year; it imported 15.2 million tons from Qatar, a year-on-year decrease of 25.3%.

In 2023, driven by China, India, and emerging markets, Asia's LNG imports steadily recovered, and the growth rate turned from negative to positive. The total import volume for the year was 264 million tons, a year-on-year increase of 3.5%, but it has not yet returned to the level of 2021. Among them, China's LNG import volume was 71.32 million tons, a year-on-year increase of 11.7%, becoming the world's largest LNG importer again; Japan's weak economy and increased use of nuclear power resulted in an LNG import volume of 66.1 million tons, a year-on-year decrease of 8.4%; South Korea's LNG import volume was 45.12 million tons, a year-on-year decrease of 4.2% due to factors such as macroeconomics and increased coal power output; India's LNG import volume was 22.59 million tons, a year-on-year increase of 11.3%; emerging market LNG import volume was 37.62 million tons, a year-on-year increase of 23.9%, with the Philippines and Vietnam both importing LNG for the first time.

In 2023, South America's LNG import volume was 6.22 million tons, a year-on-year decrease of 2.0%. The growth rate in the previous year was − 53.1%, and the decline has significantly narrowed. Among the main importing countries, Argentina's LNG import growth rate increased by 14.5% year-on-year, and the growth rate in the previous year was − 35.7%; Brazil's LNG imports decreased by 53.8% year-on-year, and the growth rate in the previous year was − 73.7%; Chile's LNG imports decreased by 1.1% year-on-year, and the growth rate in the previous year was − 23.7%.

US LNG Exports Continued to Grow, While Exports from Other Major Countries Declined

In 2023, US LNG exports jumped to the top in the world, reaching 85.92 million tons, a year-on-year increase of 9.5%. Among them, exports to Europe were 57.81 million tons, a year-on-year increase of 6.8%; exports to Asia were 21.4 million tons, a year-on-year increase of 17.5%; exports to South America were 3.47 million tons, a year-on-year decrease of 3.2%; exports to the Middle East were 760,000 tons, a year-on-year decrease of 32.7%. Qatar's LNG export volume was 79.81 million tons, a year-on-year decrease of 1.5%. Among them, exports to Europe were 15.2 million tons, a year-on-year decrease of 25.3%; exports to Asia were 59.96 million tons, a year-on-year increase of 5.3%; exports to the Middle East were 4.52 million tons, a year-on-year increase of 26.5%. Australia's LNG export volume was 79.7 million tons, a year-on-year decrease of 0.5%. Russia's LNG export volume was 31.32 million tons, a year-on-year decrease of 3.8% (Fig. 6).

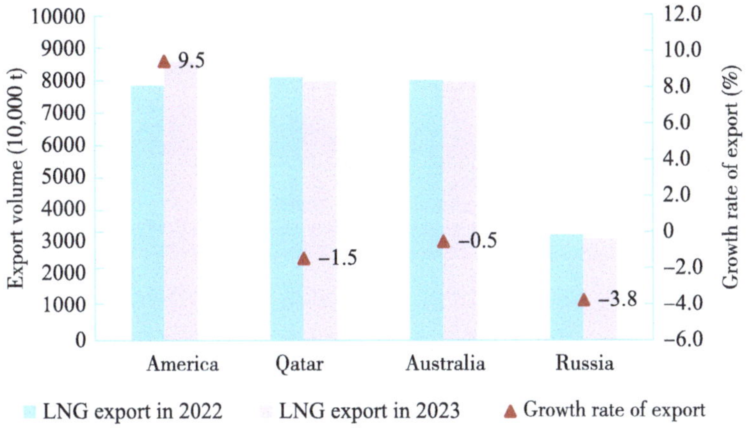

Fig. 6 Export volume and growth rate of major LNG exporting countries in 2023. *Data source* S&P Global

The Volume of Trade Contracts Signed Declines

In 2023, the global LNG market supply and demand tended to be loose, spot prices fell from high levels, and the willingness of market participants to sign long-term contracts declined. The global new LNG contract volume for the whole year was 68.96 million tons, a decrease of 29.05 million tons from the previous year, a year-on-year decrease of 29.6%. Eurasian buyers' strategies show differentiation, Europe signed multiple large LNG long-term contracts with Qatar, and new contracts in Asia are mainly concentrated in China. The LNG trade contract mainly showed the following characteristics:

The flexibility of destination clauses decreased. The Netherlands, Italy, France, China, Bangladesh, etc. have signed LNG contracts with Qatar, all using a fixed destination delivery model, with a total contract volume of 16.8 million tons/year. The proportion of flexible destination contracts fell from 83% in the previous year to 69% (Fig. 7). Long-term contracts still account for a high proportion. With increased uncertainty in the global natural gas market, the market still tends to sign stable medium- and long-term contracts to ensure stable resource volume and prices. Of the newly signed contracts, short-term contracts (less than 3 years) accounted for 3.3%, medium-term contracts (3-10 years) accounted for 16.4%, and long-term contracts (more than 10 years) accounted for 80.3% (Fig. 8).

The proportion of small and medium-sized contracts has increased. With the acceleration of the energy transition process, buyers are more inclined to sign small and medium-sized contracts for flexible resource allocation. The overall trend of the newly signed contract target volume is smaller, the proportion of contracts with a target volume of less than 1 million tonnes has increased from 33.2% in the previous year to 39.8%, the proportion of contracts with a target volume of 1 million to 2 million tonnes has increased from 33.6% in the previous year to 36.9%, and

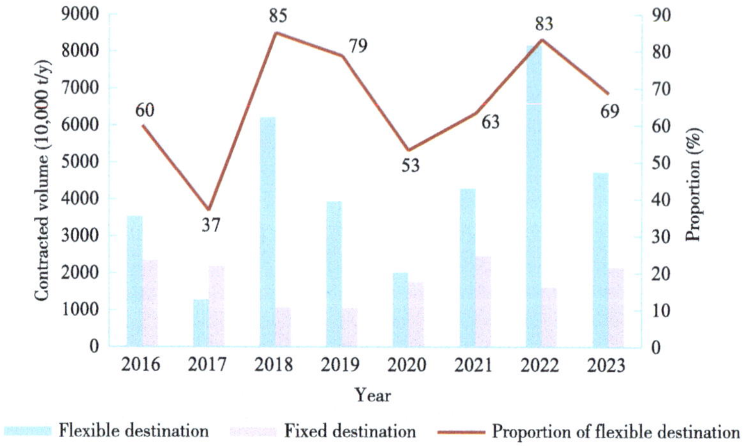

Fig. 7 Changes in global LNG trade destinations from 2016 to 2023. *Data source* S&P Global

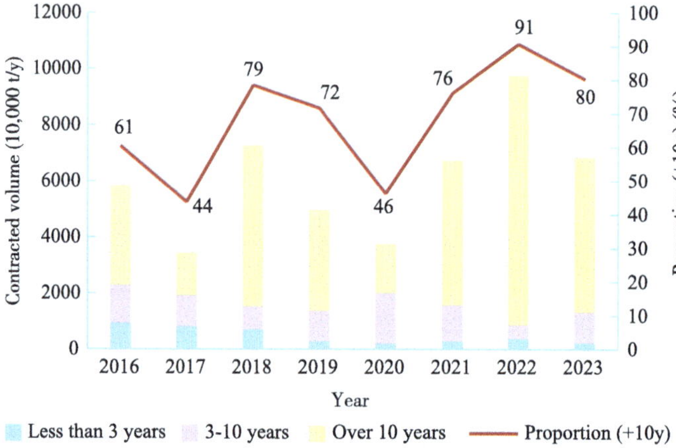

Fig. 8 Changes in global LNG contract term 2016-2023. *Source* S&P Global

the proportion of contracts with a target volume of more than 2 million tonnes has decreased from 33.2% in the previous year to 23.3% (Fig. 9).

The proportion of oil-indexed contracts has increased, with a continuous rise in the slope, and the ways of gas-indexed contracts have become more diversified. The proportion of new global LNG contracts using the Brent oil prices index is 50.0%, an increase of 16.0 percentage points from the previous year; the remaining 50.0% of new contracts are using gas-indexed prices, with indices including HH, TTF, JKM, Waha, etc. Affected by contract negotiations, formal signing and effective time, and market situation, the average slope of new oil-indexed LNG contracts has risen from 12–13% to 13–14% (Fig. 10).

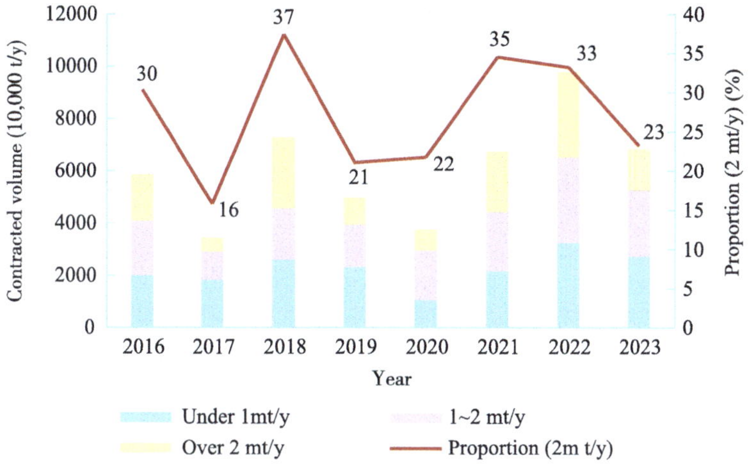

Fig. 9 Changes in global LNG trade contract target volume from 2016 to 2023. *Data source* S&P Global

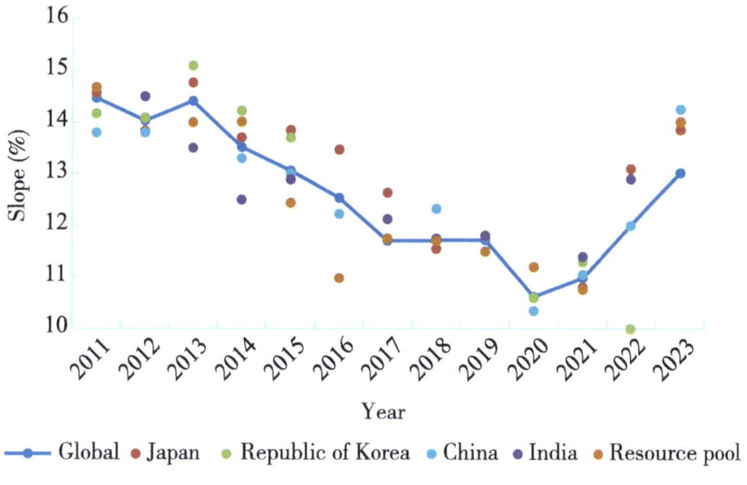

Fig. 10 Changes in the slope of global LNG trade contracts from 2011 to 2023. *Data source* Rystad Energy

Global Natural Gas Market Price Analysis

In 2023, the global natural gas market supply and demand will tend to be loose, international gas prices will fall sharply from high levels, and return to the level before the 2022 Ukraine crisis. However, due to the disturbance of supply chain risks, the "wide range high frequency" fluctuation of gas prices has increased. The price ratio between the US, Europe, and Asia has changed from 1:5.9:5.4 in the previous year to 1:5.1:5.4, the linkage between the European and Asian markets has further strengthened, and the average spot price in Asia has once again surpassed Europe.

U.S. Natural Gas Prices Fell Sharply, Returning to Below $3 Per Metric Million British Thermal Units

In 2023, the growth rate of domestic demand in the United States slowed significantly, production reached a new high, storage inventory continued to be higher than the same period in previous years, HH spot prices fell sharply, with an average price of $2.54 per metric million British thermal units, a year-on-year decrease of 60.5% (Fig. 11). In the first quarter, due to warm weather in the United States and some liquefaction project outages, gas demand was not strong, the market supply and demand were loose, and HH spot prices quickly fell from the high level at the beginning of the year to below $3 per metric million British thermal units. In the second quarter, the continued downturn in the U.S. manufacturing industry suppressed domestic gas demand, several LNG export projects were overhauled, and the demand for LNG feed gas decreased, causing HH spot prices to fall to the lowest point of the year at $1.74 per metric million British thermal units. In the third quarter, high temperatures in many parts of the United States led to a significant increase in gas consumption for electricity generation, and the impact of wildfires in Canada reduced the volume of imported pipeline gas, tightening the supply and demand fundamentals, and HH spot prices rebounded. In the fourth quarter, the conflict between Palestine and Israel increased the uncertainty of LNG market supply, and HH spot prices rose rapidly, but as market sentiment digested, coupled with relatively loose supply and demand, prices weakened again in November and December.

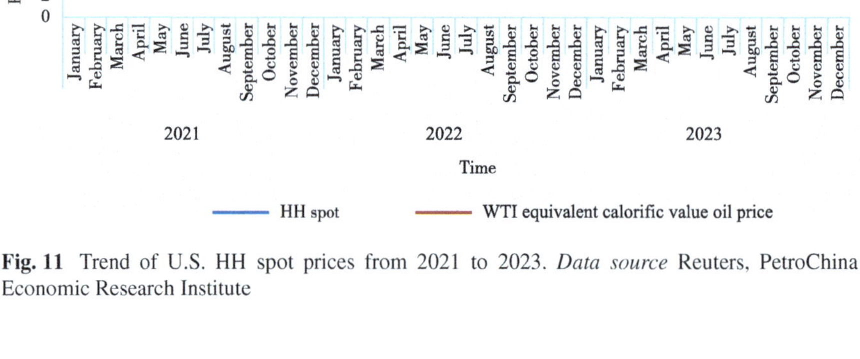

Fig. 11 Trend of U.S. HH spot prices from 2021 to 2023. *Data source* Reuters, PetroChina Economic Research Institute

European Natural Gas Prices Fluctuated and Fell, with Increased Volatility Within the Year

In 2023, Europe's natural gas demand was weak, resource supply was relatively abundant, and gas storage inventory was at a historical high for the same period, causing gas prices to plummet to 2021 levels (Fig. 12), and once again falling below the calorific value of Brent and other oil prices. The annual average spot price of TTF was $12.9 per metric million British thermal units, a year-on-year decrease of 65.8%. In the first quarter, due to warmer temperatures, the growth in heating gas was limited, and the spot price of TTF plummeted. In the second quarter, with a weak macro economy and increased use of alternative energy sources such as wind, light, and nuclear, natural gas demand continued to decline, and TTF spot prices continued to fall, once dropping to the annual low of $7.35 per metric million British thermal units. In the third quarter, high summer temperatures boosted demand for electricity generation, the maintenance time of Norwegian gas fields was extended, and news of a strike by workers on Australian liquefaction projects triggered market panic, causing TTF spot prices to rise rapidly. In the fourth quarter, multiple factors such as the Israeli-Palestinian conflict, the Red Sea crisis, and leaks in the Finland-Estonia undersea pipeline exacerbated market worries, causing gas prices to surge in the short term, but due to warmer temperatures and weak industrial production, the TTF spot price lacked upward momentum, and gas prices fell back in December.

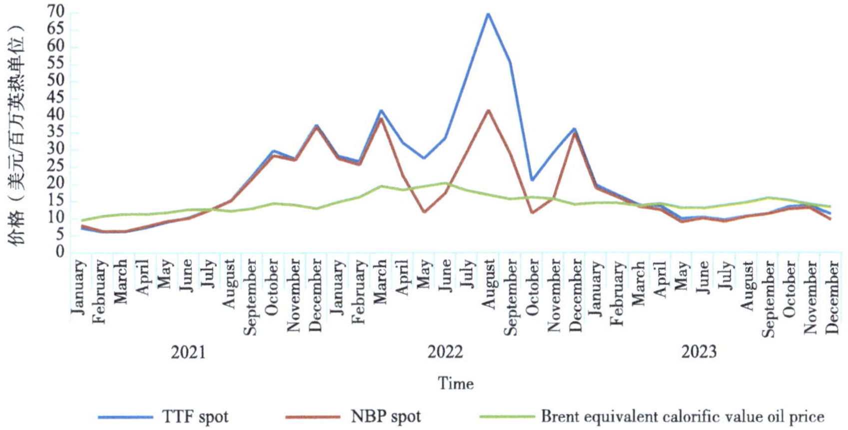

Fig. 12 Trends in European natural gas spot prices and Brent oil prices from 2021 to 2023. *Data source* Reuters, PetroChina Economic Research Institute

Northeast Asian LNG Spot Prices Fell Sharply, Showing a Wide V-shaped Trend Within the Year

In 2023, Northeast Asian LNG demand slowly recovered, the overall market supply and demand were loose, and the impact of European gas prices was superimposed, the average onshore spot price of Northeast Asian LNG was $15.8 per metric million British thermal units, a year-on-year decrease of 54.4% (Fig. 13). In the first quarter, due to warmer temperatures and a slowdown in economic growth, Northeast Asian gas demand was weak, and LNG spot prices fell rapidly. In the second quarter, the economy continued to be weak, and Northeast Asian LNG spot prices continued to fall, with the lowest daily price falling to $8.85 per metric million British thermal units. In the third quarter, high summer temperatures boosted demand for electricity generation, news of a strike by workers on Australian liquefaction projects triggered market panic, and the demand for winter stockpiling increased, causing Northeast Asian LNG spot prices to rise. In the fourth quarter, the Israeli-Palestinian conflict and the arrival of the heating season caused gas prices to fluctuate and rise, but due to warmer temperatures and a weak economy, gas prices were lower than at the beginning of the year.

In 2023, affected by the decline in gas spot prices and oil prices, the average import price of Northeast Asian LNG was $13.3 per metric million British thermal units, a year-on-year decrease of 30.8% (Fig. 14). Among them, the average price in Japan was $13.8 per metric million British thermal units, a year-on-year decrease of 24.6%; the average price in South Korea was $14.9 per metric million British thermal units, a year-on-year decrease of 28.0%.

Authors of this section: Zhang Xiaoyu, Cui Zijian Reviewer: Duan Zhaofang.

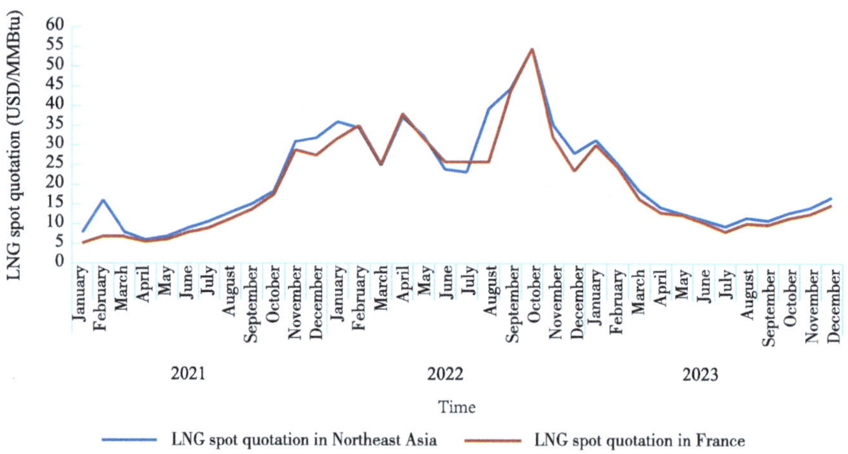

Fig. 13 Trends in Northeast Asian LNG spot quotes from 2021 to 2023. *Data source* ICIS

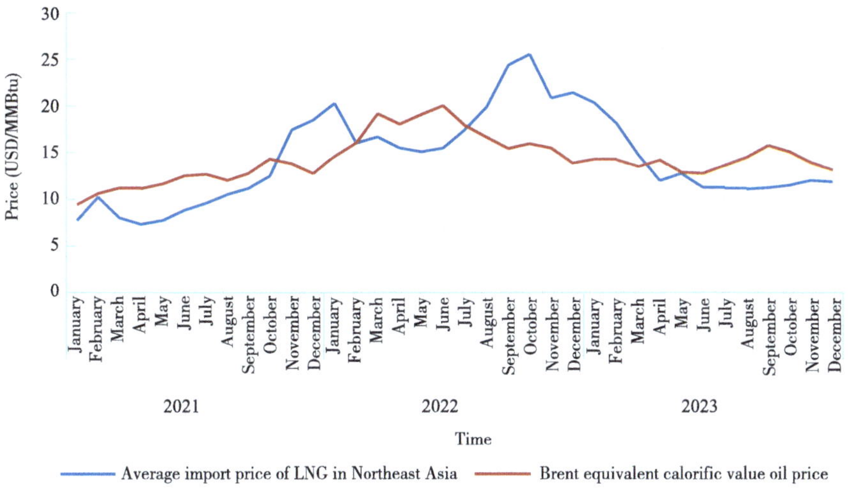

Fig. 14 Average import price of Northeast Asian LNG from 2021 to 2023. *Data source* Argus

International Natural Gas Market Outlook

In 2024, the global natural gas market demand will continue to recover, natural gas production will steadily increase, the market will continue to be in a fragile balance, and international gas prices will "fall twice and rise once". However, due to new geopolitical conflicts, extreme weather, shipping risks, and other factors, the market trend is still full of uncertainty.

Global Natural Gas Demand Outlook

Global Natural Gas Demand Grows Slightly, with the Asian Market Being the Main Growth Point

In 2024, it is expected that the global natural gas demand will be 4.02 trillion cubic meters, with a growth rate of 1.5% (Fig. 15). Looking at it by region, Europe is accelerating its energy transition, and renewable energy generation continues to grow; the transfer of manufacturing industry suppresses industrial gas use, and it is expected that the annual natural gas demand will be 456.2 billion cubic meters, a year-on-year decrease of 2.5%. Asia–Pacific natural gas consumption is 959.9 billion cubic meters, a year-on-year increase of 4.0%, among which Japan and South Korea continue to restart nuclear power and build new coal power, suppressing electricity generation gas; India's economy grows rapidly, promoting gas demand; emerging market gas-fired power generation increases promote gas demand. North America's annual natural gas consumption is 1.12 trillion cubic meters, an increase of 0.9%,

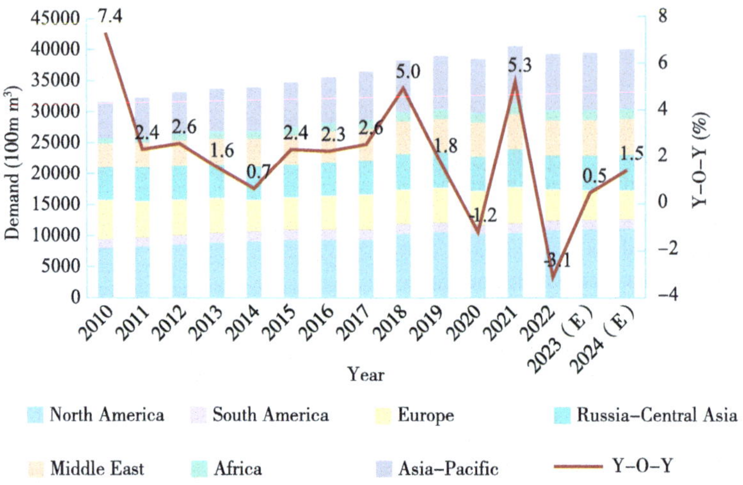

Fig. 15 Global natural gas demand by region from 2010 to 2024. *Data source* Data for 2010–2022 comes from the "Energy Institute World Energy Statistics Yearbook 2023", data for 2023–2024 comes from the PetroChina Economic Research Institute

among which the United States' residential and commercial gas grows rapidly, the use of renewable energy increases, suppressing the growth of electricity generation gas, industrial gas declines, and overall demand growth slows down. Russia's natural gas consumption growth rate has slowed down compared to the previous year, and it is expected that the annual consumption will be 429.6 billion cubic meters, a year-on-year increase of 1.2%. Central and South American natural gas consumption returns to positive growth, and it is expected that the annual consumption will be 163.1 billion cubic meters, a year-on-year increase of 1.5%. African natural gas consumption is 168.1 billion cubic meters, a year-on-year increase of 3.0%.

Global LNG Demand Growth Slows Down

By 2024, the global LNG demand is expected to be 410 million tonnes, an increase of about 4.7 million tonnes, a year-on-year growth of 1.2%. Among them, the demand for natural gas in Europe continues to decline, the import of pipeline gas remains stable, and the annual LNG import volume is expected to decrease slightly; in Asia, under the drive of China, India, and emerging markets, LNG imports continue to recover and grow. By 2024, the global gasification capacity will reach 1.135 billion tonnes/year, with an annual increase of 87.7 million tonnes/year, mainly concentrated in Germany, China, India, and emerging markets (Fig. 16).

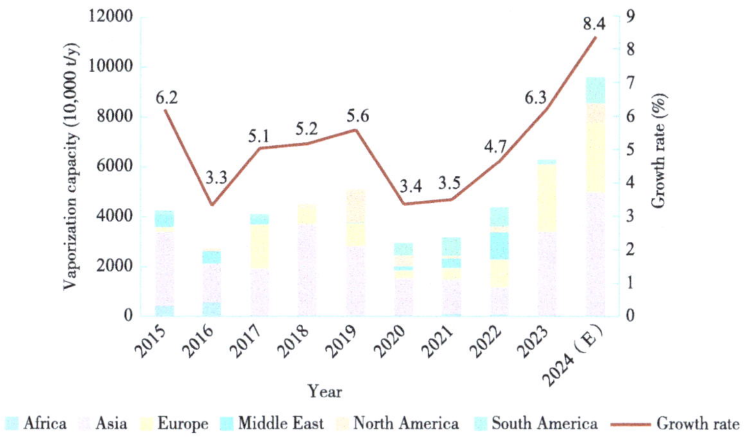

Fig. 16 Increase in global LNG gasification capacity and growth rate by region from 2015 to 2024. *Data source* S&P Global

Global Natural Gas Supply Outlook

Global Natural Gas Production Steadily Increases

In 2023, global upstream oil and gas investment will increase by about 11% year-on-year, the highest level since 2015, supporting the continuous growth of natural gas production. Due to the high volatility of international oil prices and the slow-down in energy transition, international oil companies have repositioned traditional energy, and natural gas investment continues to grow. By 2024, the global natural gas production is estimated to be about 43.7 trillion cubic meters, with a growth rate of 2.0%, and the increase mainly comes from North America and the Middle East. Among them, North America increased by about 300 billion cubic meters, and the Middle East increased by about 200 billion cubic meters; affected by the closure of the Groningen gas field in the Netherlands and other factors, the decline in European domestic production was about 60 billion cubic meters; affected by the partial commissioning of the Arctic LNG 2 project and the increase in pipeline gas supply to China, Russia's production slightly rebounded, with an increase of about 176 billion cubic meters.

LNG Liquefaction Projects Rebound, Market Uncertainty Increases

In 2024, the global new liquefaction capacity will be 28.9 million tonnes/year, with a total capacity of 489 million tonnes/year. Among them, the production capacity of the US Calcasieu Pass LNG project is 10 million tonnes/year, and the expansion capacity of the Elba Island LNG project is 600,000 tonnes/year; the production capacity of the Canadian LNG Canada project is 7 million tonnes/year; the production

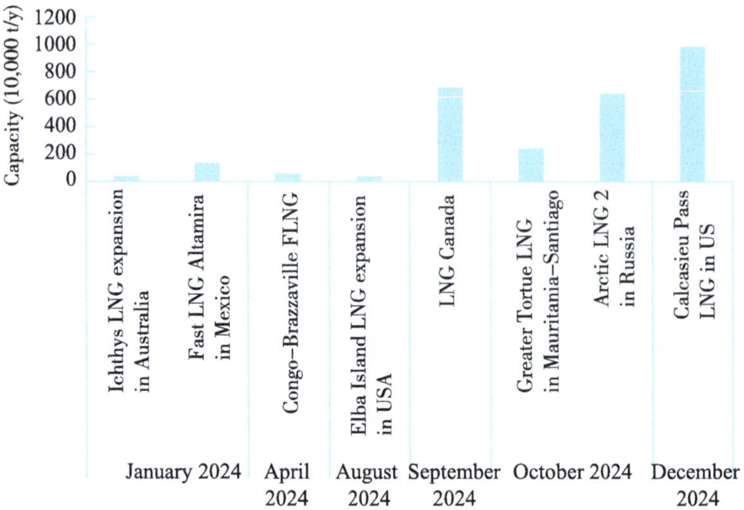

Fig. 17 Global planned LNG liquefaction projects in 2024. *Data source* S&P Global

capacity of the Russian Arctic LNG 2T1 project is 6.6 million tonnes/year; Mauritius, Senegal, Mexico, Congo (Brazzaville), Australia and others have invested in expansion capacity of 4.7 million tons/year (Fig. 17). Considering the schedule of project commissioning and reaching full capacity, and the uncertainty of the Russian Arctic LNG 2 project due to sanctions, the actual increase in liquefaction capacity for the whole year may be less than expected.

Global Natural Gas Price Outlook

In 2024, global natural gas demand is expected to increase slightly, and resources are relatively abundant. Among them, domestic demand in the United States is slowing down, but LNG exports remain high due to factors such as the new liquefaction projects, and the market supply and demand are tightening; the demand in the European market is difficult to boost, and high inventory to some extent alleviates supply-side concerns, and the market is relatively loose; the demand in the Asian market continues to recover, attracting LNG resources to flow in, and is closely linked with the European market. In 2024, it is expected that the gas prices in the three major markets will "fall in two and rise in one", and the fluctuations will be more frequent. Among them, the average spot price of European TTF is 11.0–12.0 USD/metric million British thermal units, a year-on-year decrease of 6.7–14.5%; the average spot price of Northeast Asia LNG is 12.0–13.0 USD/metric million British thermal

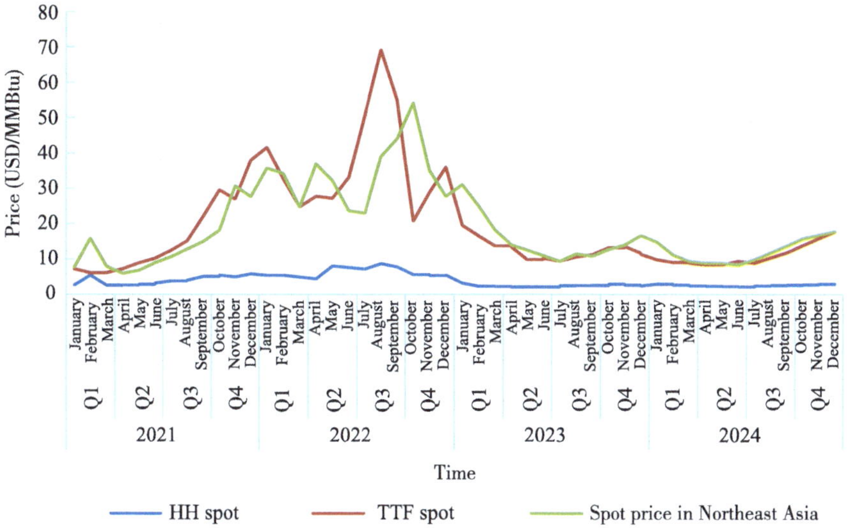

Fig. 18 Spot prices of natural gas in three markets from 2021 to 2024. *Data source* PetroChina Economic Research Institute

units, a year-on-year decrease of 17.4–23.8%; the average spot price of US HH is 2.6–3.0 USD/metric million British thermal units, a year-on-year increase of 2.5–18.2% (Fig. 18).

Author of this section: Zhang Xiaoyu Reviewer: Duan Zhaofang.

Analysis of China's Natural Gas Market

In 2023, China's natural gas consumption recovered, natural gas production maintained a growth momentum of 100 billion cubic meters for seven consecutive years, import gas resumed growth, and the degree of external dependence slightly increased. In 2024, it is expected that China's natural gas market will continue to improve, natural gas demand will maintain relatively rapid growth, domestic production will steadily increase, and it is expected to exceed 100 billion cubic meters again, and imports will grow relatively rapidly.

Review of China's Natural Gas Market

In 2023, China's natural gas consumption gradually recovered, the increase in imported gas turned from negative to positive, and the degree of external dependence slightly increased. Positive progress has been made in the end-use price linkage mechanism, pipeline construction has been steadily advanced, the newly added LNG receiving capacity has set a historical record, the operating entities have become increasingly diversified, and the construction of gas storage and peak-shaving facilities has been accelerated.

Analysis of China's Natural Gas Demand

Natural Gas Consumption Grew Relatively Rapidly at a Low Base

In 2023, affected by factors such as the relaxation of COVID-19 control, the stabilization and recovery of the macro economy, the fall in gas prices, and the sufficient supply of coal and new energy, the annual natural gas consumption was 391.7 billion

X. Qian et al., *Global Oil and Gas Industry Development Report: 2023*,
https://doi.org/10.1007/978-981-95-2375-7_6

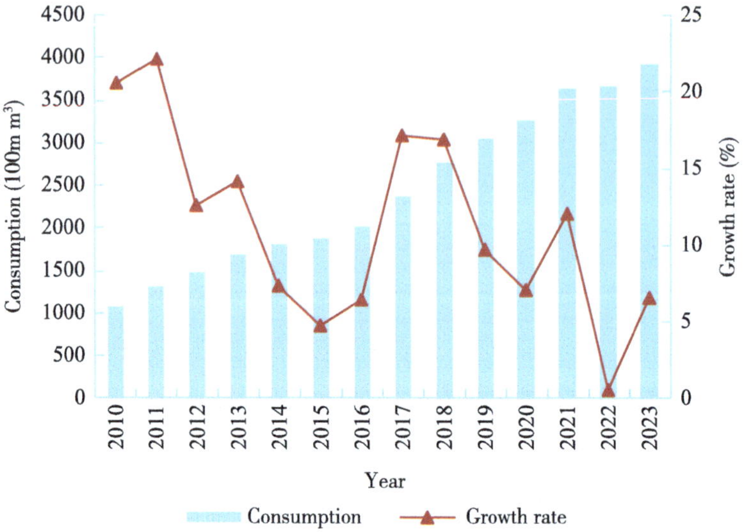

Fig. 1 Natural gas consumption in China from 2010 to 2023

cubic meters[1], an increase of 24.2 billion cubic meters, a year-on-year increase of 6.6% (Fig. 1).

Looking at it by quarter, in the first quarter, due to the combined effects of warmer national temperatures, the stable transition of COVID-19 prevention and control, and the gradual restoration of production and living order, natural gas consumption grew in a recovery manner, with a growth rate of 0.6% and an increase of 700 million cubic meters. In the second quarter, the macroeconomic operation was not as expected, the PMI index continued to be under 50, and the demand for industrial gas was impacted, but due to the insufficient output of hydropower promoting the demand for gas power supplementation, coupled with a lower base in the previous year, natural gas consumption increased by 7.4 billion cubic meters, with a growth rate of 9.4%. In the third quarter, the prosperity of the commercial service sector continued to improve, the economic advantages of vehicle LNG were significant, the consumption of city gas grew rapidly, and the national high temperature supported the use of gas for power generation, the consumption of natural gas increased by 5.3 billion cubic meters, with a growth rate of 6.3%. In the fourth quarter, the overall temperature was relatively warm, and the strong cold wave across the country in December promoted the steady increase in heating gas; the continuous recovery of the macro economy coupled with a low base in the previous year, the rapid growth of industrial and power generation gas, the consumption of natural gas increased by 10.9 billion cubic meters, with a growth rate of 10.9% (Fig. 2).

[1] The corresponding apparent consumption volume is 394.9 billion cubic meters, an increase of 25.9 billion cubic meters, a year-on-year increase of 7.0%.

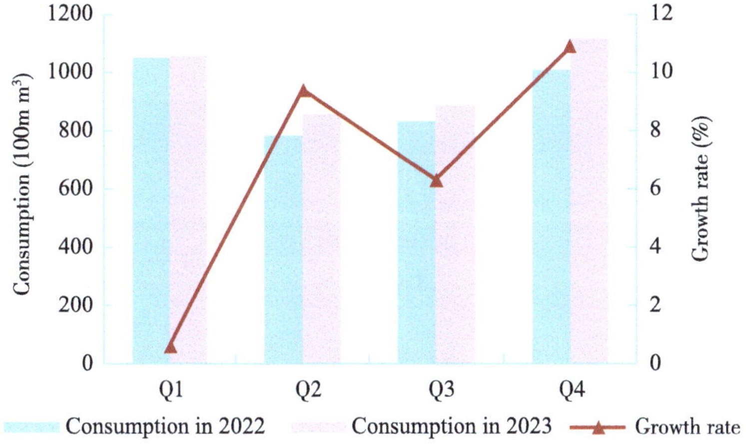

Fig. 2 Quarterly natural gas consumption and growth rate in China in 2023

The Growth Rates of All Major Gas-Consuming Sectors Have Rebounded, with Power Generation Leading the Way

In 2023, it is estimated that the consumption of city gas was 127.7 billion cubic meters, an increase of 8.2%; the consumption of industrial fuel gas was 154.3 billion cubic meters, an increase of 6.1%; the consumption of power generation gas was 71 billion cubic meters, an increase of 9.6%; the consumption of gas for fertilizer and chemical industry was38.7 billion cubic meters, a decrease of – 1.4% (Figs. 3 and 4).

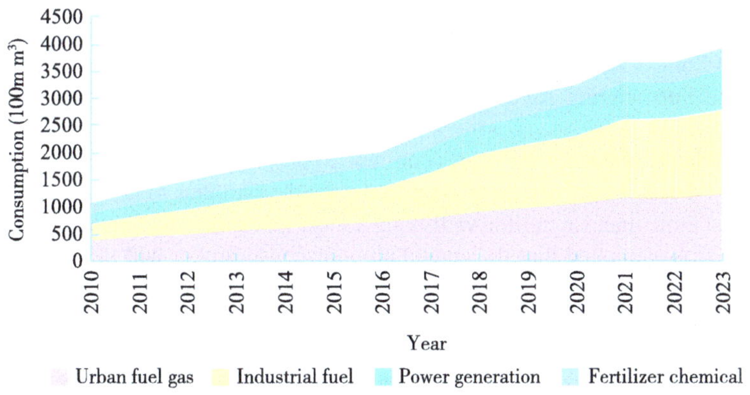

Fig. 3 Natural gas consumption by sector from 2010 to 2023

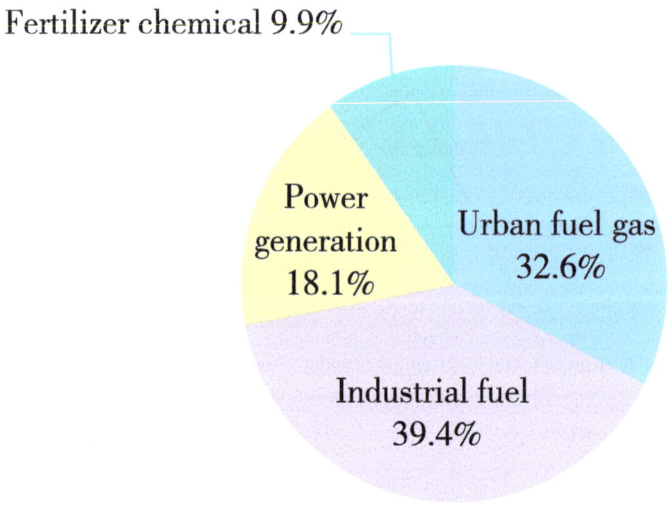

Fig. 4 Natural gas consumption by sector in 2023

Relatively Rapid Growth in City Gas Consumption

Infrastructure investment maintains strong growth, natural gas infrastructure is continuously improved, the coverage of citygas pipelines is gradually expanded, and the national gas-using population is steadily increasing. It is estimated that the citygas-using population reached 480 million, an increase of 28 million compared to 2022, and residential gas use steadily increased. The rapid recovery of the commercial service industry has led to a significant increase in gas consumption. The total social catering income in 2023 increased by 20.4% year-on-year. The consumption of heating gas steadily increased. The government has issued the "Comprehensive Air Pollution Control Action Plan for Autumn and Winter 2023–2024 in Beijing-Tianjin-Hebei and Surrounding Areas, and the Fenwei Plain", and the plain areas such as Beijing-Tianjin-Hebei are steadily and orderly promoting clean heating. Ningxia, Qinghai, Gansu and other provinces continue to promote coal-to-gas heating in key areas, and the number of heating users in the south continues to increase. It is estimated that more than 1.6 million wall-hung boilers was added nationwide throughout the year. It is estimated that the annual city gas consumption will be 127.7 billion cubic meters, a year-on-year increase of 8.2%, significantly higher than the 1.3% in the same period of the previous year.

Rapid growth in transportation gas. In terms of CNG, affected by the rapid development of new energy vehicles and other factors, the number and consumption of CNG vehicles continue to decline. In terms of LNG, domestic LNG market prices have dropped significantly, and the calorific value ratio of vehicle LNG to diesel prices has dropped to the lowest in nearly three years. The economic advantages were significant. The annual sales of LNG vehicles were nearly 200,000, a year-on-year growth rate of over 300%, driving the demand for gas to grow rapidly. It is estimated that the number of LNG vehicles in the country in 2023 was 1.04 million, a year-on-year increase of 14.3%. Guangdong and other places encourage the development of LNG ships, introduce subsidy policies, and a number of LNG ships were built and put into operation. It is estimated that there was nearly 500 LNG ships nationwide in 2023, and more than 20 LNG refueling stations. The LNG offshore refueling business continues to advance, and "ship-to-ship" LNG refueling has been realized in Shanghai, Ningbo and Guangzhou ports.

Industrial Fuel Gas Gradually Recovered

Affected by the upward trend of the macro economy and the decline in gas prices, the enthusiasm for production of industrial enterprises has rebounded, and gas use has recovered relatively quickly. The national manufacturing PMI index averaged 49.9% for the year, a year-on-year increase of 0.8 percentage points; the added value of industries above designated size increased by 4.6% year-on-year, a rebound of 1.0 percentage point from the previous year. Major gas-consuming industries such as glass and steel have increased their output by 9.8% and 5.2% respectively year-on-year. It is estimated that the industrial gas consumption in 2023 was 154.3 billion cubic meters, a year-on-year increase of 6.1%, higher than the growth rate of 0.2% in the previous year.

Rapid Growth in Power Generation Gas

Affected by the rapid growth of national electricity demand, the continuous improvement of coal supply, the rapid increase in new energy power generation, the decline in hydropower output, and the rapid growth of gas power installation, it is estimated that the power generation gas consumption in 2023 was 71 billion cubic meters, with a growth rate of 9.6%, an increase of 11.3 percentage points from the previous year. The total social electricity consumption increased by 6.7% year-on-year, an increase of 3.1 percentage points from the previous year. Coal supply has steadily increased, with an annual production increase of 2.9%, and the supply guarantee capacity has increased; new energy power generation has grown rapidly, with wind power and solar power generation increasing by 12.3% and 17.2% respectively year-on-year; hydropower generation has decreased by 5.1% year-on-year. The newly added large-scale gas-fired power generation installed capacity is 8.35 million kilowatts, and the total installed capacity is 123 million kilowatts, a year-on-year increase of 6.7%, an increase of 1.6 percentage points from the previous year.

Slight Decrease in Gas Usage for Fertilizer and Chemical Industries

The demand for gas in the chemical industry has decreased, while the demand for gas in fertilizer production remained stable. Methanol demand recovered slowly, with a loose supply–demand balance in the market. The average annual price is 2330 yuan/ton, a decrease of 10% year-on-year, and the profitability of gas companies has declined compared to the previous year. As a result, the production of methanol by gas companies has decreased by nearly 10% year-on-year. The state continues to promote the stable supply and price of fertilizers, and the demand for urea is good. Affected by the fall in coal prices and other production costs, the average annual price of urea is 2465 yuan/ton, a decrease of 8.8% year-on-year. The profits of gas companies have narrowed, and the production of urea by gas companies has remained basically stable year-on-year. It is estimated that the gas consumption for fertilizer and chemical industries in 2023 was 38.7 billion cubic meters, with a decrease of – 1.4%, lower than the 3.6% of the previous year.

Relatively Rapid Recovery of Natural Gas Consumption in Various Regions

In 2023, due to factors such as insufficient hydropower output and new power generation installations in the southeastern coastal areas, the demand for industrial and power generation gas grew rapidly, with an estimated consumption of 55 billion cubic meters (Fig. 5), a growth rate of 11.8%, accounting for 14% of China's natural gas consumption (Fig. 6). The Bohai Rim region was affected by factors such as higher temperatures during the heating season and a slowdown in the growth rate of coal-to-gas conversion, with an estimated consumption of 82.3 billion cubic meters, a growth rate of 5.0%, accounting for 21% of China's natural gas consumption. The Yangtze River Delta region continues to recover economically, with the growth rate of manufacturing investment and the added value of industries above a designated size higher than the national average, and the growth of industrial gas was relatively fast, with an estimated natural gas consumption of 65.8 billion cubic meters, a growth rate of 7.1%, accounting for 17% of China's natural gas consumption. The industrial gas recovery in the southwestern region was not as expected, and the gas consumption for fertilizer and chemical industries have decreased, with an estimated natural gas consumption of 49.6 billion cubic meters, a growth rate of 1.7%, accounting for 13% of China's natural gas consumption. The central and western regions were affected by factors such as sufficient coal supply and falling coal prices, and a slowdown in the growth rate of coal-to-gas conversion, with an estimated natural gas consumption of 53.3 billion cubic meters, a growth rate of 5.3%, accounting for 14% of the country's natural gas consumption. The rapid growth of industrial gas demand driven by new industrial users in the central and southern regions, with an estimated natural gas consumption of 40.2 billion cubic meters, a growth rate of 8.0%, accounting for 10% of China's natural gas consumption. The economy in the northwest region recovered rapidly, with rapid growth in industrial and transportation gas, with an estimated natural gas consumption of 29.3 billion cubic meters, a growth rate of 13.5%, accounting for 7% of China's natural gas consumption; the economic recovery in the northeast region is relatively slow, with stable gas consumption, with an estimated natural gas consumption of 16.1 billion cubic meters, a growth rate of 1.1%, accounting for 4% of the country's natural gas consumption.

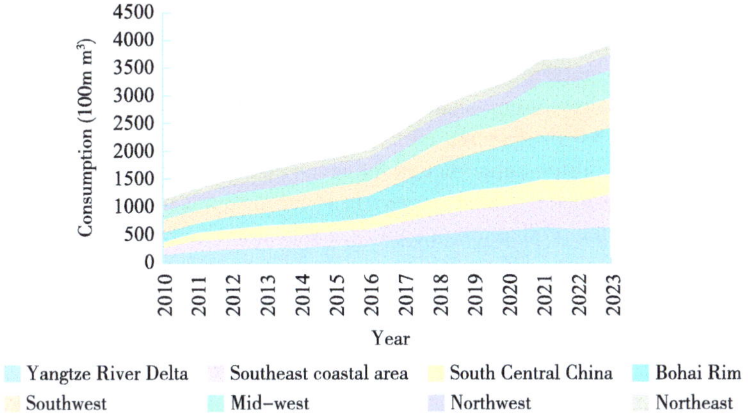

Fig. 5 Natural gas consumption by region from 2010 to 2023

Fig. 6 2023 Natural gas consumption structure by region

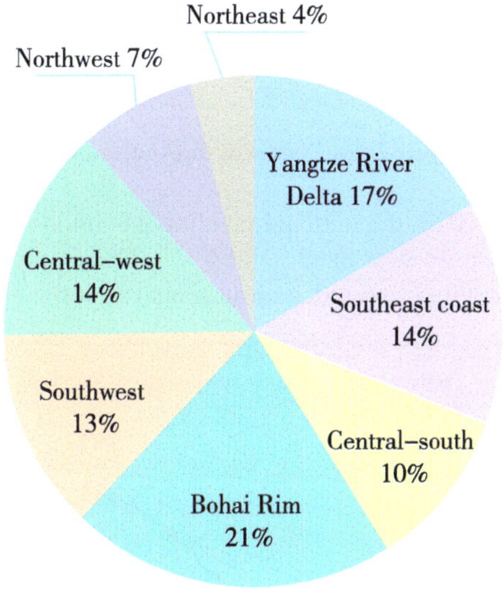

Analysis of China's Natural Gas Supply

Continuous Growth in Natural Gas Production, Significant Ballast Effect

In 2023, the oil and gas industry's "seven-year action plan" for increasing reserves and production continued to advance, with an annual natural gas production of 235.3 billion cubic meters (including shale gas, coalbed methane, and coal gas), a year-on-year increase of 5.7%, an increase of 12.6 billion cubic meters, and a continuous increase of over 10 billion cubic meters for seven consecutive years. Among them, the production of coalbed methane was nearly 14 billion cubic meters, a year-on-year increase of 20.7%; the production of shale gas was about 25 billion cubic meters, a year-on-year increase of 4.5%.

Recovery of Gas Imports, Slight Increase in Foreign Dependence

In 2023, China's natural gas import volume was 165.6 billion cubic meters, with a growth rate of 9.5%. The foreign dependence is 42.3% (Fig. 7), an increase of 1.1 percentage points from the previous year. Among them, the proportion of LNG imports was 59.4%, an increase of 1.2 percentage points from the previous year; the proportion of pipeline gas was 40.6%.

Relatively Rapid Growth in Pipeline Gas Imports

In 2023, the national pipeline gas import volume was 67.1 billion cubic meters, a year-on-year increase of 6.4%. The supply of gas from the China-Russia East Line rapidly increased, with an annual import volume of more than 22 billion cubic meters.

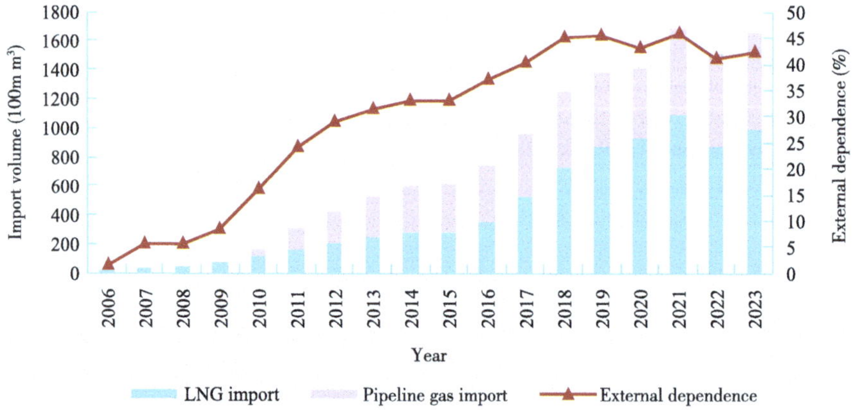

Fig. 7 China's natural gas import from 2006 to 2023

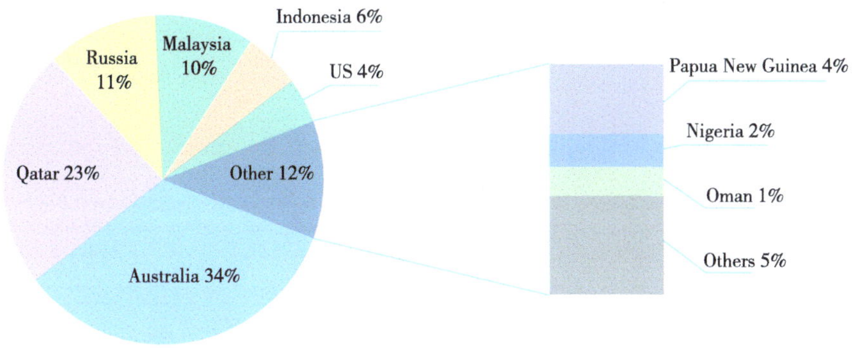

Fig. 8 China's LNG import sources in 2023

LNG Import Growth Turned Positive, but Still Lower Than 2021 Levels

In 2023, the national LNG import volume was 71.32 million tons (about 98.4 billion cubic meters), an increase of 10.3 billion cubic meters, a year-on-year increase of 11.7% (2022 was − 19.6%). Among them, spot purchases increased by about 7 billion cubic meters, a year-on-year increase of 27%. From the source, China's LNG imports mainly came from Australia, Qatar, Russia and other countries. Among them, Australia had the highest proportion of LNG, with an import volume of 24.16 million tons, a year-on-year increase of 10.6%, and the proportion has dropped from 34.5% in the previous year to 34%; Qatar's LNG import volume was 16.66 million tons, a year-on-year increase of 6.2%, and the proportion has dropped from 24.7% in the previous year to 23%; Russia's LNG import volume was 8.05 million tons, a year-on-year increase of 23.8%, and the proportion has increased from 10.3% in the previous year to 11% (Fig. 8).

The import landscape became more diversified. In 2023, there were as many as 12 LNG import companies nationwide, an increase of 4 from the previous year, including new companies such as Xintian making their first imports. The second-tier companies outside the three major oil companies (PetroChina, Sinopec, and CNOOC) imported 4.9 million tons of LNG (about 6.7 billion cubic meters), an increase of 47.4% year-on-year, accounting for 6.9% of the national LNG import volume, an increase of 2 percentage points from the previous year.

Analysis of Natural Gas Prices in China

Positive Progress in the Terminal Price Linkage Mechanism

In 2023, the state issued special guiding opinions on the natural gas terminal price linkage mechanism for the first time, and governments in various places combined

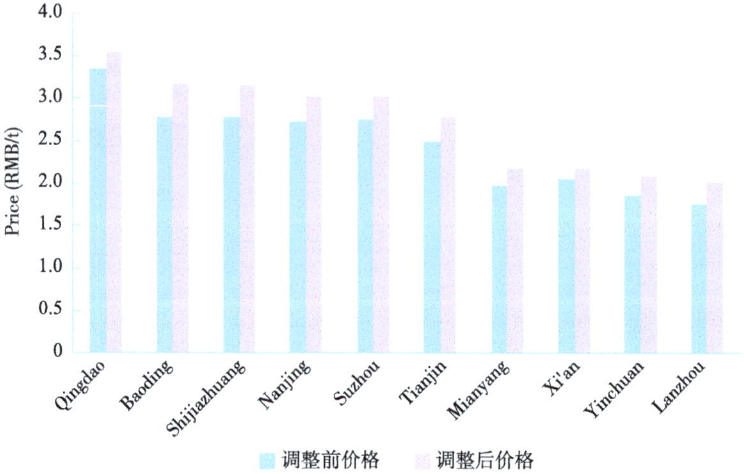

Fig. 9 Adjustment of residential gas prices in some cities in 2023

with reality to introduce or further improve the natural gas upstream and downstream price linkage mechanism from aspects such as linkage scope, start conditions, linkage cycle, and linkage procedures. Provinces such as Guizhou, Hubei, and Zhejiang require their cities to introduce corresponding linkage policies; Hangzhou, Ningbo and other places propose a price adjustment condition of a 5% change in comprehensive procurement costs; six cities in Hebei, as well as Nanjing in Jiangsu and Qingdao in Shandong, have adjusted the residential gas price, with an increase of 0.13 to 0.40 yuan/meter^3 (Fig. 9).

Domestic LNG Prices Fell Sharply

Affected by the downturn in international gas prices and relatively loose supply and demand, domestic LNG prices fell rapidly, with the annual average price falling to 4869 yuan/ton, a year-on-year decrease of 28% (Fig. 10). Looking at the trend within the year, the price first fell and then rose, with obvious seasonal characteristics. From January to August, the price continued to fall, dropping to the lowest point of the year in August; from September to December, the price quickly rebounded. Driven by the fall in LNG prices, domestic LNG consumption surged, with an annual consumption of 31.81 million tons, a year-on-year increase of 29%.

Steady Development of the Trading Center

In 2023, the Shanghai Petroleum and Natural Gas Trading Center continued to carry out pipeline gas market bidding, annual contracts, annual contract transfers, LNG capacity bidding, LNG window period monthly delivery and other transactions. The

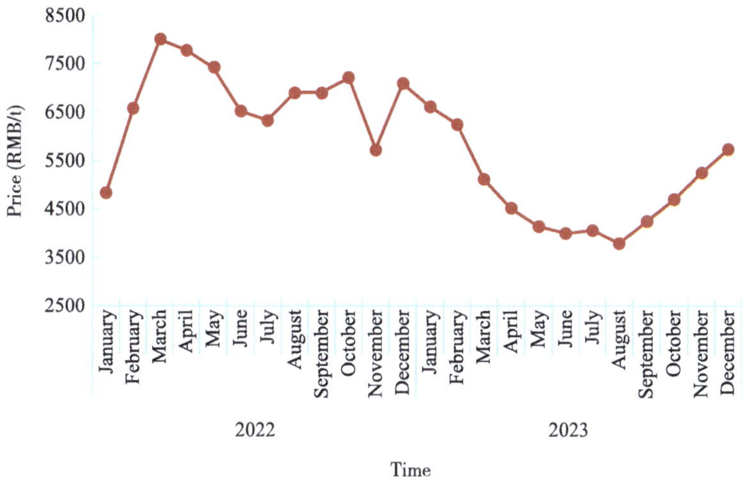

Fig. 10 Average domestic LNG market price

Chongqing Petroleum and Natural Gas Trading Center continued to carry out pipeline gas and medium and long-term warehouse receipt transactions. The National Pipeline Network Group launched the storage and transportation pass product in Shanghai and Chongqing Petroleum and Natural Gas Trading Center. The Tianjin International Oil and Gas Trading Center began trial operation at the end of the year. It is the first spot trading venue in China approved by the provincial people's government to carry out joint procurement. It is designed to launch capacity pre-sales, joint procurement, LNG order pass, warehouse receipt trading and other trading modes. In 2023, the pipeline gas unilateral trading volume of the Chongqing Petroleum and Natural Gas Trading Center was 38.2 billion cubic meters, with an average price of 2.296 yuan/meter^3; the LNG trading volume was 51,900 tons.

Development Status of Natural Gas Storage and Transportation in China

Steady Progress in Pipeline Construction

By the end of 2023, the total mileage of natural gas long-distance pipelines in China was 87,000 km. The first phase of the Mongolian Western Pipeline Project (Tianjin-Hebei Dingxing) that directly connects to Xiong'an officially put into production; the Guangxi branch line (Hengyang-Guilin) of the Qianjiang-Shaoguan gas pipeline project was successfully put into production; PetroChina's Tarim Oilfield Boz-Dabei natural gas export pipeline was successfully put into production; the Shuangtaizi gas storage pipeline project was put into production. The five main natural gas pipelines of the China-Russia East Line, West–East Gas Pipeline Line 1, Sichuan Gas Eastward Line 1, Qingning Pipeline, and Suwan Pipeline have all achieved interconnection in Jiangsu; the second phase of the Tianjin LNG terminal of the National Pipeline Network Group has been completed and successfully connected to the Mongolian Western Natural Gas Pipeline, achieving interconnection with multiple main natural gas pipelines such as the China-Russia East Line and Shaanxi-Beijing Pipeline; the Zhongkai Line and Pingtai Line interconnection project was successfully completed, providing more gas transmission channels for Wen 23 gas storage. In terms of pipeline construction, the welding of the main pipeline of the West–East Gas Pipeline Line 4 in Xinjiang has been completed, and the entire project is expected to be completed in October 2024; the Sichuan Gas Eastward Line 2 pipeline project officially started construction, and after the project is completed, it will deliver more than 200 billion cubic meters of natural gas to the areas along the line each year, and the interconnection and mutual protection capabilities of the Sichuan and Chongqing pipeline network and the national pipeline network will be further enhanced; the Nenjiang branch line of the China-Russia East Line natural gas pipeline project officially started construction. The construction of provincial natural gas pipeline networks has been accelerated. Many provinces including Shandong, Hebei, Jiangsu, Shanghai, Chongqing, and Tianjin have proposed goals to accelerate the construction of natural gas pipeline networks, improve the layout of main pipelines, and promote interconnection of networks.

The policy system for fair opening of oil and gas pipeline facilities continues to improve. The state has issued a number of policies around carrier access, industry supervision, and pipeline construction. The "Natural Gas Pipeline Facility Carrier Access and Exit Management Measures (Draft for Comment)" stipulates the conditions that companies should have to become carriers and the related functions of pipeline operating companies. The "Guiding Opinions on Energy Work in 2023" proposes to improve the operation methods related to the capacity allocation of oil and gas pipeline facilities, strengthen the supervision of the natural monopoly link of oil and gas pipelines, and deepen the fair opening of oil and gas pipeline facilities. The 2023 National Oil and Gas Pipeline Planning, Construction and Protection Work Conference proposed that the annual work focus is to implement the construction of the "National One Network" for petroleum and natural gas and major oil and gas pipeline projects.

New LNG Receiving Capacity Set a Historical Record

In 2023, 4 new LNG terminals were built nationwide, and 2 were expanded, adding a total of 18.8 million tons/year of LNG receiving capacity (Table 1). By the end of 2023, there were 28 LNG terminals in operation nationwide, with a total unloading capacity of 116 million tons/year. The tank capacity reached 18 million cubic meters, capable of storing 100 billion cubic meters of natural gas. Currently, there are more than 30 LNG terminals under construction nationwide, and the receiving capacity will exceed 210 million tons/year after completion.

Table 1 2023 China LNG terminals production situation

Project name	Additional capacity (10,000 tons/year)	Affiliated company	Status
Xintian Tangshan LNG	500	Hebei Construction Investment	In operation
Guangzhou LNG Emergency Peak Shaving Gas Source Station	100	Guangzhou Gas	In operation
Wenzhou LNG Ternimal	300	Zhejiang Energy Group	In operation
Tianjin Nangang LNG Emergency Reserve Station	500	Beijing Gas	In operation
Tianjin Nangang LNG Terminal	480	Sinopec	Expansion in operation
Jiangsu Qidong LNG Terminal	Addition of a 200,000 cubic meter tank	Xinjiang Guanghui	Expansion in operation

The proportion of LNG receiving capacity of the second-tier companies continues to increase. As of the end of 2023, the National Pipeline Network Group operates 7 terminals, with a receiving capacity of 27.6 million tons/year, accounting for 24%; the proportions of PetroChina, Sinopec, and CNOOC are 11%, 15%, and 25% respectively, with PetroChina and CNOOC's proportions declining compared to the previous year, and Sinopec's proportion rising; the proportion of the second-tier companies' LNG receiving capacity is 25% (Fig. 11), a significant increase of 10 percentage points year-on-year.

Several provinces have strengthened the price management of LNG terminal gasification services. Hainan Province issued the "Notice on Strengthening the Price Management of Imported Liquefied Natural Gas Terminal Gasification Services", Shandong Province issued the "Notice on Clarifying the Price of LNG Terminal Gasification Services", Guangxi Zhuang Autonomous Region issued the "Notice on Strengthening the Price Management of Imported Liquefied Natural Gas Terminal Gasification Services", and Guangdong Province issued the "Imported Liquefied Natural Gas Terminal Gasification Service Price Management Measures", clarifying the price of LNG terminal gasification services.

Gas Storage Peak Shaving Capacity Reaches a New Level

The construction of underground gas storage facilities is accelerating. PetroChina's Changqing Oilfield Leilongwan gas storage facility was successfully built and put into operation, the first water-flooded oil and gas reservoir renovation gas storage project—Liaohua Oilfield Ma 19 gas storage pilot project was put into operation for

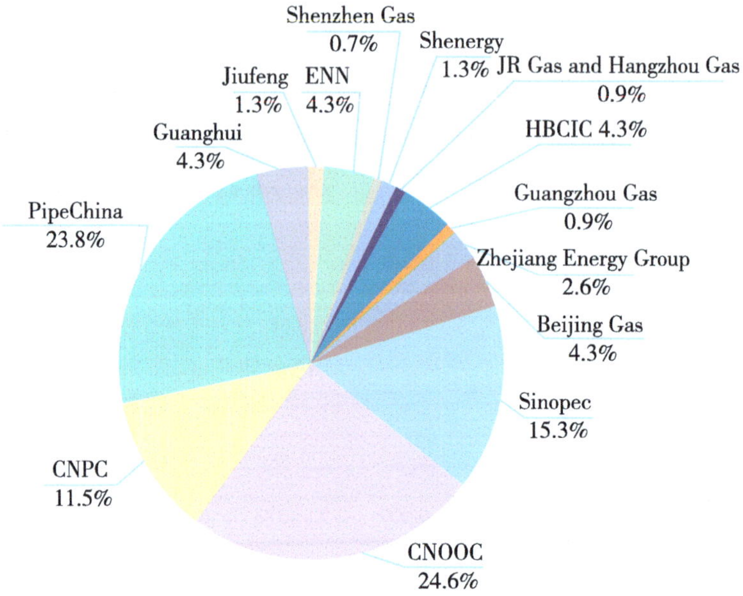

Fig. 11 2023 proportion of LNG receiving capacity of various companies

gas injection, Tuha Oilfield Wenjisan gas storage group Wenba gas storage facility was successfully tested and put into operation, Tarim Oilfield Kekoya gas storage gas injection system was launched. Sinopec's Wen 24 gas storage facility was put into operation for gas injection, and the number of Central Plains gas storage facilities increased to 6; Wen 23 gas storage facility Phase II project No. 1 well site was put into operation. By the end of 2023, there were 29 active gas storage facilities (groups) nationwide, forming a gas storage peak shaving capacity of 230 billion cubic meters, an increase of 19.8% year-on-year, accounting for 5.9% of domestic natural gas consumption, and the ability to peak shave and stabilize gas supply was further enhanced.

The construction of provincial, municipal, and enterprise-level reserve stations is progressing steadily. The first phase of the Beijing Gas Tianjin Nangang Liquefied Natural Gas (LNG) Emergency Reserve Project has been put into operation, adding 480 million cubic meters of natural gas emergency reserve capacity. The first phase of the Guizhou Gas Reserve and Emergency Peak Shaving Facility Construction Project, the Changchun City Urban LNG Emergency Peak Shaving Storage and Distribution Station, the Sichuan Province Suining Gas Storage Peak Shaving Base and other projects have been officially completed. The Guangzhou Energy Huizhou LNG Terminal is about to be put into operation, the second phase expansion project of the Shenzhen Gas Shenzhen Natural Gas Reserve and Peak Shaving Library has entered the construction stage, and the Hanas Putian LNG Project has started. In addition, many provinces (cities) have also proposed in documents such as the "Carbon Peak Implementation Plan" to vigorously promote the construction of coastal LNG terminals and supporting wharves, pipelines and other facilities, coordinate the planning and layout of gas storage facilities such as LNG terminals, and build a multi-level gas storage peak shaving system including coastal LNG storage tanks.

(This section was written by: Wu Minjie, Duan Tianyu, Sun Wenyu, Zhu Jingyi, Cui Xinyue. Reviewed by: Duan Zhaofang)

Outlook for China's Natural Gas Market

By 2024, with the continuous improvement of the domestic economy, China's natural gas consumption is expected to grow relatively rapidly. Domestic natural gas production is anticipated to keep increasing, while imported natural gas is also projected to see relatively rapid growth. Infrastructure construction is expected to advance swiftly, and the newly added LNG receiving capacity is likely to reach a new high.

Outlook for China's Natural Gas Demand in 2024

Demand Continues to Grow Relatively Fast

In 2024, the national natural gas consumption is expected to be 415.7 billion cubic meters[2], an increase of 6.1% year-on-year. The main reasons are as follows: First, the domestic economy will maintain a steady growth trend, promoting gas demand. Second, international gas prices are falling, and the economy of natural gas continues to recover. Third, in the context of building a new energy system, natural gas will continue to play an important role in energy transition. However, the stable supply of coal and the rapid growth of new energy may suppress the growth space of gas demand.

Relatively Rapid Growth in Demand for Gas in Major Industries

In 2024, city gas consumption is expected to grow steadily, with a year-on-year increase of 6.1% to 135.5 billion cubic meters; industrial fuel gas demand is expected to increase by 6.8% year-on-year to 164.8 billion cubic meters; power generation gas is expected to increase by 7.5% year-on-year to 76.3 billion cubic meters; fertilizer and chemical gas is expected to increase by 1.0% year-on-year to 39.1 billion cubic meters. In the structure of natural gas consumption, city gas consumption will account for 32.6%, industrial fuel gas consumption 39.6%, power generation gas consumption 18.4%, and fertilizer and chemical gas consumption 9.4% (Fig. 12).

[2] The corresponding apparent consumption volume is 418.9 billion cubic meters.

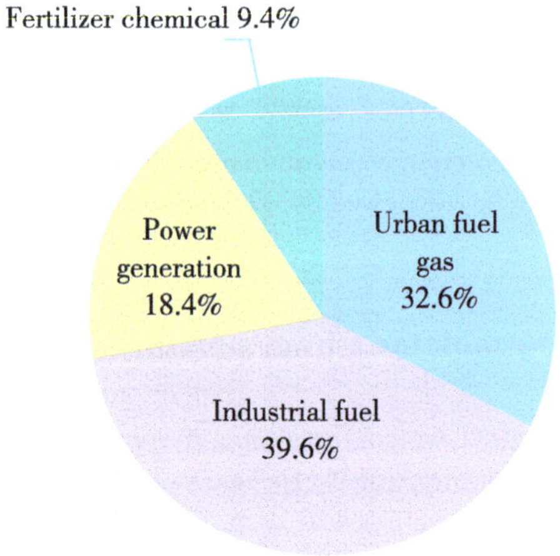

Fig. 12 Structure of China's natural gas consumption in 2024

Relatively rapid growth in city gas. Urbanization construction continues to advance, the coverage of city gas pipelines is further expanded, the popularity of city gas is steadily improved, and the gas-using population is growing rapidly. China continues to carry out comprehensive air pollution control actions, promotes clean heating according to local conditions, supports stable growth of heating gas, and the good economy of vehicle LNG and downstream logistics demand support the rapid growth of vehicle LNG demand.

Relatively rapid growth in industrial fuel gas. Under the background of economy-boosting policies, it is expected that investment will be stable and increasing in 2024, infrastructure investment will continue to exert force, manufacturing investment will grow steadily, real estate investment will narrow down, exports will steadily recover, overall boosting demand for gas products such as steel, glass, ceramics, etc., driving rapid growth in industrial gas.

The fastest growth in power generation gas. Economic growth drives power demand, and the growth rate of electricity consumption in the whole society remains stable. The demand for cooling/heating power grows rapidly during the peak summer/winter periods, and gas power has a large peak shaving space. The installed capacity of gas power is growing steadily, and it is expected that 6 million kilowatts of new gas power will be added in 2024, and the total installed capacity is expected to increase to 130 million kilowatts. It is unlikely that hydropower will continue to weaken significantly, coupled with the sufficient supply of other power sources, the demand for gas power to fill the gap is limited, and the utilization hours are basically the same year-on-year.

Slight growth in fertilizer and chemical gas. With the continuous economic recovery and improvement, the continuous advancement of policies such as "expanding domestic demand" and "dual carbon", methanol demand is expected to rebound, and the start-up rate of gas companies is expected to rise; the stable supply and price of fertilizer continue, and the good profit of gas urea supports the production enthusiasm of enterprises.

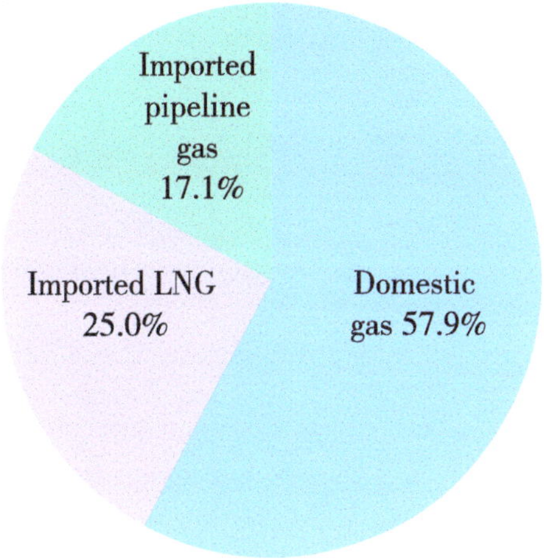

Fig. 13 Structure of China's natural gas supply in 2024

Outlook for China's Natural Gas Supply in 2024

Continuous Growth in Natural Gas Production

In 2024, domestic upstream companies will continue to actively increase reserves and production, and the annual natural gas production is expected to be 245.8 billion cubic meters, a year-on-year increase of 4.5%; domestic gas accounts for 57.9% of resource supply (Fig. 13).

Rapid Growth in Natural Gas Imports

In 2024, the annual natural gas import volume is expected to be 179.1 billion cubic meters, an increase of 8.2% year-on-year; the foreign dependence is 43.1%, an increase of 0.8 percentage points from the previous year. In terms of pipeline gas, the import capacity of the China-Russia East Line continues to increase, and the annual import volume of pipeline gas is expected to be 72.6 billion cubic meters, an increase of 8.2% year-on-year. In terms of LNG, LNG terminals such as Sinopec Longkou in Shandong and Guoxin Rudong in Jiangsu will be put into operation one after another, and the newly added LNG receiving capacity nationwide will exceed 50 million tons/year, setting a new historical record; CNOOC, Shenneng, Zheneng and other companies have started to fulfill their new long-term LNG contracts, becoming the main source of LNG import increase, and the annual LNG import volume is expected to be 77.11 million tons (about 106.4 billion cubic meters), an increase of 8.1% year-on-year.

(Authors of this section: Wu Minjie, Cui Xinyue Reviewer: Duan Zhaofang)

Development of Domestic and Foreign Oil and Gas Exploration and Development Industry

In 2023, global investment in oil and gas exploration and development continued to grow, and it is expected to continue to grow in 2024. Global oil and gas reserves slightly increased; both oil and gas production increased, and the growth rate of natural gas production was higher than that of oil. The number of global oil and gas discoveries and new discoveries in 2023 were both lower than the previous year.

Current Status and Trends of Global Oil and Gas Exploration and Development

In 2023, global investment in oil and gas exploration and development continued to grow, with a year-on-year growth of 10.6%, a growth rate lower than the previous year. Global oil and gas reserves both increased, with oil increasing by 1.3% and natural gas by 2.2%. Oil and gas production continued to grow, with oil increasing by 1.6% and natural gas by 0.8%. In 2023, there were a total of 213 oil and gas discoveries worldwide, lower than the 305 of the previous year; new oil and gas reserves were 112.4 billion barrels of oil equivalent, lower than the 192.1 billion barrels of oil equivalent of the previous year. The average discovery was 52.77 million barrels of oil equivalent, far lower than the previous year's 63 million barrels of oil equivalent.

Slight Increase in Global Oil and Gas Reserves

In 2023, the global remaining proven recoverable reserves of oil were 247.55 billion tons, an increase of 13.4 million tons from the previous year, an increase of 0.05%; the increase mainly came from North America and Latin America, which increased by 19.1 million tons and 16.4 million tons respectively. The global remaining proven

X. Qian et al., *Global Oil and Gas Industry Development Report: 2023*,
https://doi.org/10.1007/978-981-95-2375-7_7

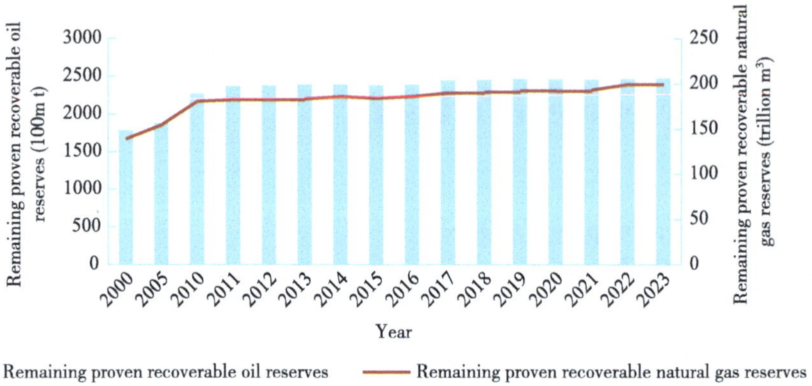

Fig. 1 Global remaining proven recoverable reserves of oil and gas from 2000 to 2023

recoverable reserves of natural gas were 191.96 trillion cubic meters, an increase of 220 billion cubic meters from the previous year, an increase of 0.11%; the increase mainly came from Indonesia and Russia, which increased by 180 billion cubic meters and 170 billion cubic meters respectively (Fig. 1). According to the analysis of oil and gas exploration investment and discovery results, it is expected that the global remaining proven recoverable reserves of oil and gas will decline in 2024.

Global Oil and Gas Discoveries and Volumes Decrease, Average Scale Shrinks Significantly

In 2023, there were 213 oil and gas discoveries worldwide, with a total of 112.4 billion barrels of oil equivalent added, and an average discovery size of 52.8 million barrels of oil equivalent; compared with the previous year, the number of oil and gas discoveries decreased, the discovered reserves decreased, and the average size decreased significantly, reaching a new low in nearly 5 years (Figs. 2 and 3). In 2023, there were only 4 major discoveries of more than 500 million barrels of oil equivalent, 3 of which were natural gas discoveries. Of these 4 major discoveries, 2 were located on land in Iran, and the other 2 were located in the deep sea of Indonesia and Guyana (Table 1).

More new discoveries of gas than oil. In the global oil and gas discoveries in 2023, there were 125 new oil fields discovered, with an additional recoverable reserve of 4.38 billion barrels; 88 new natural gas fields were discovered, with an additional reserve of 1.08 trillion cubic meters. Calculated in oil equivalent, the additional reserves of oil and gas were 4.38 billion barrels of oil equivalent and 6.86 billion barrels of oil equivalent, accounting for 39% and 61% respectively.

National oil companies are the main contributors to new discoveries. In the oil and gas discoveries in 2023, the oil and gas reserves discovered by national oil

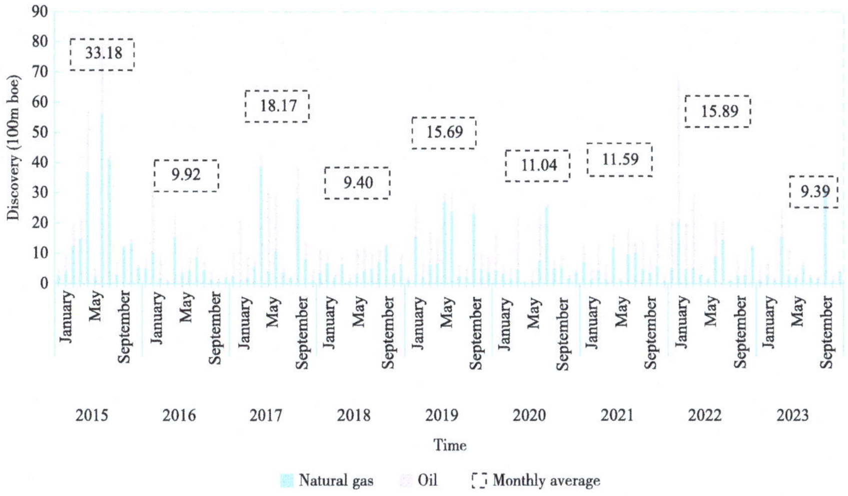

Fig. 2 Monthly statistics of global oil and gas discoveries from 2015 to 2023

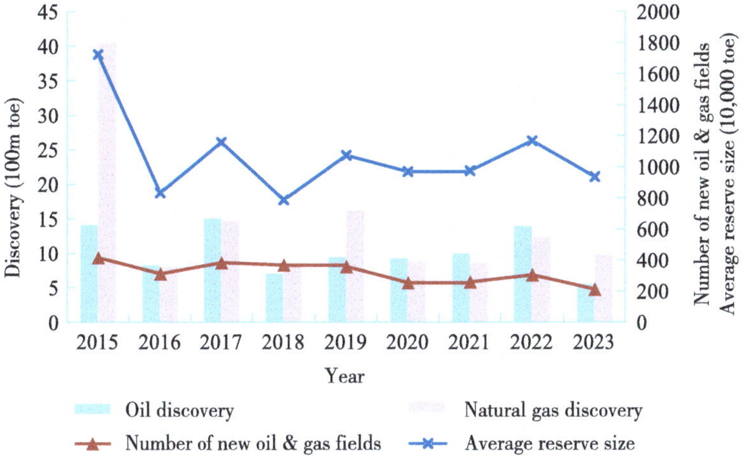

Fig. 3 Statistics of global oil and gas discoveries from 2015 to 2023. *Data source* S&P Global, 2023

companies amounted to 6.99 billion barrels of oil equivalent, accounting for 62.2% of the global discovered reserves. Independent oil companies discovered oil and gas reserves of 2.35 billion barrels of oil equivalent, accounting for 20.9% of the global discovered reserves. International oil companies discovered oil and gas reserves of 1.90 billion barrels of oil equivalent, accounting for only 16.9% of the global discovered reserves. Among the reserves discovered by national oil companies, the National Iranian Oil Company stood out, making two major natural gas discoveries, with a total discovery of 577.3 billion cubic meters of natural gas.

Table 1 Top ten global oil and gas discoveries in 2023

No.	Country	Oil and gas field	Discoverer	Onshore/Offshore	Type of oil and gas	2P reserves (Million barrels of oil equivalent)
1	Iran	Shahini 1	NIOC	Onshore	Gas	2611
2	Iran	Cheshmeh Shoor 1	NIOC	Onshore	Gas	843
3	Indonesia	Geng North 1	Eni	Offshore	Gas	607
4	Guyana	Lancetfish 1	ExxonMobil	Offshore	Oil	580
5	Namibia	Jonker 1X	Shell	Offshore	Oil and gas	429
6	Guyana	Fangtooth SE 1	ExxonMobil	Onshore	Oil	397
7	Suriname	Roystonea 1	PETRONAS	Offshore	Oil and gas	327
8	Turkey	Sehit Aybuke Yalcin	TPAO	Onshore	Oil	325
9	Australia	Canyon 2	Cypress petroleum	Offshore	Oil and gas	320
10	Indonesia	Layaran 1	MP Holding RSC Ltd	Onshore	Oil and gas	306

Data source S&P Global, 2023

Increase in Global Oil and Gas Approved and Put into Production Projects

The investment in new development projects approved for global oil and gas has increased significantly. According to Rystad's forecast, the investment in new development projects approved for global oil and gas in 2023 is about 191.04 billion US dollars, an increase of about 60% year-on-year. Among them, onshore projects are about 75.52 billion US dollars, an increase of 94.8% year-on-year; offshore projects are about 115.52 billion US dollars, an increase of 43.7% year-on-year. In 2023, North America is expected to approve new projects with an investment of about 70 billion US dollars, mainly distributed in the United States and Mexico. South America is expected to approve new projects with an investment of about 30 billion US dollars, mainly distributed in Brazil and Guyana. The Asia–Pacific region is expected to approve new projects with an investment of about 13 billion US dollars, mainly distributed in China and India. Among the major projects that have been approved for FID, there are many and large LNG projects, such as the US Rio Grande LNG Phase 1 (total investment of 11.4 billion US dollars), Port Arthur LNG (with a total investment of 13.8 billion USD) and Plaquemines LNG Phase 2 (with a total investment of 5.1 billion USD) projects, Qatargas' QatarGas T12 LNG and

QatarGas T13 LNG projects in Qatar (with a total investment of 11.5 billion USD), etc. In addition, exploration and development projects in Central and South America, such as Equinor's Pao de Acucar and Seat projects in Brazil (with a total investment of 7.9 billion USD), ExxonMobil's Uaru (Greater Liza) and Mako (Greater Liza) projects in Guyana (with a total investment of 6.7 billion USD), and Woodside's Trion project in Mexico (with a total investment of 7.4 billion USD), all indicate future growth areas for oil and gas production.

Investment in new oil and gas development projects is shrinking on land and expanding at sea. According to Rystad, the global investment in new oil and gas projects is expected to be around 117 billion USD in 2023, an increase of 19.4% year-on-year. Of this, investment in land projects is 32 billion USD (accounting for 27% of total investment), a decrease of 22.0% year-on-year; investment in offshore projects is 85 billion USD (accounting for 73% of total investment), an increase of 49.1% year-on-year. The Asia–Pacific region has the highest share of global investment in new oil and gas development projects, reaching 34%. South America, North America, the Middle East, and Africa account for 15%, 14%, 13%, and 12% respectively, Europe only accounts for 5%, and Russia only accounts for 6%.

Table 2 shows the major new oil and gas development projects to be put into production globally in 2023.

Global Oil and Gas Production Increase

In 2023, the global crude oil production is 4.45 billion tons, an increase of 1.6% year-on-year. Among them, North America has the largest increase (66.86 million tons), followed by Latin America (24.71 million tons) (Fig. 4), increasing by 4.3% and 7.6% respectively. Countries with a production increase of more than 5 million tons include the United States (64.9 million tons), Brazil (19.19 million tons), Mexico (6.74 million tons), and China (6.55 million tons). Due to the impact of the Ukraine crisis, Russia reduced production by 12 million tons.

In 2023, the global natural gas production reached 42.8 trillion cubic meters, an increase of 0.8%. Affected by the Ukraine crisis, the current Russo-European pipeline gas trade only has the flow through Ukraine and Turkey, totaling about 70 million cubic meters per day. The reduction in Russo-European pipeline gas trade has led to a continued decline in Russia's domestic natural gas production. It is estimated that the natural gas production in 2023 will be 6.42 trillion cubic meters, a decrease of more than 300 billion cubic meters from 2022. The strong demand for LNG in Europe and the recovery of LNG demand in Asia stimulate the export of US LNG and drive the growth of natural gas production. It is estimated that the US natural gas production in 2023 will be 11.525 trillion cubic meters. The shortage of natural gas in Europe also stimulates the increase in natural gas production in the Middle East and Africa. It is estimated that the natural gas production in the Middle East and Africa in 2023 will reach 7.388 trillion cubic meters and 2.638 trillion cubic meters, respectively, an increase of 58 billion cubic meters and 77 billion cubic meters (Fig. 5).

Table 2 Major new oil and gas development projects to be put into production globally in 2023

Project name	Country	Operator	Offshore/ Onshore	Water depth (meters)	Recoverable resources (Million barrels of oil equivalent)
Mero 2 (x-Libra NW) (Sepetiba)	Brazil	Petrobras	Offshore	1500–2250	300–1000
Payara (Prosperity)	Guyana	ExxonMobil	Offshore	1500–2250	300–1000
Buzios (x-Franco) V (Almirante Barroso)	Brazil	Petrobras	Offshore	1500–2250	300–1000
Marlim Revitalization 1 (Anita Garibaldi)	Brazil	Petrobras	Offshore	600–800	300–1000
Marlim Revitalization 2 (Anna Nery)	Brazil	Petrobras	Offshore	600–800	300–1000
Argos (Mad Dog Phase 2) (GC825)	USA	bp	Offshore	1500–2250	300–1000
A-2	India	Indian Oil Corporation	Offshore	600–800	30–300
Vito (MC940)	USA	Shell	Offshore	1000–1500	30–300
Lufeng 12–3	China	SK Earthon	Offshore	200–300	30–300
Sangomar Phase 1 (Leopold Sedar Senghor)	Senegal	Woodside	Offshore	1000–1500	30–300
Marmul EOR Phase 3	Oman	Petroleum Development Oman	Onshore	–	30–300
Kovyktinskoye	Russia	Gazprom	Onshore	–	>1000

In 2024, global natural gas production will continue to grow. According to EIA estimates, global natural gas production will increase by 2% in 2024, a higher growth rate than in 2023.

Exploration and Development Investment Continues to Rise

In 2023, global upstream capital expenditure in oil and gas continues to grow. Affected by factors such as the Ukraine crisis, the uneven global economic recovery, changes in the pace of energy transition, and future oil demand expectations, oil companies are still cautious about exploration and development investment. According to S&P Global research, global exploration and development investment expenditure in 2023 is $568.7 billion, an increase of $54.6 billion from 2022, an

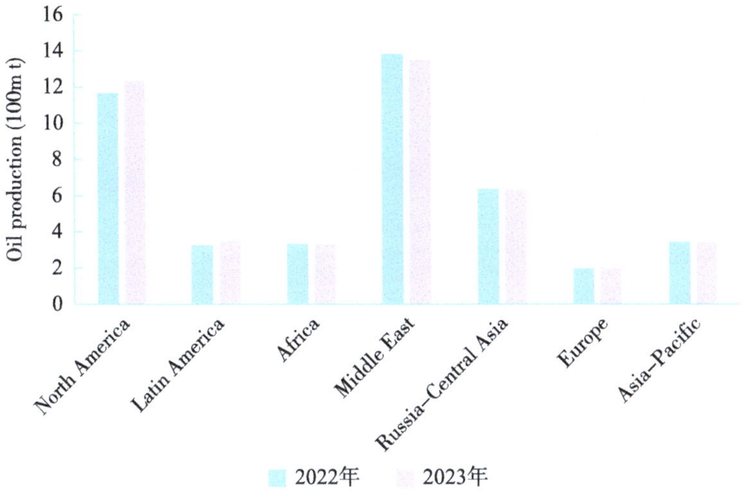

Fig. 4 Oil production in various regions in 2022 and 2023

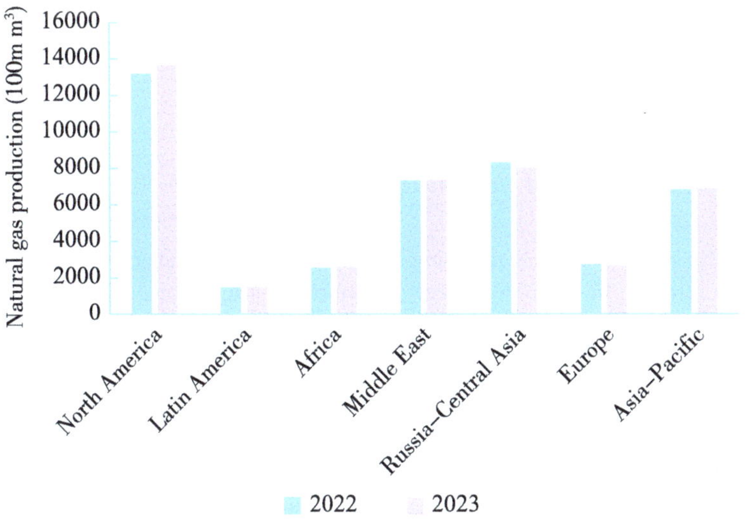

Fig. 5 Natural gas production in various regions in 2022 and 2023

increase of 10.6%; investment in the Russia-Central Asia region has negative growth, a year-on-year decrease of 4.5%.

It is expected that in 2024, global upstream capital expenditure in oil and gas will increase by $12 billion, a year-on-year increase of 2.1%. The investment performance varies from region to region. Investment in Latin America and Africa will increase by $7 billion and $5 billion respectively, with growth rates of 9.7% and 9.8% respectively. Investment in the Russia-Central Asia region will rebound, expected to

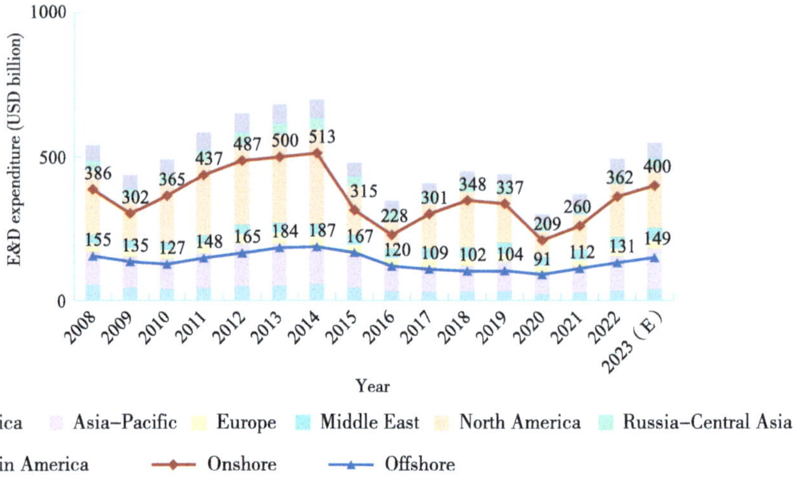

Fig. 6 Exploration and development investment in various regions around the world from 2008 to 2023. *Data source* S&P Global, September 2023

increase by $3 billion, an increase of 8.3%. Investment in the Asia–Pacific and North America regions will decrease, with decreases of 3.7% and 0.2% respectively (Fig. 6).

Growth in the Scale of the Oilfield Service Market

In 2023, the global oilfield engineering technical service market size is about $299.5 billion, continuing to maintain a large growth rate, with a year-on-year increase of nearly 10% (Fig. 7 and Table 3). Among the various sectors of oilfield services, all have seen revenue growth. Relatively speaking, the oilfield measurement and testing services and drilling and completion services sectors perform better than other sectors. The market size growth rates of the oilfield production services and geophysical equipment and services sectors are 7.5% and 8.6% respectively, lower than the average level. It is expected that in 2024, the global engineering technical service market revenue will continue to grow, but the growth rate will decline.

In 2023, due to various factors, the growth rate of the number of drilling rigs in use is low. The size of the global oilfield engineering technology service market increased by 10.2% year-on-year, the number of drilling rigs in use (4432) increased by 71 year-on-year, an increase of only 1.6%, the growth rate of the number of drilling rigs in use is far less than the growth of exploration and development investment (Fig. 8). There are two main reasons: one is that oil and gas producers in North America have reduced drilling expenditures, with 872 drilling rigs in use, a decrease of 22 year-on-year, a decrease of 2.5%; the second is that due to sanctions from the West, the number of drilling rigs in use in the Russia-Central Asia region decreased by 57, a decrease of 4.1%.

Fig. 7 Size and growth rate of global oilfield engineering technology service market from 2015 to 2023. *Data source* Spears & Associates, October 2023

Table 3 Size of global oilfield service sector market in 2022 and 2023

Technical service sector	2022 revenue (Billion USD)	2023 revenue (Billion USD)	Average annual growth (%)	Proportion of 2023 revenue to total revenue (%)
Geophysical equipment and services	64	69	8.6	2.3
Drilling and completion services	1387	1538	10.9	51.2
Oilfield production services	583	627	7.5	20.9
Logging and testing services	185	214	15.5	7.1
Oilfield engineering construction services	507	556	9.7	18.5
Total	2726	3004	10.2	100.0

Data source Spears & Associates, October 2023

It is expected that in 2024, the number of drilling rigs in use in most regions of the world will increase. Among them, the growth rate of the number of drilling rigs in use in the Middle East is expected to exceed 10%, the number of drilling rigs in use in Asia–Pacific and Europe will maintain a growth rate of more than 5%, and

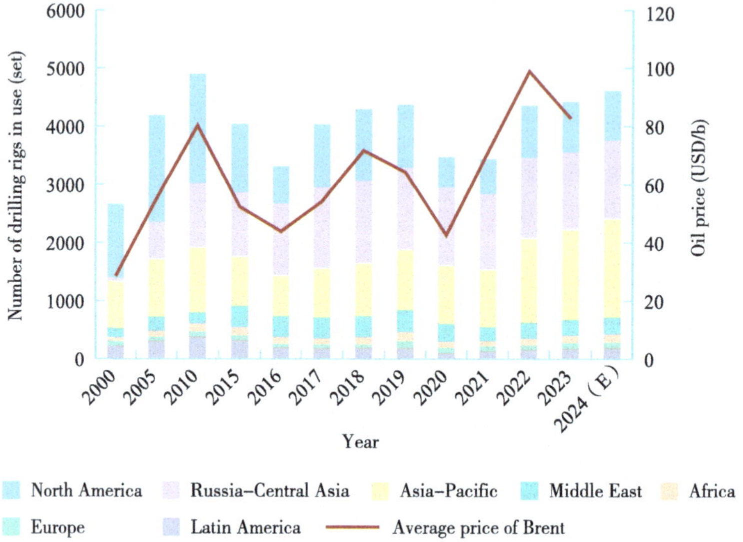

Fig. 8 Changes in the number of global drilling rigs in use and oil prices from 2000 to 2024. *Data source* Spears & Associates, September 2023

the number of drilling rigs in use in North America will decrease, with a decrease of nearly 3.0%.

New Developments in Upstream Oil and Gas Technology

Innovation and Development of Exploration and Development Theory and Technology

The global oil and gas industry has continuously made key technological advancements in the field of exploration and development. The integration of artificial intelligence with the oil and gas industry is deepening, giving birth to new technologies and new industries such as intelligent exploration and intelligent oilfields, greatly promoting the increase in oil and gas reserves, cost reduction and efficiency enhancement, and safe production. Promoting integrated collaboration throughout the process, exploration and development engineering can perform geophysical parameter calculations, geological model construction, fracture description, real-time analysis during drilling, fracture prompts during drilling, optimized fracturing design, etc., through an integrated platform, greatly improving the efficiency of exploration and development. Reservoir digital twin modeling technology, based on the integrated technology of artificial intelligence seismic inversion-reservoir description-reservoir simulation based on big data, can efficiently and dynamically approximate real reservoirs, effectively solving geological and reservoir uncertainty problems.

In the field of unconventional oil and gas, "sweet spot" prediction technology, through the combined use of geophysical methods, microseismic and core data, can determine favorable areas and undrilled target areas, becoming a key means to improve reservoir drilling encounter rate and reduce exploration costs. Three-dimensional development technology, according to the characteristics of different depth reservoirs, adopts different development technology countermeasures, through the staggered deployment of horizontal wells and the segmented implementation of vertical wells, to achieve multi-layer three-dimensional development, significantly improving the degree of resource utilization.

Facing the increasing global attention to environmental protection, as well as the challenges of environmental monitoring and sustainable development technologies, many oil companies have deployed CCUS-EOR projects and used big data, predictive analysis, artificial intelligence, automation, and 3D visualization technologies to monitor the entire process of injection and extraction in real time, to achieve the safety, accuracy, and economy of development, production, and management. In addition, the use of solar energy, wind energy, and other green energy to heat cold water to produce steam, by reducing the fuel consumption and steam usage in the thermal recovery process of heavy oil, has overcome the main obstacles of high fuel consumption, unfriendly environment, and high cost in the thermal recovery of heavy oil reservoirs, achieving the goal of reducing thermal recovery costs and significantly improving the recovery rate.

Progress in Engineering Technology and Equipment Innovation

The geophysical exploration industry has made progress mainly in green, efficient seismic acquisition, artificial intelligence data processing interpretation, and digital transformation. The progress of land node acquisition, marine node acquisition, and compressed sensing seismic acquisition processing technology is perfect; machine learning methods are steadily advancing in the application of seismic data processing interpretation, promoting full waveform inversion, full wavefield imaging, and reservoir prediction and seismic interpretation technology to a new level; leading geophysical technology service companies such as CGG and PGS have increased their efforts in digital transformation development, upgraded their multi-user data ecosystems, and launched OaaS high-performance computing and AI solutions. Geophysical technology methods continue to advance in many fields such as offshore wind energy development, geothermal resource evaluation, CCS/CCUS.

The logging industry has made breakthroughs mainly in drilling logging, imaging technology, machine learning, digitization, and integration. International oil service companies such as Schlumberger have each launched new drilling logging technology and equipment, improving the accuracy of unconventional reservoir data. Imaging technology has made breakthroughs in improving the clarity of imaging data. The introduction of machine learning for logging data interpretation, data prediction, and data inversion has improved the efficiency of logging interpretation. With the gradual promotion of the OSDU exploration and development data platform, various oil companies and oil service companies have launched their own digital platforms, and the collection, processing, interpretation, and archiving of logging data are moving towards digital transformation.

Foreign drilling technology advancements are concentrated in the fields of drilling equipment, tools, instruments, materials, and software. In the field of equipment, tools, and instruments, the intelligent pressure drilling system launched by Opla Energy Company in the United States, equipped with proprietary control software, can process data from the ground and underground in real time, significantly reducing non-productive time and speeding up significantly. Enteq Technology Company in the UK has launched a directional tool that relies on the hydraulic difference at the drill bit working face for guidance, with a length of only 1.52 or 1.83 m, strong guidance ability, and the design of the rotary guidance mode has a build-up ability of 15°/30 m. Paradigm Company in Finland has launched a downhole drill pipe tractor with a length of only 4.57 m, consisting of 6 freely rotating traction sleeves, which can apply stable drilling pressure, significantly reduce friction and stick–slip, and help extend the length of the horizontal section. In the field of materials, Schlumberger has launched a new cement-free well cementing material—geo-polymer, which can be used as a low-carbon alternative to traditional well cementing cement, helping to reduce well cementing costs and significantly reduce carbon emissions during the production process of well cementing materials, which is a breakthrough in well cementing material technology. Halliburton has launched a new nano-particle well wall sealing agent, which is a nano-composite suspension, suitable for water-based drilling fluid systems, and its performance is significantly better than traditional well wall sealing agents. In the field of software, Helmerich & Payne Company has integrated automated drilling software into automated drilling rigs. The software algorithm combined with automated sliding drilling technology can efficiently control the wellbore trajectory.

Key Focus

Global oil and gas upstream costs will continue to rise

Upstream costs include capital costs (exploration and development) and operating costs. Affected by inflation, rising raw material prices, and supply risks, the global oil and gas upstream break-even cost increased by 5.2% year-on-year in 2022, to $52.47/barrel. According to S&P Global, the global oil and gas upstream break-even cost will further increase from 2023 to 2025.

There Are Large Differences in Upstream Costs Between Regions.

Affected by factors such as geological conditions (including reserve scale, underground structure, lithology, etc., development and production technology, labor and equipment costs, tax policies, contract stability, etc.), there are large differences in oil upstream costs between regions. Among them, the costs in the Middle East and

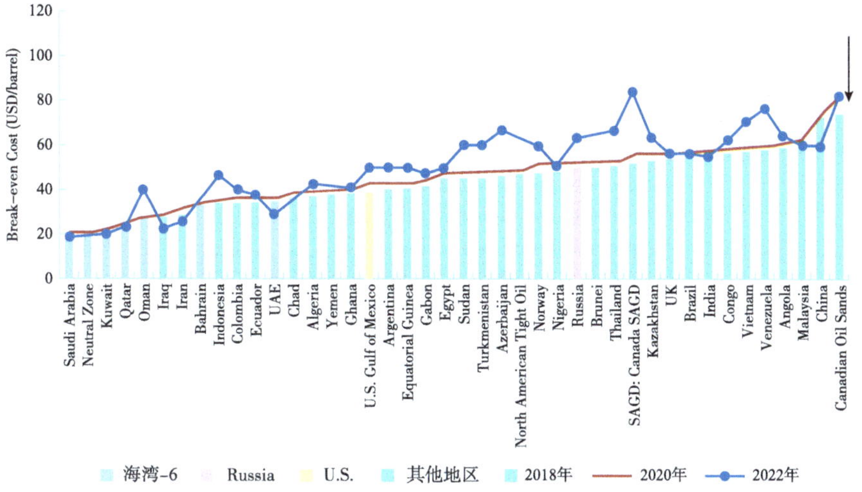

Fig. 9 Upstream cost situation of major oil producing countries worldwide

Africa are lower, and the upstream costs in Russia-Central Asia and North America are higher (Fig. 9).

The per-barrel costs of typical oil and gas projects in the Middle East, Africa, South America, Europe, Asia, North America, and the Commonwealth of Independent States (CIS) are $30.68, $52.81, $53.68, $56.86, $57.61, $61.10, and $66.35 respectively. The costs of Saudi Arabia's onshore and offshore projects are the lowest globally, at only $23.28/barrel and $24.66/barrel. The costs of onshore projects in Africa, such as those in Gabon and Nigeria, are also only $35/barrel and $36.62/barrel. The cost of Canadian oil sands projects ranks first globally, reaching as high as $104.11/barrel.

The Global Upstream Oil and Gas Costs Will Continue to Increase Before 2025.

The breakeven cost of typical global oil and gas upstream projects is on the rise. The main factor for the increase in upstream costs is the soaring inflation driving up the prices of raw materials. Secondly, the global oil and gas supply risk has increased under the Ukraine crisis, pushing up upstream costs. It is expected that between 2023 and 2025, global upstream capital costs and operating costs will increase by 9% and 13% respectively. Among them, the operating cost index in South America will increase by nearly 17%, and investment will increase by 6% ($3 billion) over three years, mainly from offshore projects in Brazil and Guyana. In Asia, due to intensified competition for labor and raw materials, the upstream cost index growth rate exceeds 12%. The indices for North American upstream capital costs and operating costs will increase by 7% and 10.5% respectively, mainly due to increased shale oil and gas production driving up labor and drilling costs. The indices for European upstream capital costs and operating costs will increase by 9% and 14% respectively, as rapid production growth will push up costs. The indices for Middle Eastern upstream capital costs and operating costs will both increase by about 11%. The indices for African upstream capital costs and operating costs will increase by 9% and 14% respectively. Increased investment in the construction and maintenance of oil and gas infrastructure for exports to Europe will further drive investment growth in the Middle East and Africa, pushing up costs. The indices for upstream capital costs and operating costs in the Commonwealth of Independent States (CIS) will increase by 7% and 10% respectively. Affected by the Ukraine crisis, Western oil and gas giants have withdrawn their investments from Russia and taken away key technologies and professionals, which may cause some projects to lack momentum for development.

(Authors of this section: Yu Gongming, Li Xiaoguang. Reviewer: Fu Lei)

Current Status and Trends of Oil and Gas Exploration and Development in China

In 2023, domestic oil and gas field enterprises made significant contributions for energy supplement, continuously developed exploration, obviously improve exploration efficiency, importantly boosted reserves, and eventually achieved significant results. A series of representativeresults in deep layer, deep water, and unconventionaloil and gas. The increasingly production of domestic oil and gas, with a focused on mature oil fields, deep oil and gas, and shale oil and gas, strengthened China's energy supply capacity and continued to play an important role in energy security.

Situation of Oil and Gas Exploration in China

Newly Found Proven Geological Reserves

At the end of 2022, in China, the remaining proven recoverable reserves of oil were approximately 3.8 billion tons, the remaining proven recoverable reserves of natural gas were 6.57 trillion cubic meters, the remaining proven recoverable reserves of shale gas were approximately 560.6 billion cubic meters, and the remaining proven recoverable reserves of coalbed methane were nearly 366 billion cubic meters.

In 2023, major exploration discoveries were made in new fields, new types, and new formations in Tarim basins, Junggar basins, and Sichuan basins. The newly found proven geological reserves of petroleum nationwide are approximately 1.3 billion tons, mainly from the Ordos Basin, Junggar Basin, Tarim Basin, and Bohai Bay Basin (including offshore); the newly found proven geological reserves of natural gas are nearly 1 trillion cubic meters, mainly from the Sichuan Basin, Ordos Basin,

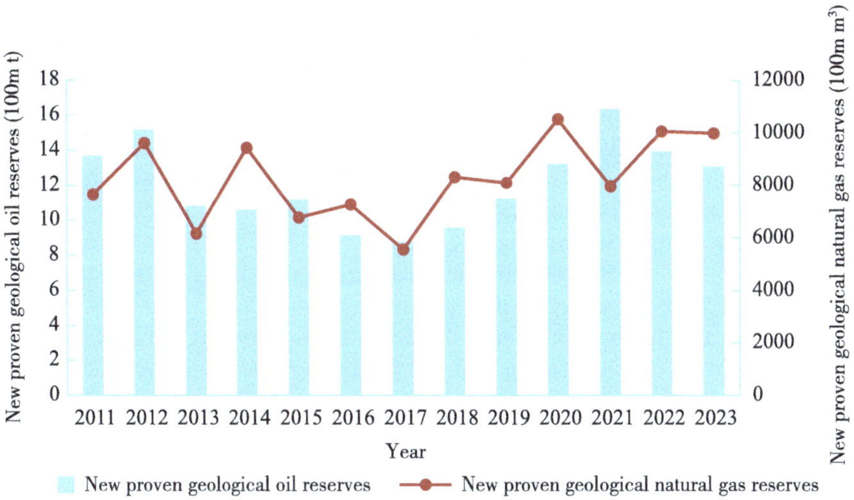

Fig. 10 New proven recoverable reserves of petroleum and natural gas nationwide from 2011 to 2023. *Data source* PetroChina, Sinopec, CNOOC, Yanchang Petroleum, etc.

and Tarim Basin (Fig. 10); the newly found proven geological reserves of coalbed methane exceed 220 billion cubic meters.

Excellent Achievements in Deep Layer, Deep Water, and Unconventional Oil and Gas Exploration

Deep and ultra-deep oil and gas have become important parts for oil and gas exploration in recent years. Significant breakthroughs have been made in large oil and gas fields such as Fuman and Shunbei in the Tarim Basin, with the discovery of high-quality reserves on a large scale. 104 ultra-deep wells deeper than 8000 m were drilled, providing strong support for the increase in oil and gas reserves and production. Xiongtan 1 well obtained high-yield oil and gas flow in the well section deeper than 6000 m, achieving the first oil and gas discovery of Cambrian Baiyun marine limestone; Tuotan 1 well made a breakthrough in Cambrian continental buried hill oil and gas exploration at a depth of 5600 m in the lower Qulitag Group. Shunbei 84 slant well, with a vertical drilling depth of 8937 m, is currently the vertically deepest thousand-ton well in Asia. The first 10,000-m scientific exploration well in China—Shendi Tako 1 well was drilled exceeding 9000 m by the end of 2023 with a designed well depth of 11,000 m; the second 10,000-m scientific exploration well—Shendi Chuanke 1 well was drilled exceeding 3600 m by the end of 2023 with a designed well depth of 10,520 m. The two 10,000-m scientific exploration wells not only promoted the progress of ultra-deep exploration technology and equipment, but also significantly increased ultra-deep oil and gas reserves.

Offshore oil and gas reserves show great potential for production growth. In 2023, four large offshore oil and gas fields were found, showing great potential for oil and gas reserves and production growth. A major breakthrough was made in deep-water exploration in the South China Sea. The Kaiping Depression in the Pearl River Mouth Basin produced more than a thousand tons of oil per day in a single test well of the Paleogene, and it is expected to become the first deep-water billion-ton light oil field in China. Two major discoveries of billion-ton oil fields were made in the Bohai Sea, further contributing to China's offshore oil and gas resources. The first is the world's largest metamorphic rock buried hill oil field—Bozhong 26-6 billion-ton oil field; the second is the first billion-ton lithologic oil and gas field in the mature area of the shallow layer of the Bohai Sea—Qinhuangdao 27-3 oil field. A major discovery of a billion-ton shale oil field was made in the Weixinan Shale of the Beibuwan Basin. Two large and medium-sized oil and gas fields, Panyu 10-6 and Huizhou 26-6 North, were found in the deep layer of the South China Sea Zhuyi Depression, with proven oil and gas reserves of about 60 million tons, marking a major strategic transition in exploration from shallow to deep layers in the Pearl River Mouth Basin.

Unconventional oil and gas exploration continues to make major discoveries. Breakthroughs were made in the deep shale gas of the Cambrian Qiongzhusi Formation and the Permian Dalong Formation in the Sichuan Basin. The Zi 201 well and Leiye 1 well both obtained high-yield gas flows, further expanding the new stratigraphic series of shale gas exploration. The Gulong shale oil in Daqing has preliminarily proven a reserve of 200 million tons; a major breakthrough in new area and new type exploration of shale oil was made in the Lijin Depression of the Jiyang Depression, confirming a billion-ton resource. Remarkable achievements were made in deep coalbed methane exploration, confirming a trillion cubic meter-level large gas area in the deep 8# and 5# coal seams of the Ordos Basin (Table 4).

The Situation of Oil and Gas Development and Production in China

The Oil and Gas Production Equivalent Has Maintained a Growth of Tens of Millions of Tons for Seven Consecutive years

In 2023, the national oil and gas production equivalent was about 400 million tons, with an average annual increase of 11.7 million tons of oil equivalent in the past seven years, forming a new peak period of production growth. Crude oil production was 209 million tons, an increase of 2.1% year-on-year. Among them, offshore crude oil production exceeded 62 million tons, shale oil production exceeded 4 million tons, and onshore deep and ultra-deep production was 11.8 million tons. Natural gas production was 235.3 billion cubic meters (including shale gas, coalbed methane, and coal-to-gas), an increase of 5.7% year-on-year, and an increase of more than 10

Table 4 Important discoveries in oil and gas exploration nationwide in 2023

No.	Category	Basin	Discovery well	Description	Operating company
1	Major breakthrough	Ordos	Nalin 1H	A major breakthrough was made in the exploration of the deep $8^{\#}$ coal seam. 6 horizontal wells in the Nalin River-Mizhi area have a daily gas production of 42,000 to 164,000 cubic meters, providing more than 120 billion cubic meters of natural gas geological reserves	PetroChina
2			Deep Coal 1	The first trillion cubic meter-level large gas field in China, the Shenfu Gas Field, was found at the 2000-m on the eastern edge of the Ordos Basin, with a proven reserve exceeding 1100 billion cubic meters	CNOOC
3			Yang Coal 1HF	A significant breakthrough was made in the exploration of Dadiniu gas field 8 $^{\#}$ coalbed methane at a depth of 2800 m, with Yangmei 1HF well obtaining a daily gas production of 104,000 cubic meters and the newly predicted reserves exceeding 120 billion cubic meters	Sinopec
4		Tarim	Xiongtan 1, Tuotan 1	A significant breakthrough was made in the exploration of the Cambrian system in the western uplift of Tarim, with a predicted reserve of 110 million tons of oil equivalent. For the first time, high-yield marine oil and gas were found in the Baiyun rock of the Lower Cambrian Qulitag Group, promoting the exploration of marine oil and gas to expand to the western part of the basin	PetroChina
5			Ketan 1	A significant breakthrough was made in the new strata of the Kuche depression, with a geological gas reserves of 1135 billion cubic meters in the Cretaceous Yagelmu Group controlled by Ketan 1 well	PetroChina

(continued)

Table 4 (continued)

No.	Category	Basin	Discovery well	Description	Operating company
6		Sichua	Datan 1	A significant breakthrough was made in the exploration of Neoproterozoic gas in the western side of the Deyang-An'yue fault, where Datan 1 well obtained 420,000 cubic meters and 820,000 cubic meters of daily gas production in Deng 2 Formation and Deng 4 Formation respectively, which is expected to form a new trillion cubic meter level reserve	PetroChina
7	Majorn breakthrough	Sichuan	Xuantan 1	A significant breakthrough was made in the exploration of natural gas in the Permian–Triassic reef beach in the Xuanhan area, where Xuantan 1 well obtained 1.53 million cubic meters and 1.09 million cubic meters of daily gas production in the upper and lower members of the Feixianguan Formation respectively, which is expected to form a 100 billion cubic meters level reserve	PetroChina
8			Zi 201	For the first time, a high-yield commercial industrial gas flow was produced from the ancient shale strata of the Cambrian system, with a daily gas production of 738,800 cubic meters and providing trillions of cubic meters of deep shale gas	PetroChina
9			Leiye 1	For the first time, a breakthrough was made in the exploration of deep marine shale gas in the Permian Dalong Group, with a daily gas production of over 420,000 cubic meters and a resource of over 1700 billion cubic meters	Sinopec

(continued)

Table 4 (continued)

No.	Category	Basin	Discovery well	Description	Operating company
10		Bohai Bay	Liye 101HF	A significant breakthrough was made in the exploration of shale oil in the Jiyang depression and Lijin depression, with Liye 101HF well clay conglomerate phase fracturing test obtaining a daily oil production of 201 tons and a new resource of 100 million tons level	Sinopec
11			Bohai 26-6-2	The Bohai South Low Uplift Bohai 26-6 oilfield achieved a new approach to explore oil fields under metamorphic rock buried hills, with an average daily production of over 270 tons of light oil and over 320,000 cubic meters of gas. The overall proven reserves reach to 200 million tons, making it the world's largest metamorphic rock buried hill oilfield	CNOOC
12		Pearl River Mouth	KP18-1-1d	The Kaiping 18-1 oil field in the Paleogene Kaiping sag obtained billions of tons of oil, with a single test well, KP18-1-1d well, obtaining a daily oil production of 12.21 million tons and a daily gas production of 14,800 cubic meters, which is expected to become the first deep-water deep-layer billion-ton light oil field in China	CNOOC
13	Important discovery	Ordos	De1	A major exploration discovery was made in the complex fault zone of the western new area, and a large oil field with geological reserves exceeding one billion tons was explored in the Hongde area, which is expected to form a reserve of one billion tons	PetroChina

(continued)

Table 4 (continued)

No.	Category	Basin	Discovery well	Description	Operating company
14			Jiping 2H	A major breakthrough was made in the exploration of the deep 5 # coal seam in the eastern edge of the Daji block, and the daily gas production of the Jiping 2H well reached a maximum of 67,000 cubic meters	PetroChina
15		Junggar	Bay Exploration 1	An important discovery was made in the natural gas exploration of the Permian Fengcheng Formation in the western depression, with predictably 75.5 billion cubic meters of gas and 15.51 million tons of condensate oil	PetroChina
16			Hu 102	An important discovery was made in the natural gas exploration of the Jurassic Kalaza Formation on the southern edge. The Hu 101 well produced 233,500 cubic meters of gas and 37.23 cubic meters of oil per day, and the Hu 102 well produced 800,300 cubic meters of gas and 107.16 cubic meters of oil per day, contributing to a new natural gas exploration on the southern edge	PetroChina
17			Basin North 1	The first discovery of Carboniferous self-generated and self-stored natural gas reservoirs in the western depression. The Carboniferous Basin North 1 well produced 526,000 cubic meters of gas and 33.6 cubic meters of oil per day, expanding the new field of natural gas exploration and showing good exploration potential	PetroChina

(continued)

Table 4 (continued)

No.	Category	Basin	Discovery well	Description	Operating company
18			Ha Mountain 5	A major exploration discovery was made in the Fengcheng Formation of the Ha Mountain foreland belt. The Ha Mountain 5 well fracture test produced 95.8 cubic meters of oil and 6926 cubic meters of gas per day. The Ha Mountain foreland belt provided a predicted oil reserve of 30.63 million tons in the original place and overthrust, and is expected to significantly increase reserves	Sinopec
19		Tarim	Shunbei 84X	A new oil and gas-rich fracture belt was found in the Shunbei area, with 84X well obtained high-yield industrial oil and gas flow, with the equivalent oil and gas reaching 1017 tons. The cumulative proven geological reserves are 307 million tons of oil and 4181 billion cubic meters of natural gas	Sinopec
20			Qia Exploration 1	An important discovery was made in the exploration of the Carboniferous-Permian in the southwestern mountain front of the tower, and the Qiatan1 well obtained a daily gas production of 79,000 cubic meters	PetroChina
21	Important discovery	Sichuan	Hongye Mao 1HF, Hongye 7HF	A major breakthrough was made in the exploration of ultra-deep shale gas in the Permian Wujiaping Formation in the Hongxing area. The Hongye Mao 1HF well obtained a daily gas production of 64,000 cubic meters from the Mao Four member, with an estimated resource volume exceeding 500 billion cubic meters and the Hongye 7HF well obtained a daily gas production of 323,500 cubic meters from the Wu Two member, both contributing to a proven reserve of 100 billion cubic meters	Sinopec

(continued)

Table 4 (continued)

No.	Category	Basin	Discovery well	Description	Operating company
22			Bazhong 1HF, Yuanlu 2HF	A major discovery of tight oil and gas was made in the new area and new strata of the Triassic-Jurassic continental series in Northern Sichuan. The Bazhong 1HF well obtained a stable oil flow of over 100 tons per day from the Jurassic river sandstone domain after fracturing; the Yuanlu 2HF well provided production enhancement for the Xu Two member of the porous reservoir, with a daily gas production of 500,000 cubic meters after fracturing	Sinopec
23		Bohai Bay	Cangtan 1	A high-quality hydrocarbon source rock was explored in the Kong Two member of the Cangdong sag during shale oil exploration. After fracturing, the Cangtan 1 well obtained a daily oil production of 118 cubic meters from the Kong Two member, providing a predicted geological reserve of over 35 million tons	PetroChina
24			Qinhuangdao 27-3	A major discovery of a billion-ton oil field was made in the shallow layers of the Bohai Sea. The overall proven reserves of the Qinhuangdao 27-3 oil field in the old area of the Shijiu protrusion exceed 120 million tons, achieving continuous oil-bearing in the entire Shijiu protrusion	CNOOC
25		Songliao	Guye 8H1	New progress was made in the pre-exploration evaluation of the Gulong shale oil, establishing the Guye 8H1 test area, confirming an oil-bearing area of over 600 square kilometers, and providing 200 million tons of proven reserves	PetroChina

(continued)

Table 4 (continued)

No.	Category	Basin	Discovery well	Description	Operating company
26			Dashen 1, North 1201	Major progress was made in the new area and new field of the Changling fault block, providing 2300 billion cubic meters of resource volume. The Dashen 1 well expanded the range of the hydrocarbon-generating depression, providing a natural gas reserve of 500 billion cubic meters; the North 1201 well found a non-typical tensional lake basin on the western slope of Changling, confirming a natural gas reserve of 1809 billion cubic meters	Sinopec
27		Northern Jiangsu	Haiye 1HF, Huaye 7HF	Shale oil discoveries were made in multiple sags in Hai'an, Gaoyou, and other places, with predicted reserves exceeding 100 million tons. The Haiye 1HF well in the Hai'an sag obtained a daily oil production of 50.7 tons after fracturing, providing a predicted reserve of 25.51 million tons. The Huaye 7HF well in the Gaoyou sag obtained a daily oil production of 52.3 tons and a daily gas production of 10,400 cubic meters after fracturing the Fu Two member, providing a predicted reserve of 52.12 million tons	Sinopec

(continued)

Table 4 (continued)

No.	Category	Basin	Discovery well	Description	Operating company
28		Pearl River Mouth	Panyu 10-6-1	The Panyu 10-6 medium-to-large oil field added over 20 million tons of proven geological reserves. The Panyu 10-6-1 well obtained a daily production of 516 cubic meters from the Enping Formation, setting a record for testing at depths greater than 3000 m in the eastern South China Sea. This is the first medium-to-large oil field discovery in the deep layer of the Panyu 4 sag, establishing a solid foundation for the exploration and development of new strata in the old oil area	CNOOC
29			Huizhou 26-6-10dSa	The HZ26-6N Enping Formation in the Huizhou sag in the South China Sea provided over 40 million tons of proven reserves, confirming a reserve of 107 million tons for the entire Huizhou 26-6 oil field, contributing to the billion-ton level of the Huizhou 26-6 oil field, and confirming the huge exploration potential of the rich sag deep layer in the old oil area	CNOOC
30		Beibu Gulf	–	A major discovery of shale oil was made in the Weixinan sag. The overall proven reserves of the Weizhou 11-6, Weizhou 11-10, Weizhou 5-11, and other oil fields exceed the billion-ton level, developing a new exploration in Weixinan and contributing to a reserve with a stable gas production of 5 million cubic meters in the Weixinan oil field group	CNOOC

billion cubic meters for seven consecutive years. Unconventional natural gas production exceeded 96 billion cubic meters, accounting for 43% of the total natural gas production, becoming an important growth pole in natural gas reserves and production. Among them, tight gas production exceeded 60 billion cubic meters, shale gas production was 25 billion cubic meters, and coalbed methane production exceeded 11 billion cubic meters. The top ten oilfields in terms of oil and gas production equivalent are Changqing Oilfield, Bohai Oilfield, Daqing Oilfield, Southwest Oil and Gas Field, Tarim Oilfield, Shengli Oilfield, South China Sea East Oilfield, Xinjiang Oilfield, Yanchang Petroleum, and South China Sea West Oilfield. Among them, the oil and gas production equivalent of Changqing Oilfield reached 66.06 million tons, ranking first in China; the oil and gas production equivalent of Bohai Oilfield exceeded 36.8 million tons, setting a new historical record, with crude oil production exceeding 34 million tons and natural gas production exceeding 3.5 billion cubic meters.

Deep and Ultra-Deep Layers Have Become New Areas for Oil and Gas Production Growth

China has efficiently built several deep oil and gas fields and has become a global leader in the field of oil and gas in deep and ultra-deep layers over 6000 m on shore. Fuman Oilfield continues to deepen the understanding of oil reservoir enrichment rules, accelerates the promotion of centralized production and scale-up, and the annual oil and gas production equivalent rapidly increases to 4 million tons, with an average annual increase of 760,000 tons. Shunbei Oil and Gas Field is anchored in the central deployment of oil and gas-rich regions, with annual oil and gas production of 1.27 million tons and 2.2 billion cubic meters respectively. The Boz—Dabei ultra-deep large gas field accelerates the production capacity construction rhythm, and the ultra-deep natural gas production in the Kuche area reaches 18 billion cubic meters, marking the new stage of large-scale development of the new trillion cubic meter gas zone in the Tarim Basin. The Anyue gas field in the Sichuan Basin continues to stabilize and increase production, the Penglai gas area has significant test mining evaluation results, and the construction of the 200 billion cubic meter super-large gas field in the central Sichuan uplift is efficiently promoted, helping Sichuan andChongqing "Gas Daqing" to accelerate progress.

Shale Oil Production Hits a New High

Shale oil exploration and development are steadily advancing, and production exceeded 4 million tons in 2023, setting a new record. The three national demonstration zones of Xinjiang Jimsar, Daqing Gulong, and Shengli Jiyang accelerated their construction, with production of 794,000 tons, 170,000 tons, and 300,000 tons respectively. The shale oil development demonstration base in the Ordos Basin has become the largest shale oil field in China, with an annual output of 2.52 million tons;

the benefit development demonstration platform of Dagang Oilfield shale oil—Cangdong Sag No. 5 platform, has formed an annual production capacity of 100,000 tons and has been built and put into production; the multi-well type test in Subei Yintong Sag has achieved commercial breakthroughs. A series of continental shale oil development results show the good prospects for the development of unconventional oil in China.

Investment and Cost in the Upstream of Oil and Gas

Investment in Oil and Gas Exploration and Development Maintains High Growth

In 2023, oil and gas field enterprises continued to enhance their exploration and development efforts, effectively optimized exploration and development deployment, improved exploration and development results, and domestic oil and gas exploration and development investment maintained a relatively high growth level. The annual exploration and development investment was about 390 billion RMB, an increase of 10% year-on-year. Among them, exploration investment was close to 90 billion RMB, and development investment was about 300 billion RMB, both of which increased by more than 10% year-on-year, both setting new historical records (Fig. 11).

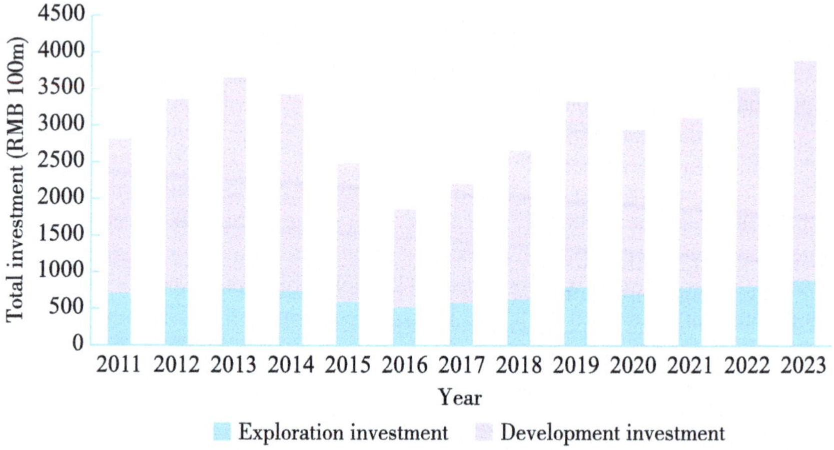

Fig. 11 Changes in national oil and gas exploration and development investment from 2011 to 2023. *Data source* PetroChina, Sinopec, CNOOC, Yanchang Petroleum, etc.

Oil and Gas Operation Costs Are Basically the Same as Last year

In 2023, various oil and gas field enterprises enhance their exploration and development efforts. While overcoming the increase in raw material prices and the difficulty of exploration and development, they adhered to benefit development, and the results of quality improvement and efficiency increase were significant. The unit oil and gas operation cost was basically the same as last year. PetroChina's oil and gas operation cost was about 13 US dollars/barrel, basically the same as the same period last year of 13.17 US dollars/barrel. Sinopec's oil and gas operation cost was 747 yuan/ton (equivalent to 14.53 US dollars/barrel, calculated at the average exchange rate of 7.01 yuan against the US dollar in 2023), a decrease of 2.4% compared with the same period last year of 765 US dollars/ton (equivalent to 14.88 US dollars/barrel); CNOOC's oil and gas operation cost was 7.78 US dollars/barrel, basically the same as the same period last year of 7.74 US dollars/barrel.

The Prosperity of the Oil Service Industry Has Rebounded

Domestic oilfield technology service companies have achieved outstanding results. In 2023, the business volume of domestic oilfield technology service companies reached a new peak, with an increase in revenue return rate and an overall positive trend in the oilfield technology service industry. According to the financial reports of the first three quarters of various companies, the net profits of most companies have steadily increased, and some companies have turned losses into profits (Fig. 12); international operations have become a focus, with companies such as Sinopec Oilfiled Services Corporation, CPECC, Offshore Oil Engineering Corporation, Yantai Jereh Oilfield Services Group, and Hilong Holdings Limited seeing an increase in overseas business contracts and a significant increase in the proportion of overseas business income.

New businesses bring more growth points to the oilfield technology service industry. Oil service companies such as Offshore Oil Development and Anruike are continuously expanding CCUS, LNG refueling, renewable energy hydrogen production, hydrogen liquefaction, pure hydrogen/hydrogen blending, lithium batteries, offshore wind power and other businesses, opening up new business markets for the oilfield technology service industry.

Continuous breakthroughs in core technologies provide important support for the increased revenue of the oilfield technology service industry. Progress has been made in the application of key core technologies such as onshore deep well drilling technology, deepwater subsea production systems, MOM oil-based drilling fluid technology, deepwater conductor rack design and installation technology, and "low permeability-marginal-heavy oil" oilfield efficient development, effectively ensuring upstream oil and gas exploration and development, and increasing the technical service income of the oilfield technology service industry.

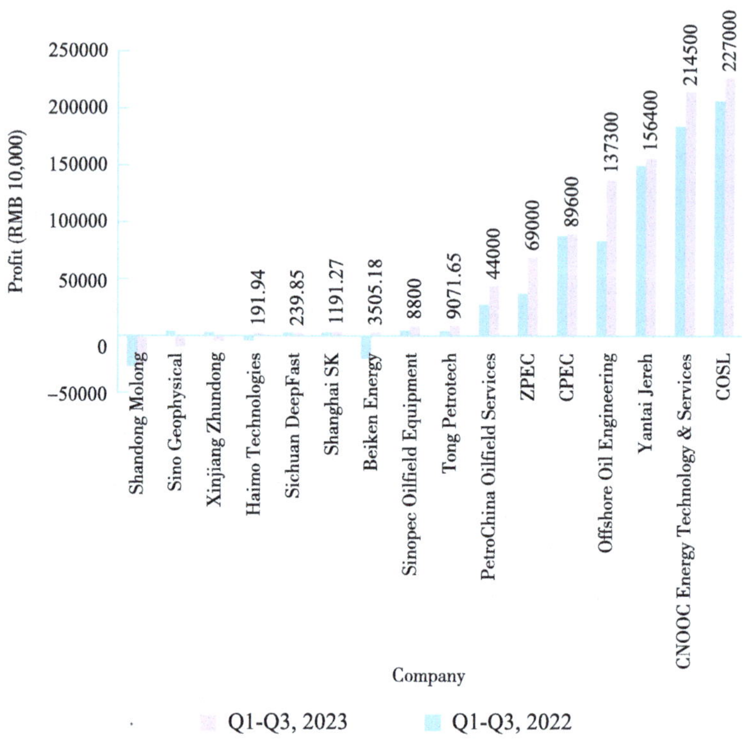

Fig. 12 Changes in profits of major domestic listed oilfield technology service companies in the first three quarters of 2023

New Progress in Upstream Oil and Gas Technology

Innovation and Development of Exploration and Development Theory and Technology

In terms of exploration theory and technology, the Tarim Basin has innovated the geological understanding of super-deep oil and gas in the superimposed composite basin based on overall research, promoting five strategic discoveries such as Ke1 well and Qia1 well; the Junggar Basin has broken through the traditional understanding of lacustrine mudstone source rocks, traditional oil and gas accumulation lower limit research ideas and traditional "from source rock to trap" oil search ideas, forming a conventional-unconventional orderly coexistence oil and gas system accumulation model, providing exploration confirmation for the concept of the whole oil and gas system; the Qiongdongnan Basin has innovatively established a strong active passive continental margin basin "continental margin transition slope control circle—low position long-term active fault control transport—late multi-stage strong injection—high position weak active fault control gathering" oil and gas enrichment accumulation model, achieving a historic breakthrough in deepwater and deep layer exploration in the South China Sea.

Innovation in development technology is mainly focused on extraction methods and downhole tools. Fire flooding + flue gas conjugate development technology further realizes oil production and carbon storage by injecting flue gas on the basis of greatly improving the recovery rate by fire flooding, achieving overall synergistic high-efficiency green development. The electric thermal molten salt heat storage steam injection technology for heavy oil fields, targeting the high silicon, high mineralization, and high chlorine content of boiler water in oil fields, can generate wet steam, realizing clean alternative energy with "electricity replacing gas" as the core. Downhole oil–water separation injection-production process technology enables efficient oil and water separation downhole, realizing injection and extraction in the same well. Fracturing process technology has formed a volume fracturing process 2.0 with "flow-limited perforation + small cluster spacing fracturing + temporary plugging and turning" as the core, promoting the large-scale application of high-performance reusable slick water system, electric drive fracturing equipment, and soluble bridge plugs.

Innovations and Developments of Engineering Technology and Equipment

Innovations in geophysical exploration technology are mainly focused on land complex area seismic data acquisition and imaging, ocean bottom node equipment development, and large-scale software development. The successful development of the marine seismic exploration production information management system has improved the efficiency of operations by managing information for different types of marine seismic acquisition projects; the independently developed marine towed-streamer acquisition system has completed seismic exploration operations in ultra-deep water areas for the first time, and the independently developed seabed node equipment has been put into use, further improving the domestically controllable marine oil and gas exploration equipment system. New progress has been made in the development and application of GeoEast integrated processing and interpretationsoftware, effectively promoting breakthroughs in oil and gas exploration and development.

Innovations in logging technology are mainly focused on imaging logging, interpretation software, perforation, and digitization. The new imaging logging instrument allows imaging measurements to be made in horizontal wells beyond the drilling tools, and it breaks the foreign technology monopoly and achieves domestic substitution; the new perforator successfully passed the field test of the American Petroleum Institute, with an average perforation depth of 2593.6 mm, breaking the world record; the logging big data platform has begun to be promoted and operated, adopting a unified data standard for further research in aspects such as machine learning and AI in the future.

Innovations in drilling technology are mainly focused on deep and ultra-deep well drilling, horizontal well drilling, deep and ultra-deep water drilling, and digital drilling. The 12,000-m ultra-deep well automated drilling rig has become an indispensable tool for China's ultra-deep oil and gas exploration and development. The high-temperature, high-inclination intelligent rotary steering drilling system has a temperature resistance of up to 175 °C and an inclination capability of 15°/30 m, filling the technical gap in this field in China. The "Longitude and Latitude Navigation" rotary geological steering drilling system, with a resistance to 165 °C high temperature, 140 GPa high pressure, and 20 times gravity acceleration, has been implemented on a large scale. Digital drilling technology has formed a new system driven by "data" to drive on-site operations, creating a new ecosystem for digital drilling.

In terms of offshore equipment, international technology blockades have been broken, achieving breakthroughs in two bottleneck technologies: gas turbines and FPSO single-point slip rings and mooring cables. The first batch of 3 sets of 2 MW domestic turbo units were successfully put into operation; the domestication of single-point liquid slip rings and mooring steel cables has achieved the first set of engineering applications of permanent mooring domestic steel cables, reaching an international advanced level; new progress has been made in engineering construction, and the independently designed and built deep-water jacket platform "Haiji-2" has been completed, becoming another significant equipment for the development of China's deep-water oil and gas resources.

Trends in Domestic Oil and Gas Exploration and Development in 2024

In 2024, the oil and gas industry will continue to deeply implement the Party's 20th National Congress's decisions on "deepening the energy revolution" and "increasing the exploration and development of oil and gas resources and the capacity to increase reserves and production", as well as the National Energy Administration's Accelerating the Action Plan for the Integration of Oil and Gas Exploration and Development and New Energy Development, to promote the intelligent and green expansion of the oil and gas industry. Oil and gas companies will further increase their exploration and development efforts, continue to exert efforts in deep and ultra-deep, deep-water, and unconventional strategic reserve and production replacement areas, it is expected that the newly added geological reserves of oil and natural gas in the country will continue to remain high, with the newly found proven geological reserves of oil exceeding 1 billion tons and the newly added geological reserves of natural gas exceeding 1 trillion cubic meters; oil production will continue to remain above 200 million tons, natural gas production will maintain rapid growth, and the newly added production is expected to continue to reach more than 100 billion cubic meters, ensuring the core demand for national oil and gas. China will lead the future development of deep oil and gas industry, make progress in the theory and major engineering technology and equipment of deep and ultra-deep oil and gas exploration and development, and promote deep and ultra-deep oil and gas to become an important area for reserve and production increase. The oil and gas industry will explore and form a new model of multi-energy complementarity and integrated development, accelerate the integration of oil and gas exploration and development and renewable energy development, and form a new pattern of integration and multi-energy complementarity between the upstream sector of the oil and gas industry and new energy industries. On the basis of stabilizing oil and increasing gas and improving the supply capacity of oil and gas resources, it will accelerate the green and low-carbon transformation and help achieve the "carbon peaking and carbon neutrality" targets.

(This section was written by: Li Xiaoguang, Zhang Jiaming, Jiao Jiao. Reviewed by: Zhang Huanzhi, Wang Hong)

Development of the Domestic and International Refining and Chemical Industry

Current Status and Trends of Global Refining and Chemical Industry Development

In 2023, the global refining and chemical production capacity continued to expand, with the growth rate slowing down. The start-up load of refining and chemical devices remained at a low level, product supply continued to grow, product demand remained weak, and growth was slow. The refining capacity in Europe and America was temporarily tight, and the refining gross profit entered the "silver period" from the "golden period" of 2022. Although the refining industry's profits fell, they still remained at a historically high level. The chemical industry was still in a downturn, market demand continued to be weak, profits continued to fall, and most chemical companies' earnings significantly decreased. Under the "dual carbon" background, the global refining and chemical industry faced a severe situation of emission reduction, and emerging large-scale refining and chemical companies accelerated the green and low-carbon transformation, intelligentization, and high-end development.

Global Refining Industry Development Status and Trends

Basic Overview of the Global Refining Industry

Refining capacity continues to grow. In 2023, the global refining capacity continued to grow, with an additional refining capacity of 86.175 million tons/year (Table 1), higher than the additional capacity of 82.5 million tons/year in the previous year. The new capacity was mainly concentrated in the Middle East and North America, with the Middle East adding 39.25 million tons/year. The United States and Japan closed some refining capacities, reducing the global refining capacity by 26.75 million tons/year. The net increase in global refining capacity was 59.425 million tons/year, lower

X. Qian et al., *Global Oil and Gas Industry Development Report: 2023*,
https://doi.org/10.1007/978-981-95-2375-7_8

than the net increase of 62.1 million tons/year in the previous year, and the expansion of refining slowed down. The total global refining capacity rose to 5.168 billion tons/year (Fig. 1).

The refining focus continues to shift eastward, and the refining status in the Eastern Hemisphere is improving. By 2023, the global share of refining capacity in the Middle East will rise from 11% in 2022 to 12%, and the combined global share of the Middle East and Asia–Pacific will remain at 46%. Africa's global share remains

Table 1 Global refining capacity increase/decrease in 2023 unit: 10,000 tons/year

Country	Company name	Location	Increased capacity
Iran	National Iranian Oil Co	Abadan, Khuzestan	1050
South Africa	Astron Energy	Cape Town	500
United States	Husky Energy	Superior, Wis	240
Argentina	Refipampa Refinery	La Pampa	72.5
Iraq	Iraq state	Karbala	700
United States	ExxonMobil	Beaumont	1150
United States	Texas International Terminals	Galveston Bay	250
Oman	Oman Refining Co., Kuwait Petroleum International	Al Duqm	1150
Bahrain	Bahrain Petroleum Co	Sitra	1025
Congo (B)	Beijing Fortune Dingheng Investment	Fouta	300
Egypt	Middle East Oil Refinery (MIDOR)	Alexandria	300
Iraq	North Refineries Co	Baiji, Salah al-Din II	350
Nigeria	IOC	Edo	180
Peru	Petroleos Del Peru	Talara, Piura	150
China	Xinhai Chemical	Cangzhou, Hebei	800
China	Jingbo Petrochemical	Danzhou, Hainan	400
Global total increase			8617.5
Country	Company Name	Location	Reduced capacity
United States	Phillips 66	Rodeo, California	600
Japan	ENEOS	Wakayama	650
Iran	National Iranian Oil Co	Abadan, Khuzestan	1275
Jordan	Jordan Petroleum Refinery	Zarka	150
Total global reduction			2675
Total global net increase in capacity			5942.5

Source PetroChina Economic Research Institute

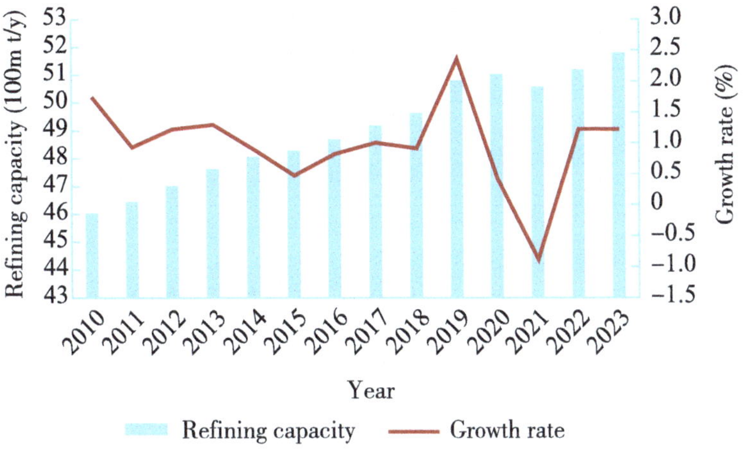

Fig. 1 Changes in world refining capacity from 2010 to 2023. *Data source* PetroChina Economic Research Institute

around 3%. The total global share of refining capacity in the Eastern Hemisphere is nearly 50%, and its global refining status continues to rise, becoming an important force in the adjustment of global refining supply. Western Europe and North America account for about 37% of the global total (Fig. 2). After continuous adjustments and reductions, the refining capacity of the two regions is temporarily tight. The refining capacity of the United States is 913 million tons/year, and China still maintains the position of the world's largest refining country (Fig. 3).

Fig. 2 Global refining pattern in 2023. *Data source* PetroChina Economic Research Institute

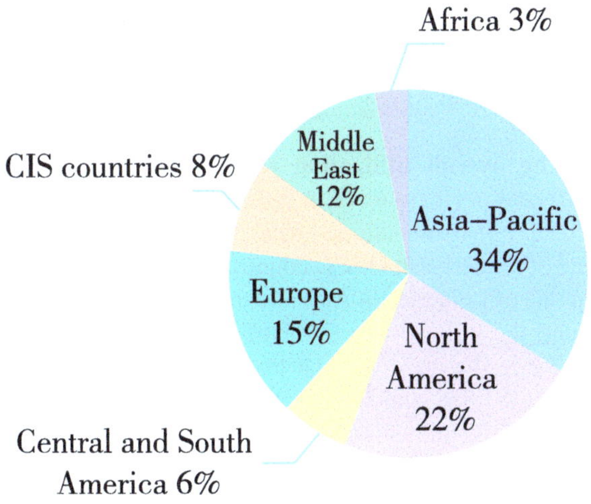

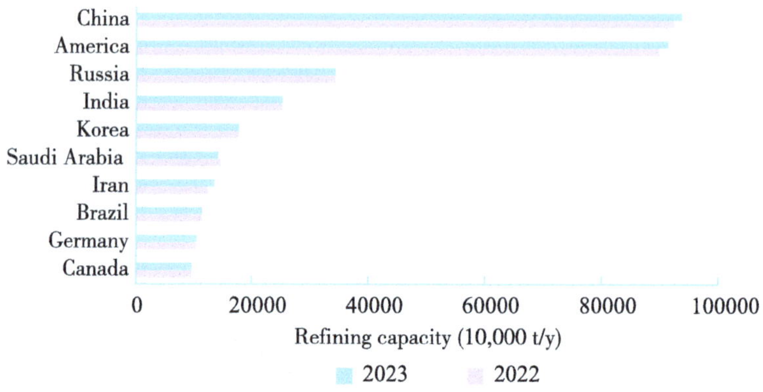

Fig. 3 Ranking of the top 10 refining countries in the world in 2023. *Data source* PetroChina Economic Research Institute

Overview of the Global Refining Industry

The total amount of crude oil processed by refineries continues to grow. In 2023, with the recovery of global oil demand, the processing volume of global refineries has increased. IEA data shows that the total processing volume of global refineries is about 82.4 million barrels/day, an increase of about 1.7 million barrels/day from the previous year, and has returned to the same level in 2019 (Figs. 4 and 5). The processing volume of refineries in the second half of 2023 is higher than that in the first half. Except for the decline in the processing volume of refineries in Western Europe and Africa, the processing volume in other regions has increased. The crude oil processing volume in the Middle East increased by 3.6% year-on-year. Affected by the recovery of oil product demand, China's crude oil processing volume increased significantly by 11%. The Asia–Pacific region outside China increased by 1.2% year-on-year, North America increased slightly by 0.5%, and Latin America increased by 2.9% (Fig. 6).

The overall utilization rate of refinery capacity continues to rise, with different performances in different regions. In 2023, the overall utilization rate of global refineries increased from 78.6% in the previous year to 80%. The utilization rate of refineries in OECD countries has declined to varying degrees, while the start-up rate of refineries in Asia and the Middle East has increased significantly. Among them, the utilization rate of the OECD Americas region slightly decreased from 87.2% in the previous year to 86.5%, OECD European countries decreased from 83.4 to 81.8%, and OECD Asia–Pacific countries decreased from 80.1 to 78.5% (Fig. 7). The utilization rate in the Middle East increased from 76.9% in the previous year to about 80%, and the utilization rate in Asia increased from 81 to 84%. The utilization rate of Chinese refineries reached a historical high, exceeding 80%, driving the overall global utilization rate higher than the previous year.

Global refining gross profit has fallen slightly but still remains at a historically high level. In 2023, global oil demand continues to recover, with demand growth

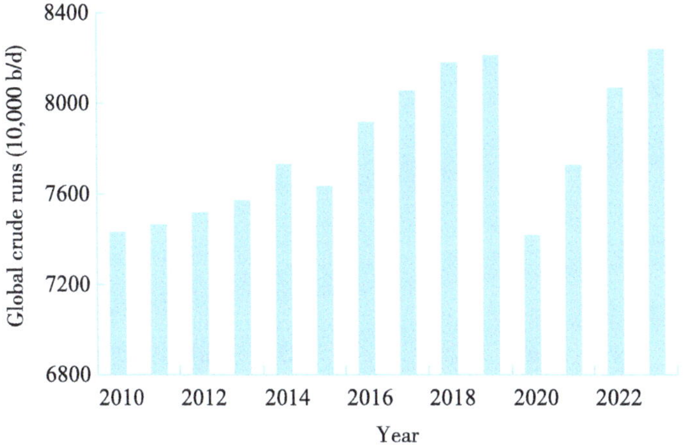

Fig. 4 Annual changes in global refinery crude oil processing volume from 2010 to 2023. *Data source* IEA

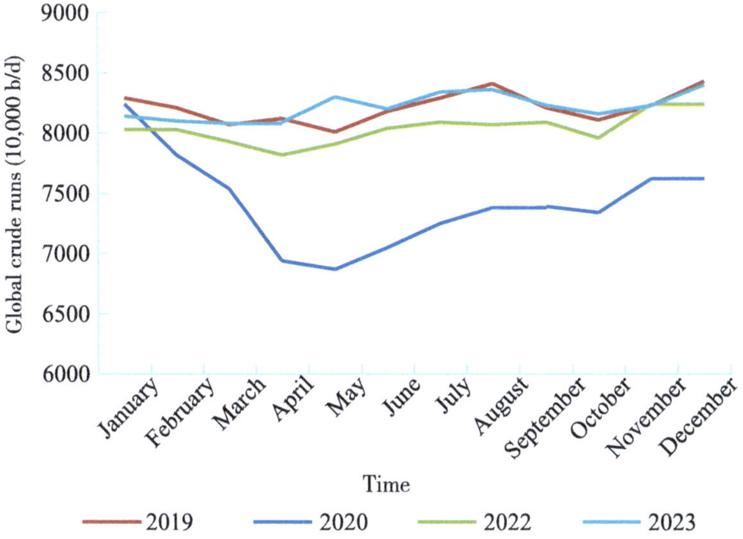

Fig. 5 Monthly changes in global refinery crude oil processing volume from 2019 to 2023. *Data source* IEA

outpacing refining capacity growth. Seasonal oil supply tightness occurs in regions such as Europe and America, and refining gross profit generally continues to maintain a high level (Fig. 8). Among them, the gross profit of medium sulfur crude oil cracking in the Gulf of Mexico in the United States fell to 29.16 US dollars/barrel, a decrease of 18% from the previous year; the gross profit of light low-sulfur oil cracking in the Mediterranean was 14.56 US dollars/barrel, a year-on-year decrease

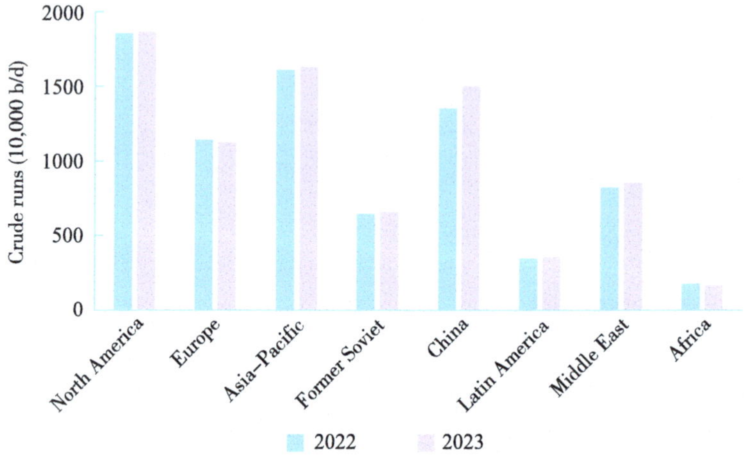

Fig. 6 Changes in crude oil processing volume of refineries in various regions of the world in 2022 and 2023. *Data source* IEA

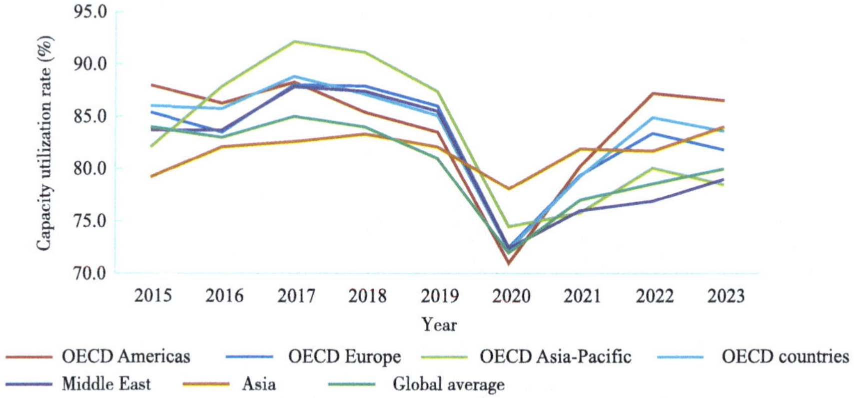

Fig. 7 Average capacity utilization rate of global refineries from 2015 to 2023. *Data source* PetroChina Economic Research Institute, IEA

of 13%; the gross profit of light low-sulfur oil cracking in Northwest Europe was 12.11 US dollars/barrel, a year-on-year decrease of 34%; the gross profit of light low-sulfur oil cracking in Singapore fell to 8.28 US dollars/barrel, a decrease of 36%. The global refining gross profit has fallen from the "golden period" of 2022, but it has maintained a historically high level, mainly due to regional imbalances in global refining capacity, restrictions on Russian oil and oil product exports, and large adjustments in the Western refining industry in the past few years. The overall tight supply of oil products globally has resulted in good profit performance for refiners, especially European and American refiners.

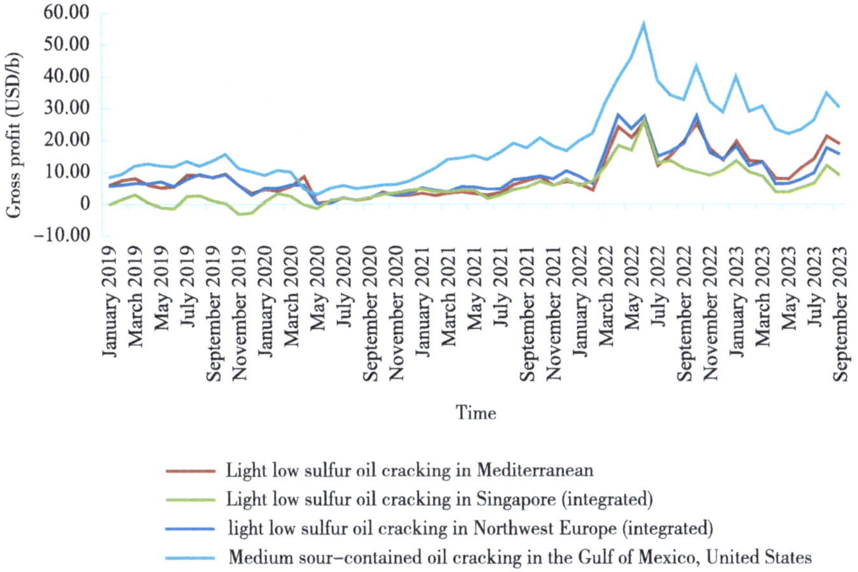

Fig. 8 Changes in global refining gross profit from 2019 to 2023. *Data source* IEA

The recovery rates of global gasoline, diesel, and fuel oil continue to decline, and only the recovery rate of aviation coal continues to increase. In 2023, major refineries around the world continue to adjust the yield of refined oil products according to changes in the refining market demand, producing more chemical raw materials, and the overall yield of major oil products fell from 84.3% in the previous year to 82.2%. The overall gasoline recovery rate fell from 30.9% in the previous year to 29.5%, the diesel recovery rate fell from 36.7% in the previous year to 35%, and the fuel oil recovery rate fell slightly from 8.4% in the previous year to 8.3% (Fig. 9). Affected by the rapid recovery of the global aviation industry, the demand for aviation coal continues to grow rapidly, and refineries increase the production of aviation coal, and the recovery rate rises from 8.3 to 9.4%.

New Developments in International Refining Technology

In the face of peak oil demand, new energy substitution, and the trend of low-carbon transformation, the refining industry is accelerating towards integration, intensification, high-end, low-carbon, intelligent, and green development. Global refining technology has made continuous progress in the processing of inferior heavy oil, the production of clean fuels, and the transformation of refining to chemicals. (1) Lightening processing technology for heavy oil and inferior oil has made breakthrough progress in fluidized bed hydrogenation technology and suspended bed

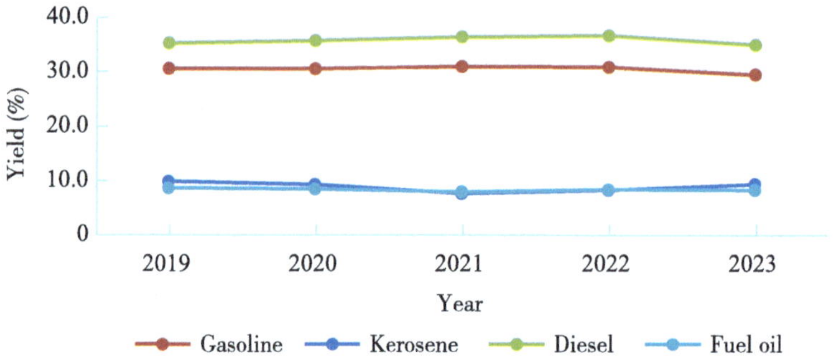

Fig. 9 Changes in the average yield of major refined oil products worldwide from 2019 to 2023. *Data source* PetroChina Economic Research Institute, S&P Global

hydrogenation technology with stronger raw material adaptability, higher conversion rate, and longer continuous operation time. (2) Many clean fuel production technologies have been developed. Gasoline is mainly desulfurization, olefin reduction, and octane number increase. Diesel is mainly desulfurization and cetane number increase. Typical clean oil technologies include catalytic gasoline desulfurization technology, catalytic gasoline olefin reduction technology, and high-octane clean gasoline component production technology. Among them, alkylation as an important technical means for the production of high-octane clean gasoline components has developed rapidly. (3) The production of high-quality low-cost chemical raw materials has become a trend for refining enterprises to transform and upgrade and develop high-quality. The process technology of "if it is aromatic, then aromatic, if it is olefin, then olefin, if it is oil, then oil, if it is chemical, then chemical" continues to improve. Major global refining companies have divested inefficient high-carbon refineries, explored "green hydrogen refining" to achieve raw material green hydrogen substitution, and tried to transform refineries into biofuel factories to produce low-carbon biofuels.

Outlook for the Development of the Global Refining Industry

Global refining capacity is expected to grow significantly. Looking at the global refining projects under construction, if they can be completed as scheduled, the global refining capacity is expected to increase by nearly 100 million tons/year in 2024, mainly from India, China, Thailand, Nigeria, Mexico, etc. Among them, India is expected to complete three refineries, and the 32 million tons/year Dangote refinery in Nigeria is expected to be completed and put into operation. It is expected that Japan and Italy will close two refineries as planned in 2024.

Refinery crude oil processing volume will still increase slightly, and refining gross profit is expected to maintain a good level. In 2024, it is expected that the global refinery crude oil processing volume will increase by about 1 million barrels/day year-on-year, reaching 83.4 million barrels/day. It is expected that the processing volume of refineries in OECD North America, Europe and Asia Pacific countries will decrease, and the processing volume of refineries in countries and regions such as China, the Middle East, and Africa will continue to grow. The global refinery capacity utilization rate is expected to remain around 80%, the tight refining capacity in Europe and America is expected to be resolved, and the refining gross profit will still maintain a high level.

The global refining industry continues to promote green and low-carbon development. The pressure to reduce emissions in the global refining industry is increasing. Under policy pressure, decarbonization and carbon reduction throughout the refining process have become an important consensus in the refining industry. Developed economies will pay more attention to the transformation of existing refineries or produce more biofuels through joint processing, improve the energy efficiency of equipment, increase the use of green hydrogen and green electricity, accelerate the process of electrification, and some refineries are exploring the implementation of carbon capture and utilization in refineries to meet policy and regulatory requirements and seek to achieve net zero emissions in refineries. The refining industry in emerging economies continues to accelerate its transformation, upgrading, and green and intelligent development process.

(Author: Xu Haifeng Reviewer: Wang Jing)

Global Ethylene Industry Development Status and Trends

Basic Overview of the Global Ethylene Industry

Global Ethylene Capacity Continues to Grow, and China's First Position Continues to Expand

In 2023, the global new ethylene capacity will be 5.323 million tons/year, an increase of 2.2% year-on-year, and the growth rate has significantly slowed down. The global total ethylene capacity will reach 228 million tons/year (Fig. 10). The new ethylene capacity mainly comes from the Asia–Pacific region, among which, China's new capacity accounts for 45.5% of the global new capacity (Table 2), and the total capacity expands to 51.95 million tons/year, continuing to consolidate the position of the world's largest ethylene country (Fig. 11). The Asia–Pacific region continues to lead the global expansion, with a total capacity of over 94 million tons/year, accounting for 42% of the global total (Figs. 12 and 13).

The Proportion of Global Ethane-Based Ethylene Remains the Largest

In 2023, the global ethylene raw material will be significantly lighter and lower-cost, the proportion of naphtha-based ethylene production will drop to 35%, and the

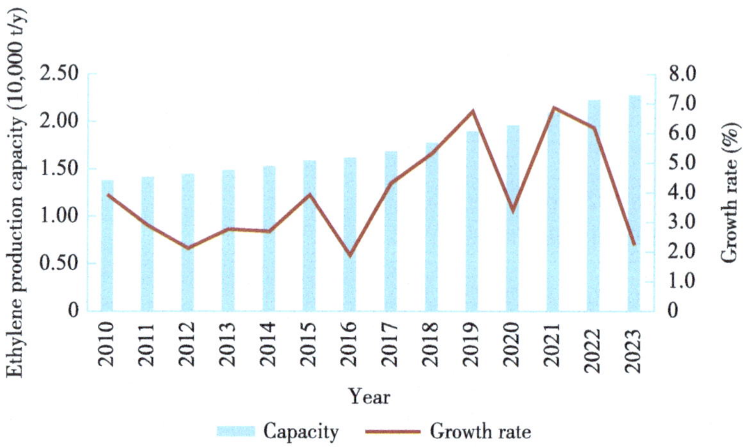

Fig. 10 Global ethylene capacity from 2010 to 2023. *Source* Wood Mackenzie

Table 2 Changes in global ethylene capacity in 2023 Unit: 10,000 tons/year

Company name	Address	New capacity
Zibo Junchen new material	Zibo, Shandong	19
Shandong Dongming petrochemical	Heze, Shandong	45
Sanjiang fine chemical	Jiaxing, Zhejiang	100
Sinopec Anqing petrochemical	Anqing	10.3
Ningxia Baofeng	Yinchuan	50
Daqing petrochemical	Daqing	18
ORLEN SA	Plock, Poland	70
HPCL-MITTAL ENERGY	Bathinda, India	120
Vietnam LONG SON	Ba Ria-Vung Tau, Vietnam	100
Total new capacity worldwide		532.3

Source Wood Mackenzie

proportion of ethane-based ethylene production will reach 39% (Fig. 14), while in 2010 these proportions were 49% and 32% respectively.

Overview of the Global Ethylene Industry

Slight Increase in Ethylene Plant Load, Slow Recovery of Ethylene Demand

In 2023, the average start-up rate of global ethylene plants slightly increased, rising from 83.1 to 83.8%, still at a historically low level. Global ethylene production was about 189 million tons, an increase of 5.5% year-on-year (Fig. 15). Global ethylene

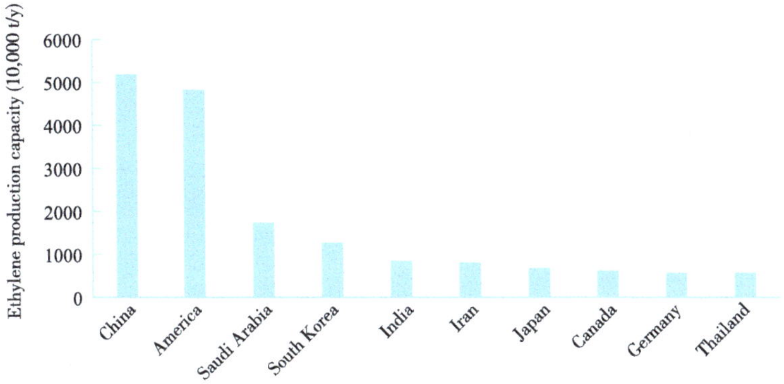

Fig. 11 Top ten ethylene producing countries in the world in 2023. *Source* PetroChina Economic Research Institute

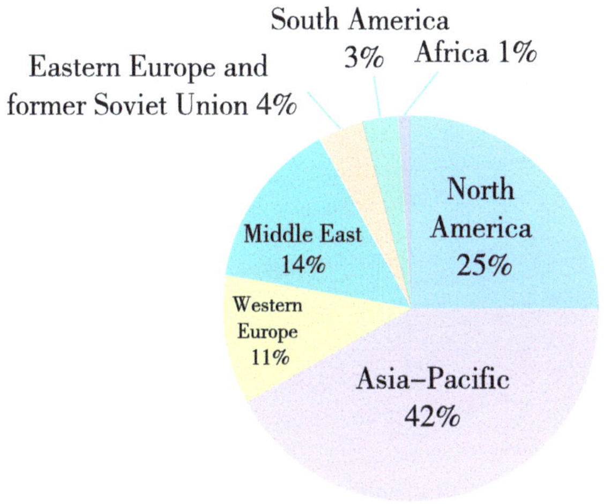

Fig. 12 Global ethylene capacity layout by region in 2023. *Source* PetroChina Economic Research Institute

demand was about 196 million tons, an increase of 5.0% year-on-year. The demand for downstream polyethylene accounted for 62%, an increase of 5.3% year-on-year; the demand for ethylene oxide, the second largest consumption area of ethylene, accounted for about 15%, an increase of 4.9% year-on-year (Fig. 16).

Overall Decline in Ethylene Prices, Ethane-Based Ethylene Still Has a Profit Advantage, Oil-Based Ethylene Profit Continues to Decline

In 2023, global ethylene prices remained relatively low, with the average spot prices of ethylene in North America, Northeast Asia, and Northwest Europe falling by 29%, 18%, and 14.6% respectively from the previous year (Fig. 17). Global ethylene gross profit performance varied, with North American ethane cracking ethylene still

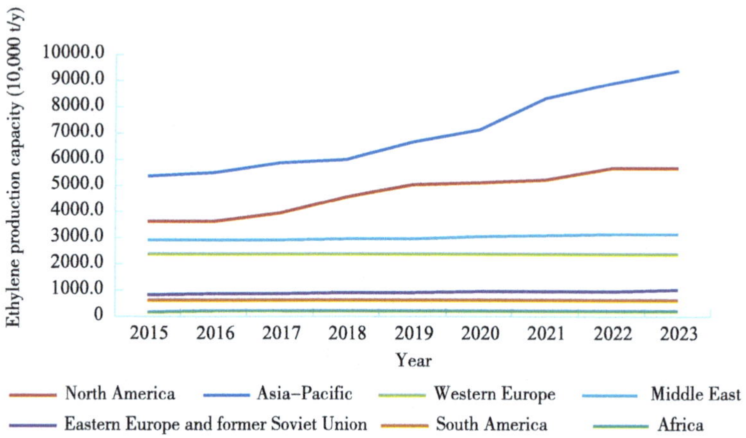

Fig. 13 Changes in global ethylene capacity by region from 2015 to 2023. *Source* PetroChina Economic Research Institute

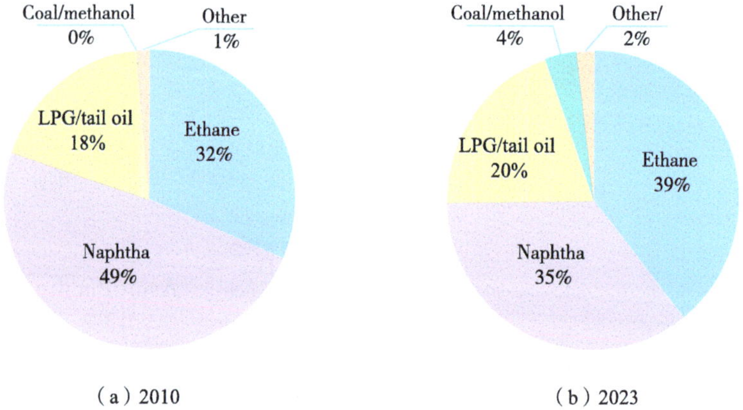

Fig. 14 Proportion of global ethylene production by different routes in 2010 and 2023. *Source* Wood Mackenzie

having an advantage, with an average gross profit of 410 US dollars/ton, an increase of 1.71% year-on-year; the performance of oil-based ethylene gross profit was poor, with the gross profit of Northwest European naphtha cracking ethylene being 283 US dollars/ton, a decrease of 37.8% year-on-year; the average gross profit of Northeast Asian naphtha cracking ethylene was 3.84 US dollars/ton, a sharp decrease of 90% year-on-year. The ethylene supply and demand in this region was imbalanced, the market continued to be depressed, and the gross profit of oil-based ethylene plants was in a loss state most of the time (Fig. 18).

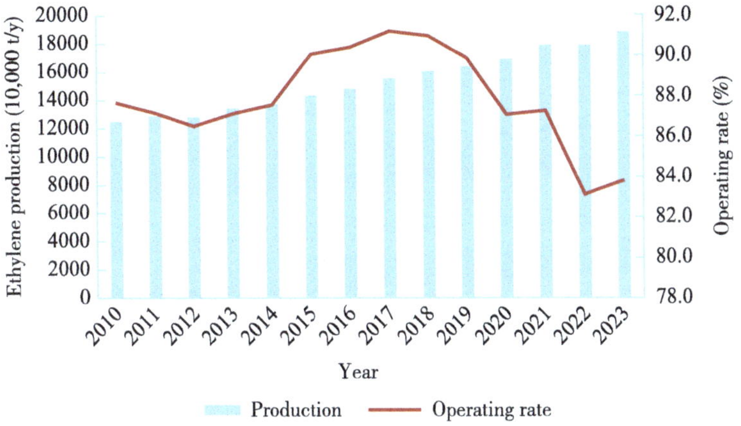

Fig. 15 Average start-up rate of global ethylene plants from 2010 to 2023. *Source* Wood Mackenzie

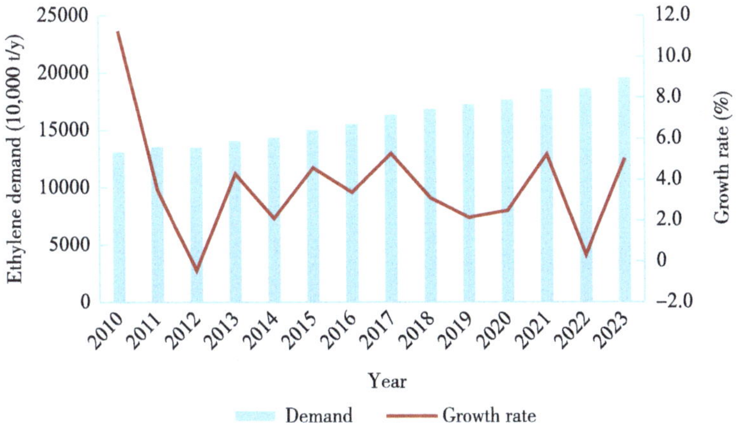

Fig. 16 Changes in global ethylene demand from 2010 to 2023. *Source* Wood Mackenzie

New Developments in International Ethylene Technology

In 2023, energy-saving renovation of ethylene production equipment is still a focus; using green electricity to reduce emissions in the production process; biomass production of ethylene technology, CO_2 reduction to produce ethylene technology, and ethane oxidative dehydrogenation to produce ethylene technology are all current research hotspots. (1) Steam cracking units are still an important source of global ethylene supply. The cracking furnace is the core equipment of the steam cracking unit and is also a high-energy-consuming device. Improving the thermal efficiency of the cracking furnace, the separation efficiency of cracking products, and seeking clean energy supply are the key paths to reduce carbon emissions from steam cracking

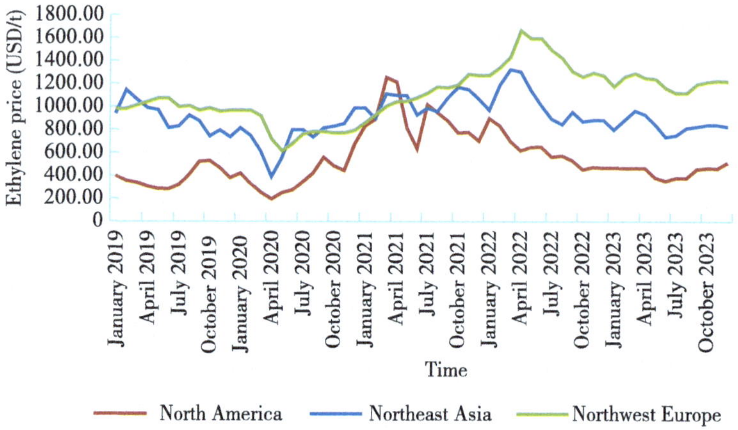

Fig. 17 Ethylene prices in the international market from 2019 to 2023. *Source* Wood Mackenzie

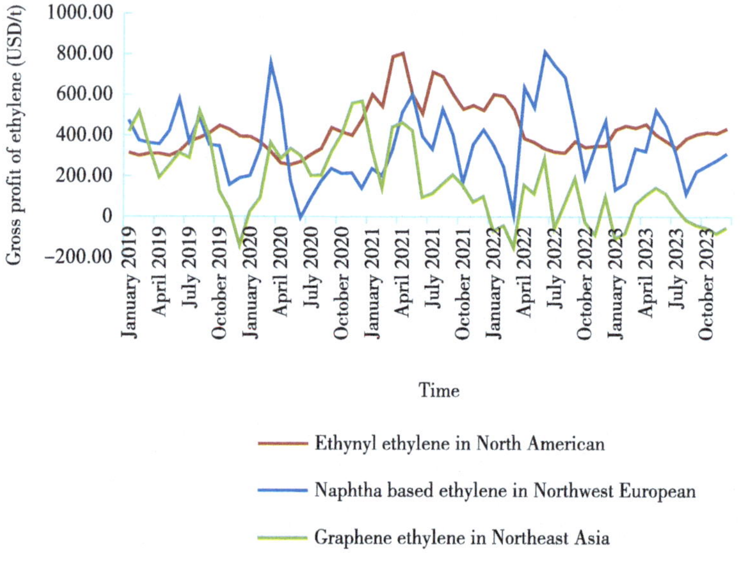

Fig. 18 Changes in ethylene gross profit in major global regions from 2019 to 2023. *Source* Wood Mackenzie

units. (2) Using biomass raw materials to produce ethylene can achieve carbon reduction at the source. The technology of producing ethylene by fermentation of first-generation grain ethanol with corn, wheat and other food crops as raw materials, and 1.5-generation non-grain ethanol with sugarcane and other economic crops as raw materials has matured. The technology of producing ethylene from second-generation cellulose ethanol with corn cobs, lignin and other cellulose materials as raw materials, and third-generation bio-ethanol with algae as raw materials is still actively being researched. (3) The production of chemicals through the reduction reaction of CO_2 is also a focus of emission reduction research. If renewable energy is used for power supply, it will greatly reduce the consumption of fossil energy in the production process. In addition, this process uses captured CO_2 as a cheap carbon resource to produce ethylene and other chemicals, realizing the effective use of CO_2. (4) Ethane oxidative dehydrogenation to produce ethylene (ODH) is an exothermic reaction with a high conversion rate. The process conditions of this process are mild, the energy consumption is low, and the research focus is on using efficient catalysts to improve selectivity.

Global Ethylene Development Trend Outlook

Global ethylene capacity continues to grow. In 2024, it is expected that the number of newly built and put into operation ethylene projects will significantly decrease, with only about 4.33 million tons of new ethylene capacity, mainly from China and the Middle East. Tianjin Petrochemical and Yulong Petrochemical will add 1.2 million tons/year and 1.5 million tons/year respectively, and the global ethylene capacity may reach 232 million tons/year.

Global ethylene demand continues to grow, the start-up rate of the device remains at a low level, and ethylene gross profit continues to rise. The global ethylene demand is expected to reach 206 million tons in 2024, an increase of 5.2% year-on-year; the start-up rate of ethylene units will rise to around 86%, still difficult to reach the high level of 90%. With the addition of less supply, the gross profit of ethylene will improve, and the price difference between naphtha and ethylene will slightly strengthen. The competition in the Asian ethylene market remains severe, with the United States continuing to maintain its ethylene exports to Asia.

The ethylene industry continues to promote green and low-carbon development. In 2024, it is still a key issue for ethylene manufacturers to achieve significant emission reductions in cracking units, and achieving significant reductions in Scope 1 and Scope 2 emissions through various means such as electrification of steam cracking furnaces is still an important way for the green and low-carbon transformation of the ethylene industry.

(Written by: Xu Haifeng Reviewed by: Wang Jing)

Current Status and Trends of China's Refining and Chemical Industry Development

In 2023, the development of domestic refining and chemical industries further differentiated. In terms of capacity, the growth of refining capacity slightly slowed down, while the capacity of the chemical industry continued to grow rapidly, with private enterprises being the main force of capacity expansion. In terms of production operation, with the recovery of refined oil demand significantly faster than the growth of chemical product demand, coupled with the significant expansion of chemical capacity, the benefits of refining improved while the overall profits of chemical products remained low, showing a temporary slowdown in the "reducing oil and increasing chemicals" trend.

In 2024, China will add 25 million tons/year of refining capacity. As the growth rate of refined oil supply exceeds the growth rate of demand, the gap between refined oil production and demand widens, and the pressure of excess refining capacity still exists, transformation and upgrading, and structural adjustment still need to be continuously promoted. The capacity investment in the chemical industry has decreased slightly, and the growth rate of demand will be slightly higher than the level in 2023, and the overall prosperity is expected to rebound slightly.

Development Status and Trends of China's Refining Industry

In 2023, China's refining capacity growth slowed down slightly. The rapid recovery of refined oil demand drove a significant improvement in the production operation of the refining industry, and several major production and operation indicators rebounded significantly. Due to the significantly faster growth of refined oil demand than the growth of chemical products and refining by-products, the "reducing oil and increasing chemicals and specialties" trend temporarily slowed down.

In 2024, China will build and expand two sets of ten-million-ton refining projects, and China's total refining capacity will reach 961 million tons/year. The pressure of excess refining capacity and structural contradictions still exist, and transformation and upgrading, "reducing oil and increasing chemicals and specialties" still need to be continuously promoted.

Characteristics of the Development of China's Refining Industry

The Growth of Refining Capacity Slows Down, and the Transformation and Upgrading Continue to Deepen

The growth of refining capacity slows down slightly. As of the end of 2023, China's refining capacity reached 936 million tons/year, an increase of 12 million tons/year from the previous year, which is the lowest increase since 2017. The new capacity mainly comes from the expansion of Xinhai Chemical and the Hainan New Materials Project of Jingbo Petrochemical (Table 3 and Fig. 19).

Diversification of operating entities and further enhancement of industrial scale. From the perspective of operating entities, the domestic competition landscape involving entities such as Sinopec, PetroChina, Sinochem, CNOOC, private

Table 3 China's new refining capacity in 2023

Group	Enterprise	Location	New refining capacity (10,000 tons/year)
Local Refining Enterprises[1]	Xinhai Chemical	Hebei	800
	Jingbo Petrochemical	Hainan	400
Total			1200

[1]Excluding enterprises owned or controlled by CNOOC, Sinochem, Sinopec, China North Industries, etc.

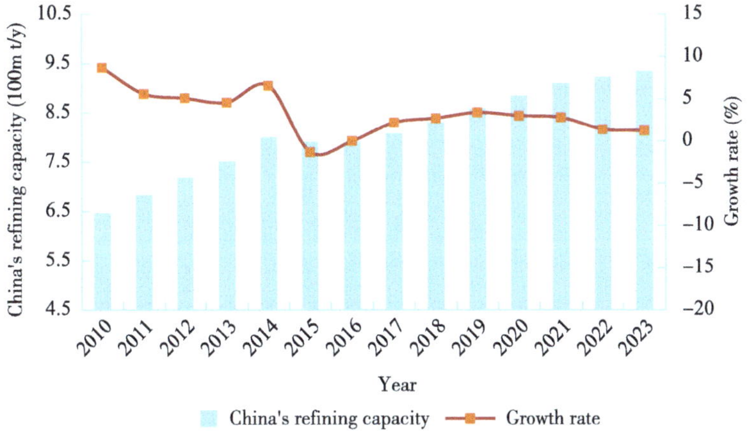

Fig. 19 Growth and growth rate of China's refining capacity from 2010 to 2023. *Data source* PetroChina Economic Research Institute

enterprises, foreign capital, and coal-based oil product companies has further developed in China. Among them, the main operating units represented by PetroChina, Sinopec, CNOOC, and Sinochem have a total refining capacity of 619 million tons/year, accounting for 66.2% of the total domestic refining capacity; private enterprises have a refining capacity of 274 million tons/year, accounting for 29.3% of the total domestic refining capacity; other operating entities account for 4.5% of the total domestic refining capacity. In terms of the number of large-scale refineries, there are 36 refineries with a capacity of over 10 million tons nationwide, an increase of 1 from the previous year, with a total refining capacity of 522 million tons/year, accounting for 55.8% of the total domestic refining capacity, an increase of 0.8 percentage points from the previous year; there are 5 private enterprises with a capacity of over 10 million tons, with a total refining capacity exceeding 100 million tons/year to 107 million tons/year. In terms of refinery scale, the average scale of refineries nationwide has risen to 4.72 million tons/year. The average scales of Sinopec and PetroChina (including foreign capital or joint ventures) refineries are 10.5 million tons/year and 8.41 million tons/year respectively (Table 4).

Table 4 Refining capacity composition of Chinese companies from 2021 to 2023

Project	2021		2022		2023	
	Capacity (10,000 tons/year)	Percentage of total (%)	Capacity (10,000 tons/year)	Percentage of total (%)	Capacity (10,000 tons/year)	Percentage of total (%)
Nationwide	91,090	–	92,350	–	93,550	–
Sinopec	27,770	30.49	27,770	30.07	27,770	29.68
PetroChina	21,000	23.05	23,000	24.91	23,000	24.59
Sinochem	5956	6.54	5956	6.45	5956	6.37
CNOOC	5200	5.71	5200	5.63	5200	5.56
Other companies	29,260	32.12	28,520	30.88	29,720	31.77
Coal-based oil companies	1080	1.19	1080	1.17	1080	1.15
Foreign companies	824	0.9	824	0.89	824	0.88

Data source PetroChina Economic Research Institute

The concentration of regional layout continues to increase. From the perspective of regional layout, East China, Northeast China, and South China are the main refining areas in the country, with a total refining capacity of 688 million tons/year, accounting for 73.6% of the total national capacity; the refining capacity in East China is 404 million tons/year, accounting for 43.2% of the total capacity. From the perspective of provincial and municipal layout, Shandong, Liaoning, Guangdong, and Zhejiang are the top four refining provinces in the country, with a total refining capacity of 495 million tons/year, accounting for 52.9% of the total national capacity. From the perspective of coastal layout, the three major refining and chemical industry clusters of the Bohai Bay, Yangtze River Delta, and Pearl River Delta have a total refining capacity of 679 million tons/year, accounting for 72.6% of the total national capacity; the ethylene capacity is 33.355 million tons/year, accounting for 64.8% of the total national capacity (Table 5).

The structure of the refining industry continues to optimize. In 2023, the significant rebound in residents' travel drove a rapid recovery in the consumption of refined oil, while the slow recovery of the manufacturing industry dragged down the demand for chemical products. The characteristic of "strong refined oil and weak chemical products" led to a short-term obstruction of "reducing oil and increasing chemicals". However, in the long-term trend, refining and chemical companies continue to promote "oil transformation and oil specialization" according to their own advantages and market demand. The supporting secondary processing and deep refining devices of refining companies have been further improved. Among them, the secondary deep processing capacity is 1.128 billion tons/year, an increase of 22.8 million tons/year compared to the previous year; the ratio to the primary processing capacity is

Table 5 Proportion of China's regional refining capacity from 2021 to 2023

Region	2021		2022		2023	
	Refining capacity (10,000 tons/year)	Percentage of national total (%)	Refining capacity (10,000 tons/year)	Percentage of national total (%)	Refining capacity (10,000 tons/year)	Percentage of national total (%)
North China	8250	9.06	8250	8.93	9050	9.67
Northeast	15,095	16.57	15,095	16.35	15,095	16.14
South China	10,930	12.00	12,930	14	13,330	14.25
East China	41,135	45.16	40,395	43.74	40,395	43.18
Northwest	8610	9.45	8610	9.32	8610	9.2
Central China	4570	5.02	4570	4.95	4570	4.89
Southwest	2500	2.74	2500	2.71	2500	2.67
Total	91,090	100	92,350	100	93,550	100

120.56%, an increase of 0.9 percentage points from the previous year (Table 6). At the same time, the deep integration of refining and chemical has become an investment trend. For existing fuel-type refineries, new generation catalytic cracking series technologies (DCC/RTC/FMP) are used to maximize the reduction of refined oil production and increase the production of low-carbon olefins and aromatic raw materials without increasing the processing volume of crude oil, such as the completion and operation of the 3 million tons/year heavy oil catalytic cracking device of Anqing Petrochemical, and the DCC catalytic cracking integrated device of Datia Petrochemical with a capacity of 3.2 million tons/year is under construction; refineries with better conditions are planning million-ton ethylene and downstream derivative projects, such as Zhenhai Refining and Chemical, Guangxi Petrochemical, Huizhou Refining and Chemical, etc. For private enterprises, some small and medium-sized refineries are transforming into world-class refining and chemical bases through mergers and integrations, and volume replacement; some large refineries are transitioning to full integration by adding to the chain; some medium-sized refineries choose to extend in certain areas with a certain foundation or their own raw materials, and strengthen their characteristic products.

Comprehensive Recovery of Industry Operations

In 2023, several major production and operation indicators of the refining industry increased significantly. Among them, the crude oil processing volume was 738 million tons, an increase of 8.6% year-on-year; the refinery start-up rate was 78.9%,

Table 6 Composition of China's crude oil secondary processing capacity from 2021 to 2023

Project	2021		2022		2023	
	Device processing capacity (10,000 tons/year)	Percentage of primary processing capacity (%)	Device processing capacity (10,000 tons/year)	Percentage of primary processing capacity (%)	Device processing capacity (10,000 tons/year)	Percentage of primary processing capacity (%)
Primary processing capacity	91,090		92,350		93,550	
Secondary processing capacity	107,386	117.89	110,506	119.66	112,786	120.56
Catalytic cracking	25,411	27.9	25,491	27.6	25,741	27.52
Delayed coking	11,491	12.61	11,971	12.96	11,971	12.8
Catalytic reforming	13,005	14.28	13,965	15.12	14,575	15.58
Hydrocracking	16,754	18.39	18,464	19.99	19,884	21.25
Hydrofining	40,725	44.71	40,615	43.98	40,615	43.41

Data source PetroChina Economic Research Institute

an increase of 5.3 percentage points year-on-year (Fig. 20). The output of gasoline, diesel, and kerosene was 444 million tons, an increase of 11.1% year-on-year; the net export volume of gasoline, diesel, and kerosene was 41.42 million tons, an increase of 25.2% year-on-year; the annual profit per ton of oil was about 360 yuan, an increase of 70 yuan/ton year-on-year. With the improvement of refining benefits, the decline in the yield of refined oil rebounded, the "reduction of oil and increase of chemical and special" phase slowed down, the annual yield of refined oil was 60.1%, an increase of 1.36 percentage points year-on-year; the yield of chemical light oil was 21.0%, an increase of 1.4 percentage points year-on-year (Table 7).

The "Ceiling" of Refining Capacity is Determined, and the Development of Integration, Green, Low-Carbon, and Intelligence Continues to Deepen

In 2023, the state continued to introduce a number of policies to guide the transformation and upgrading of the refining industry. On October 25, the National Development and Reform Commission, the National Energy Administration, the Ministry of Industry and Information Technology, and the Ministry of Ecology and Environment jointly issued the "Guiding Opinions on Promoting the Green Innovation and High-quality Development of the Refining Industry". In terms of refining capacity control, by 2025, the domestic crude oil primary processing capacity will be controlled within 1 billion tons, and the refining capacity of tens of millions of tons will account for

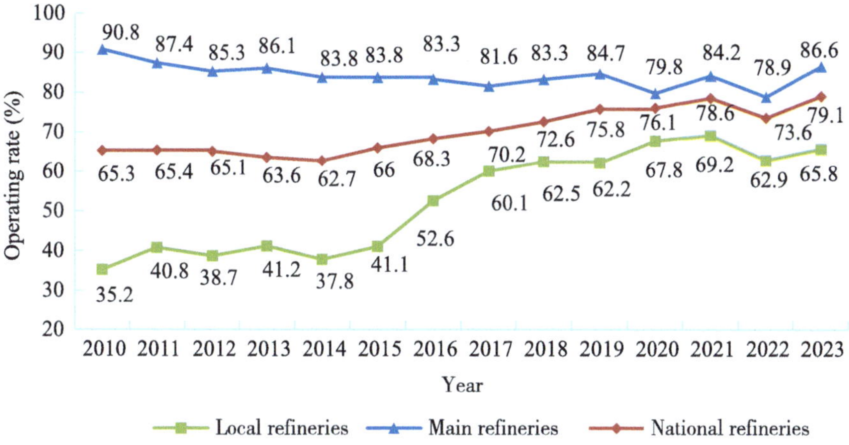

Fig. 20 Refinery start-up rate in China from 2010 to 2023. The main refineries refer to the refineries of PetroChina, Sinopec, and CNOOC

Table 7 Operation of China's refining industry from 2021 to 2023

Indicator	2021		2022		2023	
	Value	Year-on-year change	Value	Year-on-year change	Value	Year-on-year change
Crude oil processing volume (10,000 tons)	71,589	6.2%	67,995	– 5.0%	73,848	8.6%
Refinery start-up rate (%)	78.6	2.5	73.6	– 5.0	78.9	5.3
Gasoline, diesel, and kerosene production (10,000 tons)	43,276	2.5%	39,955	– 7.7%	44,401	11.1%
Gasoline, diesel, and kerosene recovery rate (%)	60.45	– 2.14	58.76	– 1.69	60.12	1.36
Diesel to gasoline production ratio	1.15	– 0.11	1.29	0.14	1.25	– 0.04
Net export of gasoline, diesel, and kerosene (10,000 tons)	3781	– 11.2%	3307	– 12.5%	4142	25.2%
Gross profit per ton of oil (Yuan/Ton)	530	290	290	– 240	360	70

about 55%; the scale of atmospheric and vacuum distillation units in newly built refineries should not be less than 10 million tons/year; and the orderly elimination and exit of atmospheric and vacuum distillation units that do not comply with national industrial policies of 2 million tons/year and below are promoted. In terms of optimizing the structure and layout of production capacity, the orderly concentration of new refining projects to petrochemical industry bases is promoted to achieve intensive and agglomerated development, and the connection with ethylene and p-xylene (PX) projects is well coordinated. In terms of green and low-carbon development, support is given to refining enterprises to accelerate the demonstration application of CCUS, effectively reduce carbon emissions; promote the integration of the refining industry and renewable energy, encourage enterprises to vigorously develop renewable energy hydrogen production; support the construction of green hydrogen refining demonstration projects, promote the replacement of green hydrogen, and gradually reduce the industry's coal hydrogen usage. In terms of intelligent development, the independent development and application of process simulation software and process optimization software are strengthened, the simulation optimization of distillation, reaction, and whole plant processes is encouraged, and the level of refined management of refining enterprises is improved; the integration of the new generation of digital technology and the refining industry is promoted, and the construction of digital and intelligent refineries is accelerated.

New Progress in Refining Technology

In 2023, the progress of refining technology was mainly reflected in product quality, environmental protection, structural adjustment of refining enterprises, and carbon reduction. The first is the key refining technology under the goal of product quality improvement. In terms of gasoline quality upgrade technology, there are lower energy consumption alkylation technology, increased production of alkylation raw material technology, low-cost benzene reduction technology, C_{9+} Technologies such as lightening of aromatic hydrocarbons; in terms of diesel quality upgrade, there are deep hydrogenation desulfurization technologies, with the focus of research and development on high-density aviation kerosene and low-cost production of low-sulfur ship fuel production technology. Second, key technologies under the continuous improvement of environmental protection goals, the sludge reduction technology for waste gas, waste liquid and solid waste pollution produced in the refining process, site pollution identification, circulating water low phosphorus corrosion and scale inhibition technology, difficult biochemical wastewater efficient pretreatment technology, etc., have initially shown results. Third, refining structure transformation technology, existing refining enterprises are developing towards the combination of oil and chemical, new refineries mainly focus on deep integration of refining and chemical. The mainstream route includes the construction of catalytic cracking/catalytic pyrolysis, hydrogenation cracking, continuous reforming and other devices, increasing the production of ethylene and aromatic raw materials, reducing the production of finished oil or not producing gasoline and diesel. Refineries with relatively good raw material quality choose crude oil direct catalytic cracking or crude oil direct steam cracking process technology. Fourth, energy-saving and carbon-reducing, cost-reducing technologies, such as molecular refining technology can provide enterprises with crude oil selection and product economic analysis model, establish a total process processing route centered on separation, greatly improve the single-pass conversion efficiency and target product yield, and reduce invalid circulation. Heat exchange network optimization, steam power system optimization, low-temperature heat efficient use and other energy efficiency improvement technologies are key links to reduce carbon emissions from refineries. The development of biofuel technology also provides support for carbon reduction.

Development Trend of China's Refining Industry

Rapid Growth of Refining Capacity Again

By 2024, the domestic net increase in refining capacity will be 25 million tons/year, with a total capacity reaching 961 million tons/year. Among them, the new capacity of 31 million tons/year mainly comes from the second phase of Zhenhai Refining and Chemical and Yulong Island Petrochemical, and Dalian Petrochemical will withdraw a set of 6 million tons/year atmospheric pressure distillation unit (Table 8). The number of domestic refineries with a capacity of 10 million tons will increase to 37, with a total refining capacity of 539 million tons/year, accounting for 56.1% of the total capacity, of which the capacity of refineries above 20 million tons/year will reach 180 million tons/year, accounting for 18.8% of the total national capacity.

① Excluding companies owned or controlled by CNOOC, ChemChina, Sinochem, China North Industries, etc.

In 2024, under the baseline scenario, in terms of main refineries, the second phase of Zhenhai Refining and Chemical will be put into operation in the second half of the year, and refineries such as Guangdong Petrochemical will make up for the loss in the first quarter of the previous year, with an expected increase in crude oil processing volume of 10 million tons. In terms of private enterprises, Yulong Island Petrochemical will be put into operation in the second half of the year, and Shenghong Refining and Chemical will make up for the loss in the first quarter of the previous year, with an expected increase in crude oil processing volume of 12 million tons; in 2023, private enterprises over-issued 17.99 million tons of crude oil quotas, if the over-issued quotas are cancelled, the expected crude oil processing volume will decrease by 9 million tons; the overall crude oil processing volume of private enterprises will increase by 3 million tons. The national crude oil processing volume is expected to be 752 million tons, an increase of 1.8% year-on-year; the output of refined oil is 451 million tons, an increase of 1.6% year-on-year; the production exceeds the demand by more than 50 million tons.

Table 8 Domestic new/eliminated refining capacity in 2024

New capacity				Eliminated capacity			
Group	Company	Location	New capacity (10,000 tons/year)	Group	Company	Location	Eliminated capacity (10,000 tons/year)
Sinopec	Zhenhai Refining and Chemical (Phase II)	Zhejiang	1100	PetroChina	Dalian Petrochemical	Liaoning	600
Local refining companies[①]	Yulong Island Petrochemical	Shandong	2000				
Total new capacity			3100	Total eliminated capacity			600
Total net increase in capacity: 25 million tons/year							

[①]Excluding companies owned or controlled by CNOOC, ChemChina, Sinochem, China North Industries, etc.

The Excess Refining Capacity is Temporarily Alleviated, but We Cannot Take It Lightly

In the next few years, while several planned domestic refining and chemical integration expansion and renovation projects are put into operation, some capacities will be replaced and cleared, the growth of refining capacity basically ends, but the problem of excess refining capacity still needs high attention and serious solutions. First, the continuous rapid development of the new energy vehicle industry makes the peak demand for refined oil imminent, and it will show a rapid decline trend after the peak consumption of refined oil; second, the "reducing oil and increasing chemical" continues to advance, the pressure of excess supply of refined oil may be alleviated, but the excess pressure has a trend of being transmitted to bulk petrochemical products; third, the elimination of most of the small refineries starts, the reduction of refined oil supply is limited, and a relatively large export needs to be maintained for a period of time in the future to balance domestic resources.

(Written by: Fei Huawei, Reviewed by: Wang Jing)

Current Status and Trends of China's Chemical Industry Development

In 2023, the domestic chemical market overall continues to expand, the market entities are diversified, and the industry development shows different characteristics. The downstream demand of the chemical market is generally poor, the loose supply situation is still severe, and the overall profit of the main petrochemical products is at a historical low.

In 2024, the capacity input of the domestic chemical market will decrease slightly, the demand growth rate will be slightly higher than the 2023 level, and the overall prosperity is expected to rebound slightly. Among them, the fundamentals of xylene are good, and the profit recovers to a historical medium–high level. The capacity input of propylene continues to be high, and the profit will still be low, and the prosperity of other products will mainly fluctuate or improve in stages.

Current Status and Trends of China's Ethylene Industry Development

Current Status of China's Ethylene Industry Development

The expansion of ethylene capacity enters a pause period, and the new capacity slows down significantly. In 2023, the domestic ethylene new capacity is 2.423 million tons/year, a total of 6 companies, far lower than the average capacity increase of 6.05 million tons/year in the past 4 years (Table 9), which is the least capacity growth since 2019, with a total capacity of 51.95 million tons/year, ranking

Table 9 Changes in China's ethylene capacity in 2023

Project/Company	Nature of the company	Process route	Location	Ethylene capacity (10,000 tons/year)
Baofeng Energy Ningdong Phase III	Private enterprise	Coal-based olefins	Yinchuan, Ningxia	50
Sanjiang Chemical	Private enterprise	Naphtha + light hydrocarbon cracking	Jiaxing, Zhejiang	100
Dongming Petrochemical	Private enterprise	Naphtha cracking	Heze, Shandong	45
Zibo Junchen New Materials	Private enterprise	Naphtha + light hydrocarbon cracking	Zibo, Shandong	19
Anqing Petrochemical	State-owned enterprise	DCC	Anqing, Anhui	10.3
Daqing Petrochemical	State-owned enterprise	Naphtha cracking	Daqing, Heilongjiang	18 (expansion)
Total				242.3

Data source PetroChina Economic Research Institute

first in the world. Among the new capacities, private enterprises have built ethylene projects to break through the upstream industry chain; state-owned enterprises actively promote the "reducing oil and increasing chemical" strategy, and extend downstream according to the company's positioning. At the same time, to accelerate the carbon peak goal, carry out equipment expansion and transformation to achieve energy saving and consumption reduction, and reach the industry energy efficiency benchmark level (Table 9).

The ethylene industry chain has a diverse range of operators, and the average scale of the devices has slightly declined. By the end of 2023, the private ethylene production capacity had reached 18.625 million tons, accounting for 35.8% of the total national capacity, an increase of 2.3 percentage points from the previous year, becoming an important participant in China's ethylene industry. The combined ethylene production capacity of Sinopec, PetroChina, and CNOOC fell slightly from 51.5% in the previous year to 49.7% (Table 10). Due to the smaller scale of the newly added coal-to-olefin and oil conversion devices, the average scale of the devices decreased slightly by 0.08 million tons/year to 0.611 million tons/year (Table 11).

Ethylene equivalent consumption continues to grow, with only a slight increase in growth rate. In 2023, China's economy continues to recover, driving the continuous increase in demand for ethylene and downstream products. It is estimated that the annual consumption of ethylene equivalent will be 61.5 million tons, an increase of 4.9% year-on-year, and the growth rate will increase by 1.3 percentage points year-on-year. However, since the release of ethylene capacity in 2022 is concentrated in the second half of the year, the actual supply scale in 2023 is still

Table 10 Composition and proportion of China's ethylene production capacity by company in 2023

Project	2005		2022		2023	
	Capacity (10,000 tons/year)	Percentage of total capacity (%)	Capacity (10,000 tons/year)	Percentage of total capacity (%)	Capacity (10,000 tons/year)	Percentage of total capacity (%)
Sinopec	552.5	70.3	1481.5	30.0	1481.5	28.5
PetroChina	217.4	27.7	872	17.6	890	17.1
CNOOC	0	0	215	4.3	215	4.1
Other ethylene companies	16	2	2384.5	48.1	2579.5	49.7
National total	785.9	100	4953	100	5195	100

Note The production capacity of Sinopec, PetroChina, and CNOOC includes foreign equity production capacity

Table 11 Composition of ethylene devices with different raw materials in China in 2022 and 2023

Raw material	2022			2023		
	Capacity (10,000 tons/year)	Production unit (Set)	Average scale (10,000 tons/year)	Capacity (10,000 tons/year)	Production unit (Set)	Average scale (10,000 tons/year)
Oil-based	3311	44	75.3	3503	48	73.0
Coal/ Methanol-based	807	27	29.9	857	28	30.6
Gas-based	835	9	92.8	835	9	92.8
Total	4953	80	61.9	5195	85	61.1

large, with an annual ethylene output of 46 million tons, a year-on-year increase of 13.5%, far higher than the demand growth, and the supply pressure further increases. The annual ethylene unit start-up rate is 91%, basically the same as the previous year, maintaining a historical low. The self-sufficiency rate of equivalent increased further to 75%, an increase of about 2 percentage points year-on-year.

Ethylene prices fell significantly, and production profits further narrowed. In 2023, due to the fall in prices of raw materials such as crude oil, coal, and ethane, and the higher growth rate of production, ethylene prices fell significantly. The average annual price of ethylene in Northeast Asia was $851/ton, a year-on-year decrease of $171/ton, which is a historically low level. Due to the continuous exacerbation of supply and demand imbalance, the profit of domestic naphtha-based ethylene units was $15/ton, a year-on-year narrowing of $10/ton, setting a new historical low; the profit of methanol-based ethylene was a loss of $70/ton, a year-on-year expansion

Table 12 Estimated new ethylene production capacity in China in 2024

Project/Company	Nature of the company	Process route	Location	New ethylene production capacity (10,000 tons/year)
Tianjin Nangang Ethylene	Joint Venture	Steam cracking	Tianjin	120
Yulong Petrochemical No.1 Ethylene	Private Enterprise	Steam cracking	Yantai, Shandong	150
Shandong New Era Polymer	Private Enterprise	Steam cracking	Zibo, Shandong	70
Total				320

Data source PetroChina Economic Research Institute

of \$100/ton; the profit of coal-based ethylene was a loss of \$100/ton, a year-on-year expansion of \$80/ton; the profit of ethylene produced using imported ethane was \$170/ton, a year-on-year increase of \$20/ton.

Development Trend of China's Ethylene Industry

In 2024, the new capacity of domestic ethylene continues to decline, and the rapid rise of capacity temporarily stops. It is estimated that the new capacity of domestic ethylene in 2024 will be 3.2 million tons/year, and the total capacity of ethylene will reach 55.15 million tons/year. Among the new capacities, all are oil-based routes, and 68% of the capacity comes from private enterprises (Table 12). In total capacity, the capacity of oil-based ethylene is 38.23 million tons/year, accounting for 69.3%, a year-on-year increase of 1.8 percentage points; the capacity of ethylene with light hydrocarbons as raw materials is 8.35 million tons/year, accounting for 15.1%, a year-on-year decrease of 0.8 percentage points; the capacity of coal/methanol-based ethylene remains 8.57 million tons/year, the proportion dropped to 15.5%, a year-on-year decrease of 1 percentage point.

In 2025, a new round of capacity deployment led by central enterprises will arrive for ethylene. From 2025 to 2027, there will still be a large number of ethylene projects put into production in China (Table 13). The new capacity is mainly from the refining and chemical integration transformation and expansion of ethylene projects by Sinopec and PetroChina, the two major groups. The expansion of private enterprises will slow down; joint ventures or wholly-owned projects in China are actively being built or promoted; non-integrated enterprises lacking naphtha raw materials are accelerating investment in light hydrocarbon cracking projects.

The ethylene industry will see a phased improvement after 2024. From the demand perspective, with the domestic consumption shifting from post-epidemic recovery to continuous expansion, it is expected that the consumption of ethylene

Table 13 Estimated new ethylene production capacity in China from 2025 to 2027

Project/Company	Nature of the company	Process route	Location	New ethylene production capacity (10,000 tons/year)	Year of completion
Yulong Petrochemical No.2 Ethylene	Private Enterprise	Naphtha cracking	Yantai, Shandong	150	2025
Sinopec Great Wall Energy Chemical	State-owned Enterprise	Coal-to-olefins	Ordos, Inner Mongolia	40	2025
Qinghai Mining	State-owned Enterprise	Coal-to-olefins	Haixi, Qinghai	26	2025
China Coal Shaanxi Yuyou Chemical	State-owned Enterprise	Coal-to-olefins	Yulin, Shaanxi	30	2025
ExxonMobil Huizhou Project	Foreign Enterprise	Naphtha cracking	Huizhou, Guangdong	160	2025
Wanhua Chemical Phase II	State-owned enterprise	Light hydrocarbon cracking	Yantai, Shandong	120	2025
Guangxi Petrochemical	State-owned enterprise	Naphtha cracking	Qinzhou, Guangxi	120	2025
Jilin Petrochemical	State-owned enterprise	Naphtha cracking	Jilin City, Jilin Province	120	2025
Lianhong New Materials	Private enterprise	Methanol to olefins	Zaozhuang, Shandong	25	2025
Dushanzi Petrochemical	State-owned enterprise	Ethane cracking	Karamay, Xinjiang	120	2025
Inner Mongolia Baofeng Coal-to-Olefins	Private enterprise	Coal to olefins	Ordos, Inner Mongolia	100	2025
Shenhua Ningxia Coal Industry	State-owned enterprise	Coal to olefins	Yinchuan, Ningxia	69.5	2025
Maoming Petrochemical	State-owned enterprise	Naphtha cracking	Maoming, Guangdong	100	2026
BASF Zhanjiang Project	Foreign enterprise	Naphtha cracking	Zhanjiang, Guangzhou	100	2026
Huajin Amec Petrochemical	Joint venture	Naphtha cracking	Panjin, Liaoning	150	2026
Fujian Zhongsha Petrochemical	Joint venture	Naphtha cracking	Zhangzhou, Fujian	150	2026

(continued)

Table 13 (continued)

Project/Company	Nature of the company	Process route	Location	New ethylene production capacity (10,000 tons/year)	Year of completion
Lanzhou Petrochemical	State-owned enterprise	Ethane cracking	Lanzhou, Gansu	120	2026
Hunan Petrochemical	State-owned enterprise	Naphtha cracking	Yueyang, Hunan	100	2026
Guangxi Huayi	State-owned enterprise	Methanol to olefins	Qinzhou, Guangxi	50	2026
Luoyang Petrochemical	State-owned enterprise	Naphtha cracking	Luoyang, Henan	100	2026
CNOOC Shell	Joint venture	Naphtha cracking	Huizhou, Guangdong	150	2027
Zhenhai Refining & Chemical	State-owned enterprise	Naphtha cracking	Ningbo, Zhejiang	150	2027
Yangba Petrochemical	Joint venture	Naphtha cracking	Nanjing, Jiangsu	100	2027
Quanzhou Petrochemical	State-owned enterprise	Naphtha cracking	Quanzhou, Fujian	100	2027

Data source PetroChina Economic Research Institute

Table 14 Ethylene supply and demand balance table from 2018 to 2024

Item	2018	2019	2020	2021	2022	2023	2024
Capacity (10,000 tons/year)	2533	3037	3518	4368	4953	5195	5515
Production (10,000 tons)	2372	2657	3137	3766	4053	4600	4900
Equivalent consumption (10,000 tons)	4799	5336	5981	5987	6138	6150	6560
Start-up rate (%)	93.6	87.5	89.2	86.2	81.8	88.5	88.8

Author Gao Zhenyu Reviewer: Wang Jing

equivalent will reach 65.6 million tons in 2024, an increase of 6.7% year-on-year, slightly higher than the level in 2023. From the supply perspective, due to the less new capacity in 2023, the actual supply growth rate in 2024 is relatively low, with an estimated production of about 49 million tons, a year-on-year increase of 6.5%, a significant decline in growth rate, the supply pressure has eased, and the profits of the ethylene industry have significantly improved. However, from the second half of 2025, with the start of a new round of capacity expansion, the profits of the ethylene industry will weaken again from 2026 to 2028 (Table 14).

Current Status and Trends of China's Propylene Industry

Current Status of China's Propylene Industry

The largest scale expansion of propylene capacity in history, with the new capacity mainly from propane dehydrogenation (PDH) units. In 2023, China's domestic propylene added capacity of 7.29 million tons/year, a total of 13 companies, is the largest year of capacity expansion in the history of the propylene industry (Table 15), with a total capacity of 63.34 million tons/year, ranking first in the world. Of the new capacity, 77% comes from private enterprises, and the propylene capacity of private enterprises has reached 17.9 million tons/year, accounting for 28.2% of the total national capacity, an increase of 2.3 percentage points year-on-year, becoming an important participant in China's propylene industry.

PDH units have become the largest source of propylene. From the perspective of raw material routes, due to the advantages of PDH units such as short process, small investment, and low energy consumption, and also considering factors such as ensuring the stability of the supply chain upstream of the industrial chain, a large number of private enterprises have invested in PDH units. In 2023, the newly added capacity of PDH units will reach 6.06 million tons/year, accounting for 83%. The total capacity of PDH is 19.7 million tons/year, accounting for 30%, which has surpassed the capacity of the steam cracking route and is the main source of propylene supply (Table 16).

The imbalance of propylene supply and demand will intensify in 2023, and the processing loss will still be severe. Driven by the rapid growth in demand for downstream propylene oxide and acrylonitrile, the consumption of propylene equivalents is expected to moderately recover in 2023, with an estimated annual consumption of 46.8 million tons, an increase of 3.6% year-on-year. The annual production of propylene is 49.52 million tons, an increase of 7.5% year-on-year, a slight increase in growth rate, and significantly higher than the demand growth rate, further expanding the supply–demand gap. In 2023, the overall start-up rate of propylene units was only 76%, continuing to drop by 4 percentage points to a historical low. In terms of profit, the gross profit of propylene production is still not good. In terms of PDH, due to the high price of propane, the PDH unit lost 750 yuan/ton, a slight narrowing of 200 yuan/ton from the previous month, and the lowest processing load dropped to less than 60%; the loss of propylene production by steam cracking was 440 yuan/ton, basically the same year-on-year; the MTO unit lost 720 yuan/ton, an increase of 30 yuan/ton from the previous month.

Development Trend of China's Propylene Industry

In 2024, the new capacity of domestic propylene is still large, continuing the rapid rising trend. It is estimated that the new capacity of domestic propylene in 2024 will be 6.065 million tons/year, continuing the rapid expansion trend, and the total capacity of propylene will reach 69.4 million tons/year. Among the new

Table 15 Changes in China's propylene capacity in 2023

Project/Company	Nature of the company	Process route	Location	Propylene capacity (10,000 tons/year)
Sanjiang Chemical	Private enterprise	Naphtha + light hydrocarbon cracking	Jiaxing, Zhejiang	27
Anqing Petrochemical	State-owned enterprise	Conventional distillation	Anqing, Anhui	30
Baofeng Energy Ningdong Phase III	Private enterprise	CTO	Yinchuan, Ningxia	66
Dongguan Juzhengyuan	Private enterprise	PDH	Dongguan, Guangdong	60
Zhejiang Huahong	Private enterprise	PDH	Jiaxing, Zhejiang	45
Jiangsu Ruiheng	Private enterprise	PDH	Lianyungang, Jiangsu	60
Quanzhou Guoheng Chemical	Private enterprise	PDH	Quanzhou, Fujian	66
Formosa Plastics	Taiwan-funded enterprise	PDH	Ningbo, Zhejiang	60
Lihuayi Chemical	Private enterprise	PDH	Dongying, Shandong	60
Ningbo Jinfar	Private enterprise	PDH	Zhejiang Ningbo	60
Guangxi Huayi New Material	State-owned enterprise	PDH	Guangxi Qinzhou	75
Jiangsu Yanchang Zhongran	State-owned enterprise	PDH	Jiangsu Taizhou	60
Binhu New Material	Private enterprise	PDH	Shandong Binzhou	60
Total				729

Data source PetroChina Economic Research Institute

capacities, oil-based and gas-based are the main ones, and private enterprises account for 78% (Table 17). Among the total capacity, the capacity of propylene production by steam cracking is 21.03 million tons/year, accounting for 31%, an increase of 1.5 points year-on-year; the PDH capacity is 21.6 million tons/year, accounting for 31%, the same year-on-year; the capacity of coal/methanol to produce propylene remains 11.88 million tons/year, the proportion dropped to 17.1%, a decrease of 1.3 points year-on-year; the capacity of refined propylene production is 16.195 million tons/year, accounting for 23%, basically the same year-on-year.

Table 16 Composition of propylene units from different raw materials in China in 2022 and 2023

Raw material	2022			2023		
	Capacity (10,000 tons/year)	Production unit (set)	Average scale (10,000 tons/year)	Capacity (10,000 tons/year)	Production unit (set)	Average scale (10,000 tons/year)
Steam cracking	1680	46	36.5	1838	50	36.8
Coal/methanol-based	1122	37	30.3	1188	39	30.5
Conventional distillation process	1438	160	9.0	1468	161	9.1
PDH	1364	28	48.7	1970	38	51.8

From 2025 to 2027, propylene will usher in a new round of capacity release dominated by private enterprises. With the transformation of refining and chemical integration dominated by central enterprises, and the unabated enthusiasm of private enterprises for investing in PDH devices, the domestic propylene capacity will still maintain rapid growth in the next few years. According to statistics, the total new capacity in the next round will reach 20 million tons/year, and the new capacity will significantly slow down after 2028 (Table 18).

The propylene industry will continue to be sluggish in 2024. From the demand perspective, the growth rate of polypropylene demand is expected to increase slightly, and the demand for propionic acid and others will improve slightly under the marginal improvement of the real estate market. It is estimated that the consumption of propylene equivalent in 2024 will be 49.1 million tons, an increase of 5.0% year-on-year, slightly higher than the level in 2023. From the supply perspective, due to the continuous launch of new capacity from 2023 to 2024, the actual supply capacity grows rapidly. It is estimated that the production of propylene in 2024 will be 53.5 million tons, an increase of 8.0% year-on-year. The situation of supply exceeding demand cannot be reversed, and the prosperity of the propylene industry continues to be sluggish (Table 19).

Current Status and Trends of China's Xylene Industry

Current Status of China's Xylene Industry

In 2023, the new capacity of PX increased, and the overall growth rate slowed down. Since the new round of PX expansion started in 2019, the domestic PX capacity has grown rapidly, with an average annual growth rate of 27.3% from 2019 to 2022. In 2023, three new PX devices were put into operation domestically, with a new capacity of 5.7 million tons/year, an average annual growth rate of 15.7%, down 10.9 percentage points from the previous year. Overall, the total domestic PX capacity

Table 17 Changes in China's propylene capacity in 2024

Project/Company	Nature of the company	Process route	Location	Propylene capacity (10,000 tons/year)
Ningxia Baoyuan New Energy	Private enterprise	PDH	Ningxia Hui Autonomous Region	30
Donghua Energy	Private enterprise	PDH	Jiangsu Province	60
Jin Guotou Petrochemical	Private enterprise	PDH	Liaoning Province	60
Guangdong Pengzun Energy	Private enterprise	PDH	Guangdong Province	30
Dongfang Hongye	Private enterprise	PDH	Shandong Province	10
Shandong Yulong Petrochemical	Private enterprise	Conventional distillation process	Shandong Province	69
Ningbo Daxie	State-owned enterprise	Conventional distillation process	Zhejiang Province	52.5
Xinhai Chemical	Private enterprise	Conventional distillation process	Hebei Cangzhou	9
Yangzi Petrochemical	State-owned enterprise	Conventional distillation process	Jiangsu Nanjing	21
Shandong Yulong Petrochemical	Private enterprise	Steam cracking	Yantai, Shandong	150
Shandong New Era Polymer	Private enterprise	Steam cracking	Zibo, Shandong	55
Tianjin Petrochemical	State-owned enterprise	Steam cracking	Tianjin City	60
Total				606.5

continued to rise in 2023, significantly increasing to 42.1 million tons/year, nearly three times that of before 2018. The new capacity all comes from state-owned enterprises. Among them, Guangdong Petrochemical's 2.6 million tons/year PX device was put into operation on February 6, Daxie Petrochemical's 1.6 million tons/year PX device produced qualified products on March 30, and Huizhou Petrochemical's 1.5 million tons/year PX device was put into operation on June 15, significantly increasing the company's total PX capacity to 2.45 million tons/year (Tables 20, 21, and 22).

The rapid growth of downstream capacity boosts PX consumption, but high prices limit its increase. PX is mainly used in PTA production, and the average proportion of PX consumption used as PTA raw material from 2018 to 2022 exceeds

Table 18 Estimated additional propylene capacity in China from 2025 to 2027

Project/Company	Nature of the enterprise	Process route	Location	Propylene capacity (10,000 tons/year)	Year of completion
Zhongfei Petrochemical	State-owned enterprise	PDH	Daqing, Heilongjiang	60	2025
Fujian Yongrong New Materials	Private enterprise	PDH	Putian, Fujian	90	2025
Wuhan United Energy	Private enterprise	PDH	Wuhan, Hubei	17	2025
Jilin Petrochemical	State-owned enterprise	Steam cracking	Jilin City, Jilin Province	60	2025
Wanhua Chemical	Private enterprise	Steam cracking	Yantai, Shandong	120	2025
BASF Zhanjiang Project	Foreign enterprise	Steam cracking	Zhanjiang, Guangdong	50	2025
Guangxi Petrochemical	State-owned enterprise	Steam cracking	Qinzhou, Guangxi	60	2025
ExxonMobil Huizhou Project	Foreign enterprise	Steam cracking	Huizhou, Guangdong	85	2025
Baofeng Energy Coal-to-Olefins	Private enterprise	Coal-to-Olefins	Ordos, Inner Mongolia	130	2025
Tongkun Petrochemical	Private enterprise	PDH	Qinzhou, Guangxi	60	2026
Yanchang Zhongran Chemical (Phase II)	State-owned enterprise	PDH	Taizhou, Jiangsu	60	2026
Xintai Petrochemical (Phase II)	Private enterprise	PDH	Zibo, Shandong	90	2026
Fujian Haiquan Chemical	Private enterprise	PDH	Zhangzhou, Fujian	90	2026
Yongrong New Materials (Phase II)	Private enterprise	PDH	Putian, Fujian	110	2026
China North Industries Group Fine Chemical and Raw Material Engineering Project	State-owned enterprise	Steam cracking	Panjin, Liaoning	80	2026
Lianhong Gerun New Materials	Private enterprise	DMTO	Zaozhuang, Shandong	65	2026
Donghua Energy (Maoming Phase III)	Private enterprise	PDH	Maoming, Guangdong	60	2026

(continued)

Table 18 (continued)

Project/Company	Nature of the enterprise	Process route	Location	Propylene capacity (10,000 tons/year)	Year of completion
Jinfaxin Materials (Phase II)	Private enterprise	PDH	Ningbo, Zhejiang	60	2027
Yuanjin New Materials (Phase II)	Private enterprise	PDH	Shaoxing, Zhejiang	75	2027

Table 19 Balance sheet of propylene supply and demand from 2018 to 2024

Project	2018	2019	2020	2021	2022	2023	2024
Capacity (10,000 tons/year)	3560	3927	4477	5000	5605	6356	6940
Output (10,000 tons)	3035	3312	3610	4297	4604	4952	5350
Equivalent consumption (10,000 tons)	3390	3694	4069	4340	4553	4680	4910
Operating rate (%)	85.3	84.3	80.6	85.9	82.1	77.9	77.1

Author Gao Zhenyu Reviewer: Wang Jing

Table 20 New PX capacity in 2023

Project	New capacity (10,000 tons/year)	Time of operation	Location
Guangdong Petrochemical	260	February 6, 2023	Guangdong
Daxie Petrochemical	160	March 30, 2023	Zhejiang
Huizhou Petrochemical	150	June 15, 2023	Guangdong
National total	570		

Table 21 Composition and proportion of China's PX capacity by enterprise from 2021 to 2023

Project	2021		2022		2023	
	Capacity (10,000 tons/year)	Percentage of total capacity Percentage (%)	Capacity (10,000 tons/year)	Percentage of total capacity Percentage (%)	Capacity (10,000 tons/year)	Percentage of total capacity Percentage (%)
Sinopec	592.5	21	725.5	20	725.5	17
PetroChina	265	9	265	7	525	12
CNOOC	94	3	94	3	404	10
Others	1925	67	2555	70	2555	61
National Total	2876.5	100	3639.5	100	4209.5	100

Note The capacities of Sinopec, PetroChina, and CNOOC include foreign equity production capacity

Table 22 Capacity scale of Chinese PX companies from 2021 to 2023

Company	2021		2022		2023	
	Production companies (Number)	Average scale of companies (10,000 tons/year)	Production companies (Number)	Average scale of companies (10,000 tons/year)	Production companies (Number)	Average scale of companies (10,000 tons/year)
Sinopec	9	66	10	73	10	73
PetroChina	3	88	3	88	4	131
CNOOC	1	94	1	94	2	202
Others	9	214	10	256	10	256
National Total	65	44	69.4	52	73.8	57

Table 23 Additional PTA capacity in 2023

Project	Additional capacity (10,000 tons/year)	Production time	Location
Hengli Huizhou	250	March 19, 2023	Guangdong
	250	July 11, 2023	
Jiatong Energy	250	June 15, 2023	Jiangsu
Hainan Yisheng	250	Q4 2023	Hainan
National Total	1000		

95%, reaching 97% in 2023. Since the rapid expansion cycle of PTA started in 2020, domestic PTA capacity has significantly increased, with an average annual growth rate of 13.5% and an average annual increase of about 8 million tons/year. In 2023, China is still in the peak period of PTA production, with a total new capacity of 10 million tons/year and an annual growth rate of 13%. The significant growth of downstream capacity greatly boosts the market demand for raw material PX, and the rise in PX prices forms a certain restriction on its consumption growth. It is expected that the annual PX consumption will increase significantly by 2.76 million tons/year to 38.05 million tons/year, an increase of 8% (Table 23).

The operating rate increased slightly year-on-year, and the output increased significantly year-on-year. In 2023, it is expected that the PX device operating rate will be 77%, a slight increase of 0.2 percentage points year-on-year, maintaining a relatively high level throughout the year; it is expected that the output will increase by 5.1 million tons/year, an increase of 19%, to 32.4 million tons. Looking at it by region, the capacity is mainly concentrated in the East China region, followed by the Northeast and South China regions, with the three regions accounting for 63%, 16.5%, and 15.3% of the total capacity, respectively (nearly 95% in total).

PX imports decrease, and the degree of dependence on foreign countries significantly decreases. With the continuous growth of domestic PX capacity, domestic PX has gradually replaced imports, and the problem of supply shortage has been effectively alleviated. In 2023, it is expected that China's PX imports will further decrease by 1.8 million tons/year (down 17%) to 8.8 million tons/year. China's PX dependence on foreign countries has significantly dropped from 61% in 2018 to 22% in 2023. In addition, since 2022, China's PX exports have significantly increased, with a cumulative export of nearly 100,000 tons to date.

The annual price fell slightly year-on-year, and the gross profit increased significantly year-on-year. In early 2023, under the transformation of COVID-19 prevention and control policies, the market outlook improved, international oil prices remained high, pushing PX prices up. However, with the gradual commissioning of new capacity and the sharp drop in international oil prices under financial pressure, PX prices began to fall, dropping to the lowest annual level of 8142 yuan/ton in June. In the second half of the year, under the boost of a significant increase in production by downstream enterprises and the bottoming out and rebound of international oil prices, PX prices resumed their upward trend. Overall, domestic PX costs fell year-on-year with the fall in international oil prices, but still maintained a relatively high level, supporting the annual PX price. The average annual PX price is expected to fall slightly by 235 yuan/ton (down 3%) to 8373 yuan/ton. In terms of gross profit, the significant increase in PX downstream demand and the continued tightening of domestic PX raw material resources due to seasonal oil adjustments in the United States significantly pushed up PX profit levels, with the average annual gross profit increasing significantly to 870 yuan/ton, 3.3 times that of 2022 and the highest since 2019. **PX futures and options are listed on domestic exchanges, opening a new chapter for the pricing mechanism.** PX futures and options were officially listed on the Zhengzhou Commodity Exchange on September 15. Overall, the listing of domestic PX futures is beneficial for upstream and downstream entities to use the hedging function of the futures market to enhance their ability to resist market and price risks. Currently, due to the concentrated distribution of domestic PX capacity, China is still a net importer of PX, and domestic trade is mainly contract-based, the growth of actual delivery of PX futures and options and their impact on the market will take time. In the long run, the market believes that PX futures prices have the potential to replace the current pricing system based on CFR China, providing a new direction for the pricing of Asian PX. 2) China's outlook for the development of the xylene industry **There is no confirmed new capacity for domestic PX, and the rapid rise in capacity will temporarily stop.** In 2024, the 3 million tons/year PX project on Yulong Island has not yet been approved. If it cannot be put into production, domestic PX capacity will still maintain 42.1 million tons/year. **Downstream capacity has grown significantly, significantly pushing up PX demand.** In 2024, the new PTA units to be put into operation in China include: Yizheng Chemical Fiber (3 million tons/year), which has postponed its production plan from the fourth quarter of 2023 to the first quarter of 2024, and Ningbo Taiwan Chemical (1.5 million tons/year); scheduled to be put into operation at the end of 2024, Shenghong Group (2.4 million tons/year). It is estimated that the total new PTA capacity for the whole year will be 6.9 million tons/year, an increase of 7.7%. The continuous growth of PTA capacity will significantly drive PX demand, and it is expected that the annual PX consumption will continue to increase by 4.5 million tons/year to 42.5 million tons/year, an increase

Table 24 PX supply and demand balance table 2018–2024

Item	2018	2019	2020	2021	2022	2023	2024
Capacity (10,000 tons/year)	1412	2047	2627	2877	3640	4210	4210
Production (10,000 tons)	1040	1561	2127	2214	2731	3242	3452
Consumption (10,000 tons)	2610	2900	3390	3534	3547	3936	4486
Operating rate (%)	74	76	81	77	75	77	82

of 11.8%. **PX imports have rebounded slightly, and the degree of dependence on foreign countries still maintains a low level.** The significant increase in downstream demand will stimulate companies to increase their operating rates. It is expected that the PX device operating rate will increase in 2024, with an annual average of up to 82%; PX production will increase by 2.1 million tons/year to 34.5 million tons/year. Due to the significant increase in domestic PX demand, but no new capacity has been put into production, coupled with the continued tightening of PX raw material resources due to seasonal oil adjustments in the United States, the gap between PX supply and demand will increase year-on-year, and imports will increase accordingly. It is estimated that PX imports will increase by 1.7 million tons/year to 10.5 million tons/year. China's dependence on foreign PX only increased slightly by 2 percentage points year-on-year, to 25%. **International oil prices remain at a medium-to-high level, and downstream demand has grown significantly, keeping PX prices at a medium-to-high level.** In 2024, as international financial pressure gradually eases and geopolitical factors rise again, it is expected that medium-to-high international oil prices will continue to affect PX cost prices. There is no new capacity for PX, and downstream PTA is still in a rapid expansion cycle, which will continue to increase PX demand and provide some support for PX prices. It is expected that PX prices will fluctuate widely in 2024 and maintain a medium-to-high level. In terms of gross profit, the explosive growth of downstream PTA, the continued existence of seasonal oil adjustment demand in the United States, and no new capacity for PX, the supply–demand gap expands to support PX gross profit. In addition, it is expected that the price difference of refined oil cracking will be relatively weak, and the oversupply situation of PTA will intensify, and the profit of the industrial chain will further concentrate on PX, which also supports the fluctuation of PX gross profit at a medium-to-high level (Table 24).

From 2025 to 2027, the new capacity of PX will continue to grow. It is expected that the new capacity in 2025 may exceed the level in 2023, and then the growth rate will slow down (Table 25).

Table 25 New PX capacity from 2025 to 2027

Project	New capacity (10,000 tons/year)	Commissioning time
Yulong Petrochemical	300	2025
Huajin Petrochemical	200	2025
Jiujiang Petrochemical	320	2025
Gulei Petrochemical	320	2026–2027
Total	1140	2025–2027

Author: Li Yunfei Reviewer: Wang Jing

Table 26 Monthly capacity of major domestic synthetic resin products from 2018 to 2023

Year	Capacity (10,000 tons/year)					
	Polyethylene	Polypropylene	Polyvinyl chloride	Polystyrene	ABS resin	Total capacity
2018	1813	2244	2470	364	400	7291
2019	1906	2446	2550	369	422	7692
2020	2256	2816	2713	385	427	8596
2021	2661	3131	2771	455	477	9494
2022	2871	3394	2796	515	552	10,127
2023	3131	3942	2932	615	752	11,371

Current Status and Trends of China's Synthetic Resin Industry

Current Status of China's Synthetic Resin Industry

The capacity of synthetic resin is growing rapidly, and the capacity of polyolefins and acrylonitrile–butadiene–styrene copolymers (ABS) is soaring. In 2023, China's synthetic resin capacity maintained rapid growth, with a total capacity of 114 million tons/year, an increase of 12.44 million tons/year, an increase of 12.3%, which is the largest scale of capacity expansion in the past five years. Among them, polyolefins and ABS have the largest scale of new capacity, accounting for 64% of all new capacity. In terms of products, the capacity of polyethylene is 31.31 million tons/year, with a new capacity of 2.6 million tons, the capacity of polypropylene is 39.42 million tons/year, with a new capacity of 5.48 million tons/year, a total of 8.08 million tons/year; the capacity of polyvinyl chloride (PVC) is 29.32 million tons/year, with a new capacity of 1.36 million tons/year. The capacity of ABS resin is 7.52 million tons/year, with a new capacity of 2 million tons/year. The capacity of polystyrene (PS) is 6.15 million tons/year, with a new capacity of 1.46 million tons/year, a capacity elimination of 460,000 tons/year, and a net increase in capacity of 1 million tons/year (Table 26).

The development of the five major general-purpose synthetic resin products has its own characteristics. The growth rates of polyethylene and polypropylene capacity are clearly differentiated. With the completion of integrated refining and chemical projects and the relative difficulty of obtaining ethane raw materials, the growth rate of ethylene capacity has significantly slowed down. At the same time, private enterprises have made great efforts to enter the PDH industry to connect the upstream and downstream industrial chains, and the downstream generally matches polypropylene devices, leading to the continued rapid expansion of polypropylene capacity. All PVC devices put into operation in 2023 are ethylene-based, and the proportion of ethylene-based PVC capacity has increased to 25%. Moreover, all new capacities are in coastal areas rich in ethylene raw materials. Among them, Fujian Wanhua and Huayi Qinzhou are both full-industry chain constructions with low production costs and energy consumption. ABS resin leading companies continue to consolidate their traditional advantageous positions, and companies such as LG Chem, Ningbo Taihua, and Jilin Petrochemical continue to expand. Home appliance and automobile modified plastic companies have also participated.

The downstream demand for synthetic resin is not good in 2023. It is estimated that the domestic plastic product output in 2023 will be 73 million tons, a year-on-year decrease of 6.0%. The main reasons are: first, the slow recovery of industry after the epidemic. Since 2023, China's economy has shown that the service industry recovers faster than the industry. The manufacturing PMI is significantly lower than the service PMI, and the recovery speed of industrial production is slower, which drags down the demand for synthetic resin. Second, the lack of consumer confidence. Due to the generally pessimistic attitude of plastic production enterprises towards the future market, they mainly purchase for rigid demand and actively reduce raw material inventory. Third, the lack of overseas demand, exports have significantly fallen, and the export amount of plastic products in the first 11 months has fallen by 5.0% year-on-year, and the growth rate has dropped sharply by 15 percentage points.

The imbalance of supply and demand has intensified, and the profits of synthetic resin production are generally not good in 2023. The situation of supply exceeding demand in China's synthetic resin industry has further intensified, production profits have fallen sharply, and the long-term profitable situation has been broken. It is estimated that the loss of oil-based polyethylene is 1000 yuan/ton, a year-on-year narrowing of 430 yuan/ton, and the profit of imported ethane-based polyethylene is 2200 yuan/ton, a year-on-year increase of 800 yuan/ton. The profit of oil-based polypropylene is 200 yuan/ton, a year-on-year narrowing of 600 yuan/ton, and the loss of imported propane-based polypropylene is 150 yuan/ton, a year-on-year narrowing of 350 yuan/ton. ABS has a loss of 200 yuan/ton, a year-on-year sharp drop of 1300 yuan/ton. The production profit of PVC (ethylene method) is 200 yuan/ton, a year-on-year decrease of 800 yuan/ton. PS has a loss of 55 yuan/ton, a year-on-year decrease of 100 yuan/ton.

The phenomenon of structural surplus of synthetic resin is obvious, and the product structure is not entirely reasonable. After many years of capacity deployment in the synthetic resin industry, the overcapacity of general products has become increasingly prominent. By 2023, the start-up rates of polyethylene, polypropylene, polyvinyl chloride, polystyrene, and ABS resin devices will drop to 81.0%, 80.0%, 74.2%, 67.0%, and 82.0% respectively (Table 27), basically dropping to the lowest

Table 27 Start-up rates of major domestic synthetic resins from 2018 to 2023

Year	Start-up rate (%)				
	Polyethylene	Polypropylene	Polyvinyl chloride	Polystyrene	ABS resin
2018	87.4	93.2	81.0	73.0	94.6
2019	94.0	91.4	78.5	78.8	97.0
2020	90.1	90.7	76.5	78.4	99.5
2021	86.1	91.3	78.6	72.3	95.6
2022	84.9	88.8	74.0	68.1	86.5
2023	81.0	80.0	74.2	67.0	82.0

level in history, further exacerbating the overcapacity of general synthetic resins. In addition, some general engineering plastics also show signs of overcapacity, such as polyoxymethylene (POM), polycarbonate (PC), polybutylene terephthalate (PBT), etc., all of which have seen a significant drop in prices. The dependence on imports for high-end synthetic resins is still high, with self-sufficiency rates for metallocene polyethylene and metallocene polypropylene only at 11% and 6.5% respectively. High-end synthetic resins such as polyolefin elastomer (POE) resin, polyvinyl butyral (PVB) film for laminated glass, electronic grade epoxy resin, and polyvinylidene fluoride (PVDF) separator for power batteries are completely dependent on imports, becoming a key factor restricting the high-quality development of the synthetic resin industry.

The rapid development of the new energy industry has led to a high-speed increase in the demand for ethylene–vinyl acetate copolymer (EVA) and POE. With the continuous advancement of the "dual carbon" goal, China's photovoltaic industry is developing rapidly, and it is expected that the installed capacity of photovoltaics will exceed 520 million kilowatts by the end of 2023. In addition, China is the world's largest exporter of photovoltaic modules, accounting for more than 80% of the global market. The consumption of photovoltaic film has also increased significantly, and the demand for the main raw materials EVA and POE has been growing rapidly for several years. By 2023, the consumption will reach 3.3 million tons and 920,000 tons respectively, an increase of 20% and 31% year-on-year, making them one of the fastest-growing synthetic resin products.

Outlook for the Development of China's Synthetic Resin Industry

In 2024, China's synthetic resin industry will further increase its new capacity. It is estimated that by 2024, China's synthetic resin production capacity will increase by up to 15.5 million tons per year, with a total capacity of 129 million tons, an increase of 13.6%. Among them, polypropylene has the fastest capacity expansion, with an increase of 8.4 million tons per year and a total capacity of 48.54 million tons per year; polyethylene has an increase of 2.5 million tons per year and a total capacity of 34.01 million tons per year; ABS has an increase of 1.7 million tons per year and a total capacity of 9.66 million tons per year; PS has an increase of 1.48 million tons per year and a total capacity of 7.41 million tons per year; PVC has an increase of 1.5 million tons per year and a total capacity of 30.65 million tons per year.

The growth rate of synthetic resin demand is slowing down overall, but there are still structural opportunities. Although the growth rate of domestic synthetic resin demand has shifted from high-speed growth to medium-speed growth, there are still development opportunities in some sub-sectors. First, the demand for packaging film from e-commerce continues to grow steadily; second, with the acceleration of agricultural modernization, the demand for high-end agricultural film is considerable; third, China will maintain its status as a major exporter for a long period of time, and the export potential of plastic products is still large; fourth, the demand for high-end film materials from new materials and high-end consumer goods is growing rapidly; fifth, there is a large growth space for the demand for building substitute materials and engineering substitute materials.

It is expected that the profit of synthetic resin will rebound slightly in 2024. Although the new production capacity of synthetic resin will break through the historical high again in 2024, the supply pressure will increase without decreasing, and the terminal demand is difficult to grow significantly. However, from the perspective of actual demand, since the middle of 2022, downstream enterprises of synthetic resin have voluntarily reduced raw material inventory, and the inventory growth rate of rubber and plastic enterprises announced by the National Bureau of Statistics has almost dropped to the lowest level in history, and is expected to enter an upward stage in 2024. Coupled with the fact that the inventory of manufacturing industries in Europe and the United States has dropped to a historical low, which is conducive to the growth of plastic product exports, it will bring a huge increase in demand. Therefore, it is expected that the profit of synthetic resin will improve slightly year-on-year in 2024.

The synthetic resin industry insists on promoting high-quality development and enhancing industrial competitiveness and profitability. First, we must have firm confidence in development and grasp the direction of development. The whole industry should learn and use policies well, and implement the "Fourteenth Five-Year" Development Guidance for the Plastic Processing Industry, the "Fourteenth Five-Year" Scientific and Technological Innovation Guidance for the Plastic Processing Industry, and the Outline for the Green Development of China's Plastic Processing Industry. Second, we must insist on innovative development. We must adhere to the "five transformations" direction of scientific and technological innovation, namely functionalization, lightweighting, precision, ecologicalization, and intelligentization. Focus on innovation priorities. Strive to meet the needs of the people's better life, the country's major strategic needs, and the real needs of the industry, and tackle key problems in "bottleneck" technology, core technology, key common technology, import substitution technology and products. Strengthen the collaborative innovation of the industrial chain and promote the advanced nature of the industrial base and the modernization of the industrial chain. Third, we must insist on digital empowerment. Take digital transformation as the strategic fulcrum of

Table 28 China's new synthetic rubber production capacity in 2023

Variety	Project/ Company	Nature of the company	Process route	Location	Increased capacity (10,000 tons/year)
Styrene-butadiene rubber	Zhejiang Petrochemical	Private enterprise	Solution polymerization	Ningbo, Zhejiang	6
Butadiene rubber	Zhejiang Petrochemical	Private enterprise	Solution polymerization	Zhejiang Ningbo	10
	Changhong High-tech	Private enterprise	Solution polymerization	Zhejiang Ningbo	2
SEBS	Hainan Refining	State-owned enterprise	Solution polymerization	Hainan Danzhou	5
SBS	Hainan Refining	State-owned enterprise	Solution polymerization	Hainan Danzhou	12

Data source PetroChina Economic Research Institute

industrial transformation and upgrading. Play the driving role of the digital economy and make reasonable use of data. Drive product innovation with data analysis and strengthen the application of information technologies such as artificial intelligence in product production, after-sales service, product design, and market marketing.

(Author: Gao Zhenyu Reviewer: Wang Jing)

Development Status and Trends of China's Synthetic Rubber Industry

Development Status of China's Synthetic Rubber Industry

The production capacity of synthetic rubber continues to grow rapidly, with the total capacity exceeding 7 million tons per year. In 2023, the domestic synthetic rubber production capacity increased by 350,000 tons per year, and the total capacity reached 7.202 million tons per year. The increased capacity mainly comes from styrene-butadiene rubber and butadiene rubber. From the main body, the new capacity of state-owned enterprises accounts for 49%, and private enterprises account for 51% (Table 28).

The synthetic rubber business is diversified. In 2023, the synthetic rubber production capacity of Sinopec and PetroChina will reach 3.43 million tons/year, accounting for 47.6% of the total national capacity, and the production capacity of private synthetic rubber will account for 52.4% of the total national capacity (Table 29).

The rapid recovery of the automotive industry drives the steady growth of synthetic rubber demand. In 2023, the country introduced a series of automotive industry policies to stimulate automobile consumption. The development momentum of the new energy vehicle industry remained strong, and China became the world's

Table 29 Composition and proportion of synthetic rubber enterprises in China in 2023

Project	2021		2022		2023	
	Capacity (10,000 tons/year)	Percentage of total capacity (%)	Capacity (10,000 tons/year)	Percentage of total capacity (%)	Capacity (10,000 tons/year)	Percentage of total capacity (%)
Sinopec	179	28	190	27.7	207	28.7
PetroChina	126.5	19.8	136	19.9	136	18.9
Other enterprises	334.2	52.2	359.2	52.4	377.2	52.4
National total	639.7	100	685.2	100	720.2	100

Note Sinopec's capacity includes foreign equity capacity

largest automobile exporter, all of which supported the demand for synthetic rubber. Data released by the China Association of Automobile Manufacturers showed that domestic automobile production and sales both broke through 30 million units, an increase of 11.6% and 12% respectively year-on-year; exports were 4.91 million units, an increase of 57.9%. Among them, the production and sales of new energy vehicles were 9.587 million units and 9.495 million units respectively, an increase of 35.8% and 37.9% respectively, and exports were 1.203 million units, an increase of 77.6%. The apparent consumption of synthetic rubber for the whole year was about 4.9 million tons, an increase of 4.0% year-on-year.

The listing of butadiene rubber futures and options. On July 28, butadiene rubber futures and options contracts were officially listed on the Shanghai Futures Exchange. Butadiene rubber is one of the two largest types of synthetic rubber (the other is styrene-butadiene rubber). After the listing of butadiene rubber futures, as a new pricing model, futures tools gradually penetrated into daily transactions, playing a role in assisting pricing. At the same time, after the commodity is given financial attributes, the impact of macroeconomic changes on butadiene rubber will be more intuitive.

Profit differentiation in the synthetic rubber industry, improved profitability of styrene-butadiene rubber. In 2023, due to the significant fluctuations in international oil prices and naphtha prices, the violent fluctuations in the price of raw material butadiene, and the differentiated performance of different types of cars, the price trend of synthetic rubber differentiated. Styrene-butadiene rubber, as the main application field of small cars, led by new energy vehicles, had a smaller price drop and improved profits; the rapid deployment of butadiene rubber capacity, supply growth significantly faster than consumption growth, a larger price drop, profit narrowing; due to the significant increase in domestic production and imports of nitrile rubber, the increase in supply, profits significantly narrowed (Fig. 21).

The price gap between natural rubber and synthetic rubber has widened. In 2023, due to abnormal weather disturbances in domestic and foreign natural rubber producing areas (abnormal phenological interference in Thailand's producing

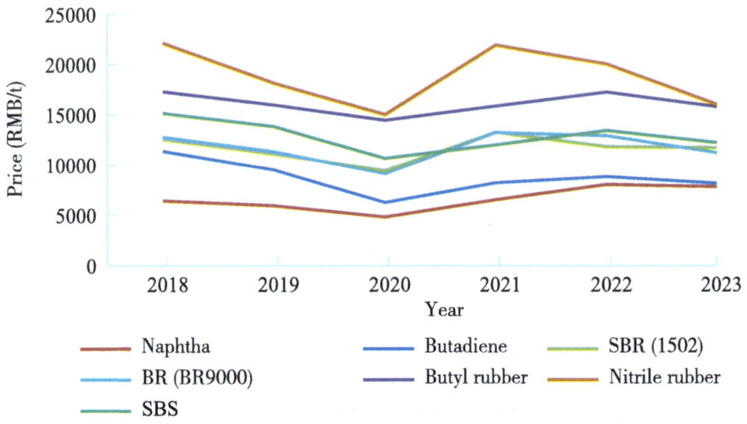

Fig. 21 Price trend of synthetic rubber from 2018 to 2023

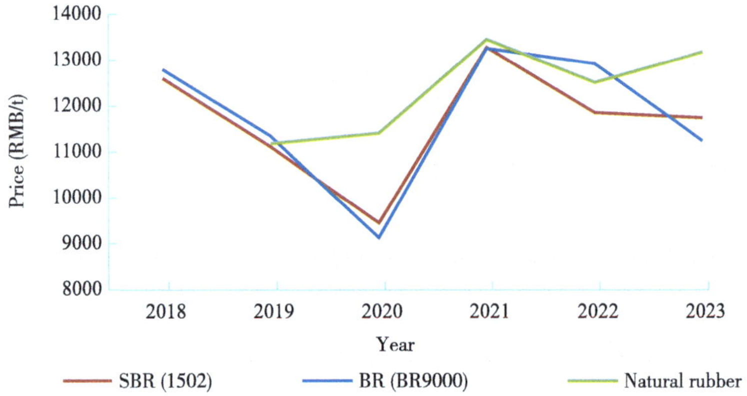

Fig. 22 Comparison of the price trends of natural rubber and synthetic rubber from 2018 to 2023

areas affecting latex output, excessive rainfall in Vietnam's producing areas affecting normal rubber collection work), frequent adjustments to Thailand's rubber policies (the establishment of a special rubber management working group when Tamanan, the Minister of Agriculture and Cooperatives of the new government, took office, the main work is to investigate, raise prices and crack down on illegal rubber), the price of natural rubber rose, diverging from the trend of synthetic rubber (Fig. 22).

Outlook for the Development of China's Synthetic Rubber Industry

Domestic synthetic rubber production capacity will continue to grow rapidly. According to the project construction situation, it is estimated that the new synthetic

Table 30 New synthetic rubber capacity in China in 2024

Type	Project/Company	Nature of the company	Process route	Location	New capacity (10,000 tons/year)
Styrene-butadiene rubber	Yulong Island Petrochemical	Private enterprise	Solution polymerization	Shandong	6
Butadiene rubber	Yulong Island Petrochemical	Private enterprise	Solution polymerization	Shandong	15
	Transfar Synthetic Materials	Private enterprise	Solution polymerization	Zhejiang	12
Styrene–Butadiene–Styrene Block Copolymer (SBS)	Shanghai Petrochemical	State-owned enterprise	Solution polymerization	Shanghai	14
	Baihao Sheng	Private enterprise	Solution polymerization	Anhui	5
	Fuhua Luhua	Private enterprise	Solution polymerization	Fujian	4
	Jinhai Chenguang	Private enterprise	Solution polymerization	Zhejiang	3.5
Hydrogenated Styrene–Butadiene–Styrene Block Copolymer (SEBS)	Shanghai Petrochemical	State-owned enterprise	Solution polymerization	Shanghai	5
	Xiangjiang Petrochemical	Private enterprise	Solution polymerization	Fujian	2
	Dongying Yike	Private enterprise	Solution polymerization	Shandong	3
Styrene–Isoprene-Styrene Block Copolymer (SIS)	Shanghai Petrochemical	State-owned enterprise	Solution polymerization	Shanghai	6
	Hainan Refining	State-owned enterprise	Solution polymerization	Hainan	3

Data source PetroChina Economic Research Institute

rubber production capacity in China in 2024 will be 785,000 tons/year, and the new capacity will mainly come from private enterprises (Table 30).

The synthetic rubber industry continues to expand, intensifying market competition. From 2024 to 2026, several synthetic rubber projects will be put into production domestically, with the increased capacity mainly coming from styrene-butadiene rubber and polybutadiene rubber. As the development of the automotive industry shows a slowing trend, the growth rate of synthetic rubber demand will also slow down. However, capacity continues to expand, and the synthetic rubber industry will show the following trends: First, the import gap will gradually decrease, and the market will shift from the current supply shortage to oversupply; Second, as the market participants become more diversified, market competition will intensify, especially in the polybutadiene rubber market (Table 31).

Table 31 Estimated additional synthetic rubber capacity in China from 2025 to 2026

Variety	Project/Company	Nature of the company	Process route	Location	Increased capacity (10,000 tons/year)	Year of production
Styrene Butadiene Rubber	ShenHua Chemical	Private enterprise	Emulsion polymerization	Jiangsu	22	2025
Styrene Butadiene Rubber	Guangxi Petrochemical	PetroChina	Solution polymerization	Guangxi	12	2025
Styrene Butadiene Rubber	ShenHua Chemical	Private enterprise	Solution polymerization	Jiangsu	5	2026
Polybutadiene Rubber	Jilin Petrochemical	PetroChina	Solution polymerization	Jilin	12	2025
Polybutadiene Rubber	Sinopec Zhongke Refining and Chemical	Sinopec	Solution polymerization	Guangdong	10	2025
Polybutadiene Rubber	Tarim Petrochemical	PetroChina	Solution polymerization	Xinjiang	10	2026
Ethylene Propylene Diene Monomer	Jilin Petrochemical	PetroChina	Solution polymerization	Jilin	4	2025
Ethylene Propylene Diene Monomer	Yanan Energy Chemical	Yanchang Petroleum	Solution polymerization	Shaanxi	2.5	2025
SBS	Guangxi Petrochemical	PetroChina	Solution polymerization	Guangxi	8	2025
SEBS	Huizhou Li Changrong	Private Enterprise	Solution polymerization	Guangdong	2	2025
SIS	Zhejiang Henghe	Private Enterprise	Solution polymerization	Zhejiang	8	2025

Data source PetroChina Economic Research Institute

(Author: Fei Huawei, Reviewer: Wang Jing)

Development Status and Trends of China's Synthetic Fiber Industry

Development Status of China's Synthetic Fiber Industry

The production capacity of synthetic fibers has increased year-on-year, mainly concentrated in the production of polyester filament. In 2023, both polyester staple fiber and polyester filament of synthetic fiber products are in the accelerated production stage, with an estimated increase in production capacity of 5.405 million tons/year; excluding the 180,000 tons/year filament production capacity of Zhejiang Jinxin Chemical Fiber Company eliminated within the year, and a total of 400,000 tons/year of staple fiber production capacity of companies such as Jiangsu Desai Chemical Fiber, the production capacity of synthetic fibers increased by 4.825 million tons/year year-on-year. Among them, the production capacity of polyester filament has significantly increased, with an increase of 1.1 times the previous year, mainly from leading companies (Table 32). The significant increase in the production capacity of leading companies, the industry has entered the "oligopoly competition" pattern, which will accelerate the exit of backward production capacity and increase industry concentration.

The trend of operation rate is diverging, and the output has increased. In 2023, domestic open policies boosted the downstream demand for synthetic fibers. The operation rate of staple fiber fluctuated widely between 70 and 84%, and the operation rate of filament rebounded from the low point at the end of the previous year. Both had an average annual operation rate at a medium-to-high level, with staple fiber higher than filament. Against the backdrop of a significant increase in new capacity and fluctuating operation rates, the annual output of synthetic fibers increased year-on-year.

The domestic demand for synthetic fibers weakened, and foreign trade diverged. In 2023, the raw silk procurement of downstream textile enterprises was mainly based on low-price rigid demand, and the demand for staple fiber and filament both fell significantly year-on-year. The foreign trade situation of the two was severely divergent, with "staple fiber decreasing and filament increasing". Affected by the poor global economy, textiles and clothing, which are non-essential goods, performed poorly. Yarn and weaving foreign trade orders were significantly insufficient, and there was a phenomenon of some orders flowing back to Southeast Asian countries. In terms of filament, China's comprehensive opening of epidemic prevention and control to a certain extent promoted the export of polyester filament market, coupled with the transfer of the textile industry in recent years, Southeast Asian countries are unable to match polyester, and the import volume of raw material filament is increasing day by day, which overall boosts the foreign trade volume of filament.

The price of synthetic fibers remained relatively high, still in the negative gross profit margin range. The prices of polyester staple fiber and filament mostly follow the changes in the cost end, that is, the impact of international crude oil prices on them is quite significant. In 2023, the fluctuation of international oil prices at a medium-to-high level formed a certain support for the price of synthetic fibers. However, the weak downstream demand was bearish on the price of synthetic fibers.

Table 32 Increase in synthetic fiber production capacity in 2023

Project	Increased capacity (10,000 tons/year)	Production time	Product
Xinfengming	22.5	January 2023	Polyester staple fiber
Suqian Yida	30	May 2023	
Yizheng Chemical Fiber	23	August 2023, October 2023	
Sichuan Jixing	20	December 2023	
Subtotal	95.5	–	
Tongkun Moyang Project	30	February 2023	Polyester filament
	30	May 2023	Polyester filament
	30	July 2023	Polyester filament
	30	September 2023	Polyester filament FDY
Xinfengming Xintuo	40	February 2023	Polyester filament
	40	August 2023	Polyester filament
Tongkun Xinjiang Project	30	March 2023	Polyester filament
Tongkun Hengchao	60	March 2023	Polyester filament
Hengyi Hengming	20	March 2023	Polyester filament
Jiatong Energy	30	March 2023	Polyester filament (bright)
	30	June 2023	Polyester filament (bright)
	30	October 2023	Polyester filament
Hengli Hengke	25	April 2023	Polyester filament
Wujiang Shenghong	20	April 2023	Bicomponent filament
Total filament nationwide	445	–	Polyester filament
Total	540.5	–	Polyester staple fiber, Polyester filament

It is estimated that the price of polyester staple fiber fell by 414 yuan/ton year-on-year, to 7351 yuan/ton; the price of polyester filament fell by 285 yuan/ton year-on-year, to 7538 yuan/ton. Under the "double empty" scenario of high upstream raw material prices and weak downstream operation, the gross profit of synthetic fibers continued to be suppressed, even at a loss, and the gross profit of the two main products both weakened significantly year-on-year. However, the divergence in foreign trade caused a gap in the decline of their gross profits. It is estimated that the gross profit of polyester staple fiber fell by 108 yuan/ton year-on-year, to − 129 yuan/ton; the gross profit of polyester filament slightly increased by 18 yuan/ton year-on-year, to − 166 yuan/ton.

Table 33 Additional capacity of synthetic fibers in 2024

Project	Additional capacity (10,000 tons/year)	Product
Jiatong Energy	30	Polyester filament
Rongsheng Shengyuan	50	
Xin Fengming Xintuo	40	
Xinjiang Zhongtai	25	
Xinjiang Zhongtai	25	Polyester staple fiber
Sichuan Jixing	30	
Suqian Yida	30	
Tongkun Hengyang	30	
National Total	260	

Outlook for the Development of China's Synthetic Fiber Industry

The additional capacity of synthetic fibers increased year-on-year, but the increase was significantly reduced. It is expected that in 2024, the additional capacity of synthetic fibers will total 2.6 million tons/year, 48% of the previous year, most of which comes from filament enterprises (Table 33).

The slowdown in global economic growth will continue to drag down the demand for synthetic fibers, and both domestic and foreign demand will be hindered. In 2024, downstream textile companies will still maintain a "low demand level". In terms of foreign trade, the demand in the European and American clothing market will recover, but the further slowdown in the global economy will continue to limit the increase in demand. After the BIS certification of Chinese companies in India, the largest exporter of filament, expires, it will take time to re-certify, and the export of filament in 2024 will also be significantly hindered.

The "high cost + low demand" model continues to suppress gross profit, and prices will maintain a relative median level. In 2024, it is expected that international oil prices will still maintain a medium-to-high level, and the price of synthetic fibers will "rise with the tide", maintaining a relative median. However, the high cost and the slowdown in demand growth will continue to suppress the gross profit of synthetic fibers, which will fall compared to the same period.

From 2025 to 2026, the additional capacity of synthetic fibers will continue to grow, but it will be difficult to return to the high growth level of 2023. The main source of growth is filament companies (Table 34).

In the coming years, with the continuous increase in domestic petrochemical production capacity, the production capacity of a large number of petrochemical products will tend to be saturated and even surplus. The complexity, severity, and uncertainty of the external environment continue to rise. The competition among market entities mainly composed of central enterprises, private enterprises, and foreign capital in China will continue to intensify. The petrochemical industry will face severe challenges, and some enterprises will encounter difficulties in operation.

Table 34 Additional capacity of synthetic fibers from 2025 to 2026

Project	Additional capacity (10,000 tons/year)	Year of production	Product
Jiuhua Blue	65	2025	Polyester filament
Tongkun Shuyang	120	2025	
Yuxin Phase II (Tongkun)	30	2025	
Anhui Youshun (Tongkun)	30	2025	
Hengli Hengke	60	2025	
Tongkun Hengchao Phase III	60	2026	
Yuxin Phase III (Tongkun)	40	2026	
Zhangzhou Gulei (Tongkun)	200	2026	
Tongkun Hengyou	30	2026	
Guowang Gaoke	200	2026	
Sichuan Nengtou	28	2025	Polyester staple fiber
Tongkun Hengyang	30	2025	
Total	893		

However, in the long run, China's economy and petrochemical industry have great resilience and potential, and the long-term positive fundamentals have not changed. In this regard, the domestic petrochemical industry should firmly believe in development, adhere to demand-oriented, use digitalization, high-end, greening, and internationalization as means, strive to break through technical bottlenecks, vigorously develop mid-to-high-end products, achieve high-quality green and effective development, and continuously move towards the goal of high technology level, low resource consumption, low "three wastes" emissions, and good economic benefits.

(Writer: Li Yunfei Reviewer: Wang Jing)

Overseas Investment Environment

Overview of the Overseas Investment Environment

Analysis and Outlook of the World Geopolitical and Security Situation

The Ongoing Ukraine Crisis and a New Round of Israeli-Palestinian Conflict Have Intensified Geopolitical Tensions

The Ukraine crisis has made the situation in the Russia-Central Asia region more complex, with escalating gamesmanship among various parties. Issues such as power struggles, military disagreements, and social dissatisfaction within Russia have become prominent. In June 2023, the Wagner mercenary group staged a rebellion, which was quelled in the short term, but the risk of domestic turmoil remains in the future. Western countries such as the United States and Europe are escalating sanctions against Russia, limiting Russia's income from oil and gas exports, and restraining Russia's power; on the other hand, they are continuously enhancing military aid to Ukraine, causing the Ukraine crisis to continue to ferment. In addition, the deterioration of Russia-Europe relations, the strengthening of cooperation between the United States and Europe in the fields of energy, food, and other commodities, has strengthened the binding of interests between the United States and Europe. The Ukraine crisis has brought the risk of great power confrontation to its highest level since the Cold War, and the issue of nuclear weapons use has become a focus of international attention.

The new round of Israeli-Palestinian conflict has disrupted the trend of easing in the Middle East driven by Saudi-Iranian reconciliation. In 2023, China promoted the resumption of diplomatic relations between Saudi Arabia and Iran, and the reconciliation between Saudi Arabia and Iran led Middle Eastern countries to start

X. Qian et al., *Global Oil and Gas Industry Development Report: 2023*,
https://doi.org/10.1007/978-981-95-2375-7_9

improving relations with Iran and Syria, and the situation in the Middle East showed signs of easing. However, the outbreak of a new round of Israeli-Palestinian conflict has brought new variables to the gradually stabilizing situation in the Middle East. As the Israeli-Palestinian conflict escalates, the flames of war have spread to countries in the region such as Lebanon and Syria, and even obstructed Red Sea shipping, making the security situation in the Middle East more severe.

The risk of "camping" in the world political structure has increased, but the trend of "re-globalization" is irreversible. After the Ukraine crisis, Western countries such as the United States and Europe have imposed multiple rounds of sanctions on Russia. Developing countries such as China and India have remained neutral towards Russia, but the United States has been constantly pressuring them on the grounds of their involvement in Russian business. After the Israeli-Palestinian conflict, the Islamic world supports Palestine, while Western countries such as the United States, France, Italy, Germany, and the United Kingdom have expressed firm support for Israel, and some southern countries, including China, have remained neutral. In geopolitical hotspots such as the Ukraine crisis and the Israeli-Palestinian conflict, the United States leads and forcibly promotes the division of camps based on values, limiting the development of countries with different values and political camps such as China and Russia in key areas such as high-tech, and the risk of global division into different camps based on ideology is increasing. However, some of the "camping" measures of the United States and Western countries have prompted the transfer of supply chains and industrial chains to southern countries such as Vietnam, Indonesia, and Mexico, forming a broader globalization network. "Re-globalization" will follow objective economic laws, not be transferred by the subjective will of the United States and Western countries, and form a broader, more representative, and fairer globalization system, which will be the mainstream trend of future development.

The U.S. Policy Towards China Shows a "Dualistic" Trend

Sino-US strategic competition has entered a new stage of "fighting while talking", high-level communication has been resumed, and there are signs of a rebound and stabilization in bilateral relations. Since May 2023, high-level officials from China and the United States have visited each other, held talks, established working groups in the economic field, etc., to strengthen communication and exchanges. The relationship between the two sides has come out of the cold and tense situation caused by the "balloon incident" and "Tsai Ing-wen's visit to the United States", showing signs of easing. However, the United States' positioning of China as the primary competitor has not changed, and it frequently carries out political activities in countries around China to build a "small circle" to contain China. It has launched the "Indo-Pacific Economic Framework (IPEF)", established a comprehensive strategic partnership with ASEAN, continued to promote the "Indo-Pacific Strategy", held special meetings with leaders of India, South Africa, and Brazil, and upgraded the "C5 + 1" dialogue mechanism established with the five Central Asian countries since 2015 to the head of state level, and proposed to support and expand the US-Central Asia partnership, to "shape" China's peripheral strategic environment through a dense alliance network.

The United States is building a "small yard high wall" to strengthen restrictions on China in the field of technology. In November 2020, the U.S. policy report "How to Respond to China's Challenge: A New Strategy for U.S. Technology Competition" initiated a new policy towards China to impose restrictions in the field of technology, and the "small yard high wall" strategy gradually became the technology strategy towards China. **Advancing legislation and rule-making against China.** The U.S. Senate and House of Representatives passed a bill to cancel China's developing country status, launched the 2024 Fiscal Year "National Defense Authorization Act" emphasizing confrontation with China, and on October 17, 2023, the Bureau of Industry and Security (BIS) of the U.S. Department of Commerce announced new export control rules for advanced computing chips and semiconductor manufacturing equipment, limiting China's ability to purchase and manufacture high-end chips.

Escalating sanctions against China on the grounds of Iran and Russia. On October 6, the United States listed 42 Chinese companies on the export control "Entity List" on the grounds of supporting Russia's military and defense industry. On October 18, the U.S. Treasury Department imposed sanctions on 11 individuals, 8 entities, and 1 vessel, including companies and individuals from mainland China and Hong Kong, on the grounds of participating in Iran's missile and drone manufacturing. On November 2, sanctions were again imposed on foreign entities on the grounds of involvement with Russia, including 3 Chinese companies.

Traditional and Non-traditional Security Challenges Are Intertwined

Tensions continue to rise in some regions. In addition to the Ukraine crisis and the Israeli-Palestinian conflict, India and Pakistan have repeatedly clashed in the Kashmir region, with ongoing tensions and frequent border conflicts. The meeting between Russian and North Korean leaders and their strengthening of military cooperation have provoked strong reactions from the United States and South Korea, leading to increased tensions on the peninsula. The United States' frequent intervention in the South China Sea issue, as well as its arms sales and military aid to Taiwan, have made the situation in the South China Sea and the Taiwan Strait more severe and complex, disrupting peace and stability in the Asia–Pacific region.

Political instability continues in many countries. There is a resurgence of military coups in Africa with a spreading trend. Mali, Guinea, Chad, Sudan, Niger, and Gabon have successively experienced military coups, causing concern in Rwanda and Cameroon. Frequent armed conflicts between different factions in South Sudan and the Democratic Republic of the Congo, the conflict in Sudan continues to expand and become protracted, making the security situation in Africa complex and severe.

Since the general election in Ecuador, several politicians have been attacked, there is division within the government over the election, and the future political situation faces significant uncertainty. Myanmar has been in a state of chaos since the military coup in 2021, with the military constantly extending the state of emergency and postponing elections, and frequent battles between the military and opposition armed groups, posing a high domestic security risk.

The situation in high-incidence areas of terrorist attacks is severe. Data from the "2023 Global Terrorism Index Report" released by the Australian think tank Institute for Economics and Peace (IEP) shows that the top ten countries facing the threat of terrorism include Afghanistan, Burkina Faso, Somalia, Mali, Syria, Pakistan, Iraq, Nigeria, Myanmar, Niger, concentrated in the Middle East, Africa, and South Asia. Terrorism in the Sahel region of Africa has increased dramatically, and the border region between Mali, Burkina Faso, and Niger has become a hotbed of terrorist organizations. South Asia remains the region with the worst average score on the 2022 Global Terrorism Index. After the United States withdrew from Afghanistan, Afghanistan and Pakistan have become the hardest-hit areas of global terrorism.

Global Economic Situation and Outlook

The Global Economic Recovery Process is Slow and Uneven

In 2023, the global economic recovery was better than expected, but due to factors such as tight monetary policy, unilateralism and protectionism, and geopolitical conflicts, the endogenous momentum of the economy was insufficient, and the economic recovery showed a weakening trend. The report released by the International Monetary Fund (IMF) in October indicates that the global economic growth rate is expected to slow from 3.5% in 2022 to 3.0% in 2023 and 2.9% in 2024, far below the average level of 3.8% from 2000 to 2019. The World Trade Organization (WTO) and the Organization for Economic Cooperation and Development (OECD) also have lower economic growth forecasts for 2024, at 2.5% and 2.7% respectively. Developed economies have become the main reason for the downturn in global economic growth, while the economic growth prospects of emerging markets and developing economies are relatively stable. In October, the preliminary value of the Eurozone Manufacturing Purchasing Managers Index (PMI) fell to 43, and the preliminary value of the service PMI fell to 47.8, both lower than the previous value and expected value, and consistently below the boom-bust line. Economic activity in the Asia–Pacific region remains on a growth trajectory and is expected to become a major driver of the global economy. In September, the Asian Manufacturing PMI was 50.8%, up 0.1 percentage points from August, continuing to rise slightly and remaining above the 50% boom-bust line.

Inflation Falls but Will Remain High in the Short Term

In 2023, with the tightening of monetary policy and the decline in commodity prices, the global inflation rate significantly dropped from its high point in 2022. The IMF predicts that the global inflation rate will gradually and steadily decline, from 8.7% in 2022 to 6.9% in 2023 and 5.8% in 2024; the decline in core inflation is slower than the overall inflation rate, falling from 6.4% in 2022 to 6.3% in 2023 and 5.3% in 2024. The IMF believes that in 2023, about 93% of economies will have an annual inflation level above the target range, and in 2024, 89% of economies will still have an inflation level above the target range. Global inflation will not fall back to near the target level until 2025, and global inflation will remain at a relatively high level in the short term.

Divergence in the Economic Situation of Major Resource Countries

In 2023, due to multiple factors such as high international oil and gas prices, high inflation pressure, "OPEC+" production cuts, and domestic instability in some resource countries, the economic situation of major resource countries shows a divergent trend. The economic situation in Russia and Brazil is relatively good, with Russia's economic growth rate estimated to increase from − 2.1% in 2022 to 1.6% in 2023, and Brazil's economic growth rate in 2023 is 2.5%. The economic growth rate in the Middle East is declining, mainly due to the slower-than-expected economic growth in Saudi Arabia, affected by the "OPEC+" production cuts, the regional economic growth rate in 2023 is 1.8%. The economic growth rate in Africa is also declining, affected by the deterioration of the domestic security situation and other factors, Sudan's economy has significantly declined, with an estimated growth rate of − 18.2% in 2023.

Characteristics and Trends of Policies and Laws in Major Resource Countries

Russia-Central Asia's Major Resource Countries Tighten Cooperation Policies to Protect Local Interests

Affected by the direct and spillover effects of the Ukraine crisis, oil and gas-related revenues in countries such as Russia and Kazakhstan have declined, affecting government fiscal revenue and economic development. These countries have successively tightened oil and gas cooperation policies to alleviate government fiscal pressure and protect local interests. The Russian government has changed its method of taxing oil companies, changing the tax base for mining tax and export tax from the original Ural crude oil price to the Brent crude oil international benchmark price minus a fixed discount, to fill the gap in oil export revenue. Kazakhstan, on the one hand, ensures domestic oil and gas consumption by increasing oil export transportation fees, extending the ban on refined oil exports, and requiring natural gas produced by the Tengiz project to supply the domestic market. On the other hand, it has established a national interest protection working group to strengthen the supervision of Production Sharing Agreements (PSAs). In addition, as Kazakhstan's political and economic reforms deepen, the government's supervision of corporate compliance operations has become stricter, and it has filed multiple lawsuits against the Kashagan project and the Karachaganak project, alleging non-compliance in environmental protection and contract execution.

Latin America and Asia–Pacific Regions Actively Promote Energy Transition

The Lula government in Brazil attaches great importance to the development of clean energy, striving to develop a green economy by reforming the tax system and reducing tax rates. It plans to reduce the cost of renewable energy such as solar energy to make it affordable for low-income people and ordinary consumers, and hopes that Petrobras, the Brazilian national oil company, will pay attention to renewable energy investment. Petroecuador, the national oil company of Ecuador, announced its joining of the Latin American Oil and Gas Methane Association to promote the country's decarbonization and sustainable development plan. Australia has introduced the "Safeguard Mechanism (Credit) Amendment Act (2023)", which restricts carbon emissions from industrial facilities and projects, requiring large industrial facilities that have already been put into operation to reduce carbon emissions by 4.9% per year before 2030; new LNG projects must maintain net zero emissions from the start; and actively deploy CCUS projects. Countries such as Indonesia and Malaysia strongly support the development of CCUS. Indonesia has issued MEMR 2/2023 decree, supporting the addition of CCS and CCUS schemes in upstream oil and gas projects. Malaysia has provided a series of tax relief policies for CCS projects. Myanmar has exempted tariffs on some imported photovoltaic products to encourage domestic solar energy utilization.

(Authors of this section: Duan Yixuan, Jin Huandong Reviewer: Fu Lei)

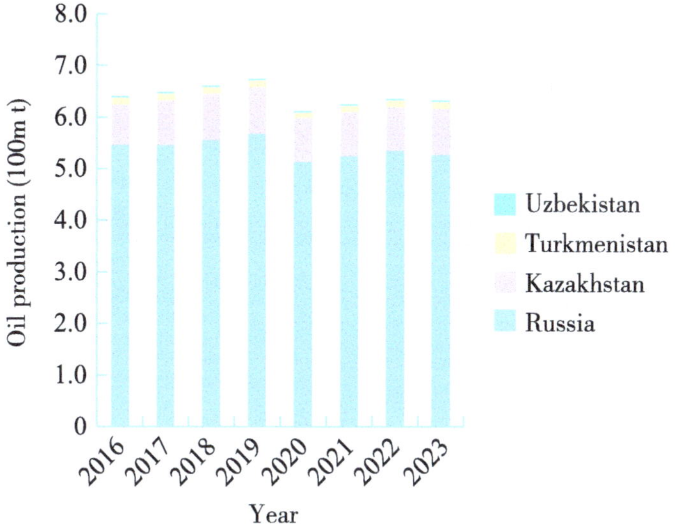

Fig. 1 Oil production in the Russia-Central Asia region from 2016 to 2023. *Data source* PetroChina Economic Research Institute

Russia-Central Asia Region

Development of the Oil and Gas Industry in the Region and Major Countries

Oil and Gas Production and Trends

In 2023, affected by multiple factors such as "OPEC+" production restrictions, the Ukraine crisis, and Western sanctions, the oil production in the Russia-Central Asia region decreased by 1.2% year-on-year to 626 million tons; the contraction of Russian gas exports led to a 3.6% decrease in the region's natural gas production to 805.9 billion cubic meters (Figs. 1 and 2).

Oil and Gas Exports Show a Downward Trend

In 2023, the oil export volume of the Russia-Central Asia region was 298 million tons, a year-on-year decrease of 3.5%; the natural gas export volume was 177.6 billion cubic meters, a year-on-year decrease of 25%. Among them, Russia's oil export volume dropped sharply to 225 million tons, a year-on-year decrease of 7%; Kazakhstan's oil export volume was 71.7 million tons, a year-on-year increase of 10%. In 2023, Russia's oil supply to China exceeded 100 million tons, making it China's largest source of crude oil imports.

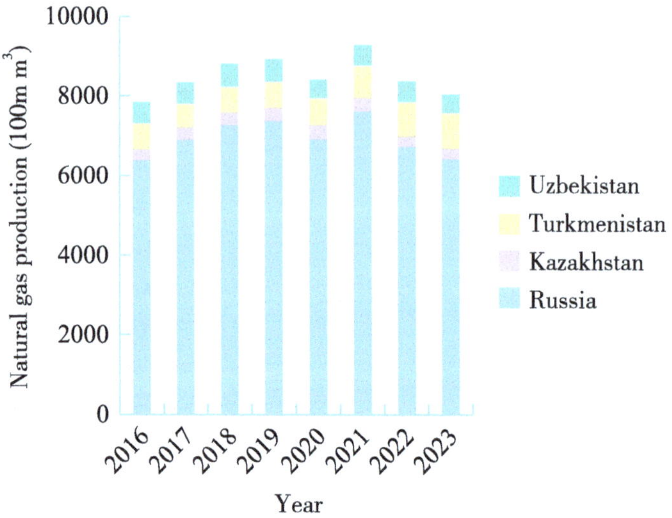

Fig. 2 Natural gas production in the Russia-Central Asia region from 2016 to 2023. *Data source* PetroChina Economic Research Institute

The natural gas exports from the Russia-Central Asia region have dropped significantly. Russia exported 1,217 billion cubic meters of natural gas, a significant decrease of 34% year-on-year, mainly due to a significant reduction in natural gas supply to Europe. In Central Asia, due to increased demand for natural gas in the region, Kazakhstan and Uzbekistan both reduced their exports, but Turkmenistan's natural gas exports increased significantly by 13.8%, reaching 489 billion cubic meters. Russia's natural gas exports to China continue to grow, with pipeline natural gas exports of 227 billion cubic meters in 2023, an increase of 46% year-on-year. Central Asia exported 412 billion cubic meters of pipeline natural gas to China, slightly less than in 2022.

Investment Environment in Regions and Major Countries

Comparison of Investment Environment in Important Oil and Gas Resource Countries in the Region

The investment environment in Russia is deteriorating. Politically, the Ukraine crisis is difficult to resolve, and Russia's confrontation with the West is intensifying. By the end of 2023, the war is still at a stalemate, neither Russia nor Ukraine can gain an overwhelming advantage, making it difficult to reach a reconciliation. In June 2023, the "Wagner Incident" occurred in Russia and was quickly quelled, which did not affect the Russian political situation, but it also reflected the increase in internal power struggles, military disagreements, and social dissatisfaction in Russia under the influence of the Ukraine crisis. **In terms of policy**, Russia frequently adjusted its policies in 2023, firstly to restrict foreign capital withdrawal, and secondly to increase government fiscal revenue. In terms of restricting foreign capital withdrawal, it depreciates the assets of foreign companies withdrawing from Russia and establishes temporary management agencies. If Russian overseas assets are seized, the agency will take over the assets of the relevant country in Russia. In terms of increasing government fiscal revenue, the Russian government will levy a windfall tax of up to 10% on some large companies in 2024, applicable to companies with an average pre-tax profit of more than 1 billion rubles since 2021, but excluding small and medium-sized enterprises, single agricultural taxpayers, and oil, natural gas and coal companies. **Economically**, after the outbreak of the Ukraine crisis, the West imposed unprecedented sanctions on Russia in terms of intensity and breadth. In 2023, the list of technologies for restricted exports to Russia was continuously expanded, while imports of energy, ores and other raw materials from Russia were banned, and the prices of Russian crude oil and refined oil exports were restricted, affecting Russia's fiscal revenue and the ruble exchange rate. The Russian government has implemented relevant measures to stabilize the exchange rate, requiring some large exporters to forcibly settle foreign exchange in the domestic market to stabilize the ruble exchange rate. The World Bank estimates that Russia's economic growth rate in 2023 will be 1.6%.

The investment environment in Central Asia is complex and changeable. Politically, the political situation in Central Asia is generally stable. After Tokayev was elected President of Kazakhstan with a high vote in the early elections held in November 2022, he further stabilized the domestic situation and consolidated his regime through comprehensive political and economic reforms. Uzbekistan's President Mirziyoyev was re-elected in July 2023 with 87.05% of the vote, ensuring the continuity of Uzbekistan's comprehensive reforms and demonstrating a distinctive governance style of "pro-people, enterprising, pragmatic, and open". **In terms of policy**, Kazakhstan has increased oil export pipeline fees, mining taxes and value-added tax rates to increase oil and gas industry income. Uzbekistan, in order to stabilize the situation, has reduced the corporate income tax, business tax, corporate property tax and land tax rates in Karakalpakstan by 50% to increase the investment attractiveness of the region; in addition, it provides tax incentives for renewable energy producers nationwide, exempting them from major taxes for 3–10 years. **Economically**, driven by the recovery of international trade and tourism, the entry of Russian companies and remittance income, the Central Asian economy has begun to recover. The European Bank for Reconstruction and Development predicts that Central Asia's economy will grow by 5.7% in 2023 and reach 5.9% in 2024. **In terms of security**, Russia's main energy is used to deal with the Ukraine crisis, and its influence on security affairs in Central Asia has declined. Armed conflicts continue to occur between Tajikistan and Kyrgyzstan due to border issues. The unstable security situation in Afghanistan also impacts the Central Asian region, especially the security situation in Tajikistan (Table 1).

Table 1 Investment environment in major countries in Russia-Central Asia Region

Country	Politics	Economy	Policy
Russia	The crisis in Ukraine has led to a complete breakdown in relations between Russia and the West, damaging its international political influence, increasing domestic contradictions, and increasing the pressure on Putin's administration	A large amount of foreign capital is flowing out, the domestic financial market is experiencing turbulence, the volatility of the ruble exchange rate is intensifying, and the risk of economic recession is increasing	Deduct foreign enterprise assets, restrict foreign capital withdrawal; increase taxes on large companies to make up for the fiscal deficit
Kazakhstan	The government has completed its restructuring, President Tokayev has replaced the leaders of major departments, eliminating the political influence of the former president; a new parliament has been formed, with the ruling party winning a majority of seats	Although oil and gas revenues have declined, Kazakhstan's retail, wholesale trade, and construction industries have seen significant expansion	Increase oil export pipeline fees, plan to increase mining tax and value-added tax rates
Uzbekistan	Mirziyoyev was re-elected as president, comprehensive reforms continue, and his ruling position is more stable	After the epidemic, the retail, catering, and tourism industries have recovered, driving economic recovery; policies to stimulate exports and optimize imports have been adopted, and the total foreign trade volume continues to expand	Introduce preferential policies to increase the attractiveness of foreign investment; provide tax relief for the development of renewable energy

Regional Foreign Cooperation Bidding and Asset Acquisition Situation

Kazakhstan continues to expand foreign cooperation. The National Oil Company of Kazakhstan (KMG) continues to promote its IPO, actively seeking partners to advance some exploration and development projects to increase the company's oil and gas production and improve its financial situation. In September, the National Oil Company of Kazakhstan transferred 50% of the shares of Kalamkas Khazar Operating Company to Lukoil of Russia.

Main Risk Points of Investment in Key Resource Countries in the Region

Political risk. After the outbreak of the Ukraine crisis, the West has intensified sanctions against Russia, and Russia faces high diplomatic conflicts and geopolitical risks.

Economic risk. The risk of economic recession in Russia is increasing, a large amount of foreign capital is flowing out, the domestic financial market is experiencing turbulence, the volatility of the ruble exchange rate is intensifying, and the risk of debt default is increasing. Kazakhstan's oil and gas export revenue is a major source of fiscal revenue, and fluctuations in international oil prices and the control of crude oil exports by Russia will affect its economic development.

Business environment risk. Western sanctions make it difficult for Russia to guarantee oil and gas income. To make up for the fiscal deficit, the Russian government may continue to increase taxes in the oil and gas industry. To prevent foreign capital from exiting, Russia strengthens capital control measures and forcibly nationalizes some Western joint-stock companies. Kazakhstan tightens labor policies and gradually reduces the number of foreign workers.

The Situation of International Oil Companies in Regional and Major National Energy Cooperation

Western companies completely withdraw from Russia. Since 2023, Western oil companies have accelerated their exit from the Russian market. ExxonMobil and Shell have successively withdrawn from the Sakhalin 1 and 2 projects, and Japan continues to cooperate for energy security reasons. Schlumberger, TechnipFMC and other oil service companies have also successively withdrawn from projects in Russia. The Russian government chooses to let domestic companies take over the oil and gas projects that Western companies have withdrawn from, such as approving Novatek to acquire Shell's 27.5% stake in the Sakhalin 2 project for 94.8 billion rubles (about 1.2 billion US dollars). In addition, in response to the behavior of unfriendly countries seizing Russian assets, the Russian government temporarily takes over its projects in Russia. At present, Russia has temporarily taken over 83.73% of the shares of the Russian Unipro division held by Germany's Uniper SE, and the shares of Fortum held by Fortum Russia B.V. and Fortum Holding B.V.

Russia's westward exports are blocked, and it accelerates the promotion of eastward natural gas exports. Since the Ukraine crisis, Russia has successively shut down the Yamal-Europe, Nord Stream 1 and other natural gas pipelines for export to Europe. At the same time, it has accelerated the advancement of eastward natural gas export pipelines. Gazprom plans to supply natural gas to China from Sakhalin Island via the Far East Line natural gas pipeline in early 2027, with an annual agreed supply of 10 billion cubic meters. Russia and Mongolia will sign a cooperation document for the Eastern Union Natural Gas Pipeline Project (Mongolian section of the China-Mongolia-Russia Natural Gas Pipeline) in the first quarter of 2024, and construction will begin as early as the first half of 2024. In October, the leaders of China and Russia proposed that they hope the China-Mongolia-Russia natural gas pipeline project will make substantial progress as soon as possible.

The diversification of oil and gas exports is accelerating. Kazakhstan is continuously promoting the construction of the trans-Caspian transport corridor, hoping to export crude oil through this corridor. In addition, Kazakhstan proposed that if the situation in the Black Sea further deteriorates due to the Ukraine crisis, the crude oil exported from Russia through Kazakhstan may be diverted from the Novorossiysk port in the Black Sea to the Ust-Luga port in the Baltic Sea. Turkmenistan plans to cooperate with Azerbaijan and Turkey to open a new route for natural gas exports to Europe. In 2023, the natural gas alliance between Russia, Kazakhstan, and Uzbekistan made some progress. On January 18 and 24, Russia signed a natural gas cooperation roadmap with Kazakhstan and Uzbekistan, respectively, and started supplying Russian natural gas to Uzbekistan through Kazakhstan on October 7.

Outlook for 2024

The Ukraine crisis continues to ferment, and the regional situation is difficult to improve. In 2024, the Ukraine crisis will be difficult to resolve, the strategic goal of the West to continue to suppress Russia will not change, sanctions against Russia will continue to strengthen, and the situation in the Russia-Central Asia region remains severe.

Russia's economy is difficult to improve. In the future, the normalization of Western sanctions against Russia will further reduce Russia-Europe trade, Russian oil and gas revenues will be difficult to guarantee, a large amount of foreign capital will flow out, causing the ruble exchange rate to fluctuate, the risk of debt default will increase, and the risk of economic recession will increase. The World Bank predicts that Russia's economic growth rate will drop to 1.3% in 2024 and to 0.9% in 2025.

Oil and gas production may decline. Western companies are still constantly exiting the Russian oil and gas industry, and Western sanctions against Russia also hinder the entry of new foreign capital. In the future, the problem of insufficient investment in the Russian oil and gas industry will become prominent, and the country's oil and gas production capacity and export volume may decline. It is estimated that in 2024, the oil and gas production in the Russia-Central Asia region will slightly decrease to 600 million tons and 800 billion cubic meters respectively. Europe further reduces Russian oil and gas imports, and Russia's eastward exports are limited by insufficient infrastructure in the short term. It is expected that in 2024, the region's oil and gas export volume will still show a downward trend.

Russia will accelerate the development of the Asia–Pacific market. Russia's westward oil and gas exports continue to decrease, and it is actively seeking a breakthrough. According to the latest revised national energy development plan by the Russian government, by 2030, Russia plans to increase its energy export share to the Asia–Pacific market from the current 49–76%.

Central Asia will continue to promote the diversification of export strategies. The Ukraine crisis is difficult to improve in the short term, and Central Asian resource countries will continue to seek oil and gas export channels bypassing Russia, including promoting eastward oil and gas export channels to China, and strengthening cooperation with Turkey, Azerbaijan and other Caspian countries to open new westward export channels.

(This section is written by: Feng He, Zhang Jing, Liang Meng, Peng Yingying, Reviewed by: Fu Lei)

Middle East Region

Development of the Oil and Gas Industry in the Region and Major Countries

Oil and Gas Production and Trends

Oil production declines, natural gas production grows. In 2023, the upstream capital expenditure of regional oil and gas resources countries was 54.9 billion US dollars, an increase of 13.4% year-on-year. The growth mainly came from Gulf Cooperation Council countries such as Saudi Arabia, the United Arab Emirates, Qatar, and Kuwait. In terms of oil production, affected by the "OPEC+" production limit, the production of countries such as Saudi Arabia, Iraq, the United Arab Emirates, and Kuwait declined, and the regional oil production was 1.35 billion tons, a year-on-year decrease of 2.2%. Iran's oil production is not subject to production limit agreements, and its production has increased by 11% year-on-year, making it the country with the largest increase in regional production. In terms of natural gas production, regional countries have increased their efforts to develop natural gas, and the natural gas production is 738.8 billion cubic meters, a year-on-year increase of 2.4%. The increase in production mainly comes from Saudi Arabia, Kuwait, and Qatar (Figs. 3 and 4).

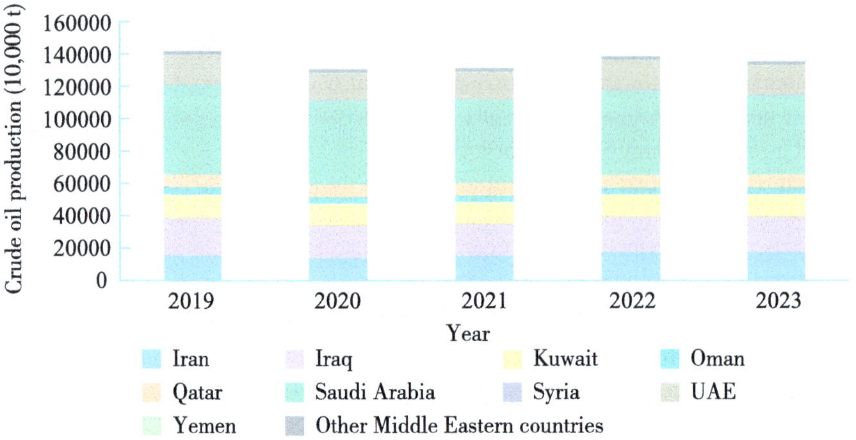

Fig. 3 Crude oil production in the Middle East from 2019 to 2023. *Data source* OPEC

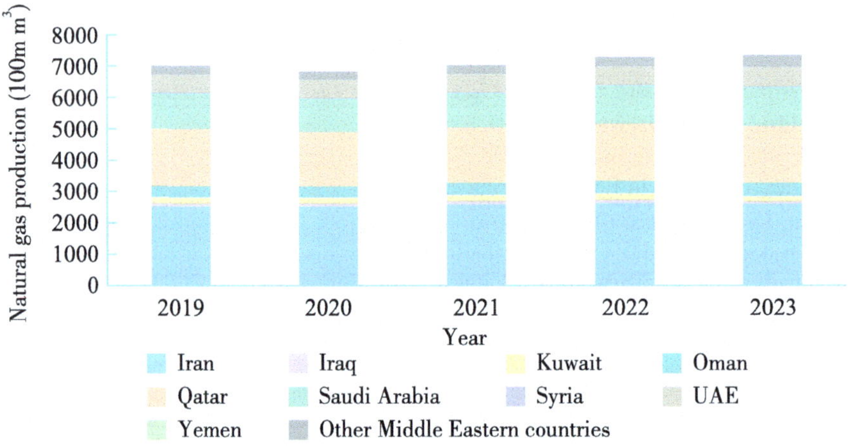

Fig. 4 Natural gas production in the Middle East from 2019 to 2023. *Data source* OPEC

Oil and Gas Import and Export and Trends

Oil exports decrease, mainly to Asian countries. Affected by the "OPEC+" production limit and the increase in regional consumption, in 2023, the regional oil export volume was 890 million tons, a year-on-year decrease of 6.7%. The decrease in export volume mainly came from Iraq, Saudi Arabia, and Kuwait. Among them, the suspension of Iraq's northern export of 450,000 barrels/day of crude oil through the Turkish Ceyhan port is the main reason for the country's decline in oil exports; Saudi Arabia's export volume has decreased due to the "OPEC+" production limit and voluntary production cuts; Kuwait's Al Zour refinery has been put into operation, and domestic consumption has increased, leading to a decrease in exports. In terms of export destinations, regional crude oil is mainly exported to Asia, with a small amount exported to Europe. The proportions of exports to Asia by Saudi Arabia, Iraq, the United Arab Emirates, Kuwait, Oman, and Qatar are 85%, 76%, 96%, 97%, 98%, and 99% respectively, mainly to countries such as China, Japan, South Korea, India, and Singapore.

Natural gas export volume remains the same as the previous year, with Asian countries being the main export destinations. In 2023, the regional natural gas export volume was 179 billion cubic meters, basically the same as in 2022, with exports mainly from Qatar and Oman. Among them, 80 and 96% of Qatar and Oman's natural gas exports flow to Asia, mainly including countries such as China, India, South Korea, and Japan.

Investment Environment in the Region and Major Countries

Comparison of Investment Environment in Important Oil and Gas Resource Countries in the Region

The tendency of regional countries to focus inward has increased. In 2023, the tendency of countries in the Middle East to focus inward has increased. The governments of major countries attach importance to stabilizing the political situation, revitalizing the economy, and improving people's livelihoods, and independently adjust their domestic and foreign policies according to their own interests and needs, presenting a new situation of reconciliation, cooperation, and development. The reconciliation between Saudi Arabia and Iran has brought positive radiation effects to regional security, stability, and development, and Syria's return to the Arab League has also become an important symbol of the continued "reconciliation wave" in the Middle East. Many countries in the region have joined multilateral organizations representing the interests of developing countries. Iran has officially joined the Shanghai Cooperation Organization, and Saudi Arabia, Iran, and the United Arab Emirates have officially become members of the BRICS countries, actively participating in shaping the international system and becoming an important force in the "Global South".

The Israeli-Palestinian conflict stirs up the regional situation. Since October 2023, a new round of Israeli-Palestinian conflict has continued, with obvious spillover effects, becoming the biggest variable in regional security. The conflict has caused a serious humanitarian disaster in Gaza and temporarily halted the reconciliation process between Israel and Arab countries. At the same time, regional pro-Iran political forces such as Hezbollah in Lebanon, Syrian and Iraqi militia forces, and Houthi forces in Yemen have launched attacks on Israeli and American targets, and the spillover effects of the conflict continue to ferment. Among them, the Houthi forces in Yemen have launched multiple attacks on ships passing through the Red Sea. In response, the United States and the United Kingdom have launched air strikes on many places including the Yemeni capital Sana'a, escalating the crisis in the Red Sea.

The overall economic growth rate of the region slows down. In 2023, driven by the investment of the governments of the Gulf Cooperation Council (GCC) countries, non-oil industries in countries such as Saudi Arabia and the United Arab Emirates have grown strongly, and the economies of the GCC countries have maintained growth. However, affected by the "OPEC+" production limit, regional oil export income has decreased, and the new round of Israeli-Palestinian conflict has continued, leading to foreign investment withdrawing from the regional capital market. The economies of some already troubled Middle Eastern countries have added insult to injury, and the overall economic growth rate of the region has slowed down, with economic growth in 2023 being 1.8%.

The development of the oil and gas industry is moving forward with energy transformation. In 2023, the regional resource countries are actively developing clean and low-carbon energy businesses while developing the oil and gas industry, promoting energy transformation. The United Arab Emirates hosted the 28th United Nations Climate Change Conference (COP28), during which it released the "Oil and Gas Decarbonization Charter" (OGDC) with Saudi Arabia, aiming to achieve net-zero emissions in the oil and gas industry by 2050. The Abu Dhabi National Oil Company (ADNOC) announced that it will achieve the net-zero emission target five years ahead of schedule in 2045 and achieve the methane zero-emission target in 2030 (Table 2).

Regional Foreign Cooperation Bidding and Asset Acquisition Situation

Iraq signs the fifth round of oil and gas bidding contracts. In February 2023, the Iraqi government signed exploration and development agreements for six oil and gas fields with three energy companies, including Chinese companies, all of which were winning projects in the fifth round of oil and gas bidding after the 2018 Iraq war. Among them, the Chinese company Continental Oil & Gas Co., Ltd. will explore and develop two oil and gas fields in Diyala Province in eastern Iraq and Maysan Province in southern Iraq; another Chinese company, United Energy Group, will explore and develop an oil and gas field in Basra Province in southern Iraq; the UAE's Crescent Petroleum Company will explore and develop three oil and gas fields in Diyala and Basra provinces.

Regional new energy bidding is mainly concentrated in Oman. In 2023, Oman launched the second round of green hydrogen project bidding, introducing three green hydrogen projects in the Dhofar region, bringing investment opportunities to foreign companies.

Table 2 Investment environment of major countries in the Middle East

Country	Politics	Economy	Policy
Saudi Arabia	The political situation is stable, and power is smoothly transferred to the Crown Prince and Prime Minister. Diplomatic autonomy is increasingly strengthened, and while valuing security and strategic relations with the United States, relations with Russia and China are enhanced. Regional diplomacy is generally easing	Voluntary additional production cuts are implemented, oil export revenue declines, overall economic growth slows, non-oil industries grow strongly, GDP growth in 2023 is 4.1%, and it is expected to grow by 4.4% in 2024	The "Vision 2030" is progressing orderly; plans to increase oil and gas production, vigorously develop the petrochemical industry; plans to launch a greenhouse gas credit quota plan in early 2024, allowing companies to offset their emissions by purchasing credits from voluntary reduction or elimination of greenhouse gas emission projects; proposes to exempt companies that relocate their regional headquarters to Saudi Arabia from corporate tax for 30 years
Iran	The political situation is basically stable, and parliamentary and expert assembly elections (responsible for electing the supreme leader) will be held in March 2024. The Iran nuclear talks have not made progress, seeking to strengthen cooperation with other regional countries and international organizations. Domestic attacks by extreme terrorist organizations have intensified tensions with neighboring countries, and domestic security risks have risen	Strive to increase oil production, expand oil exports, vigorously develop non-oil and gas industries, and curb the impact of inflation and sanctions. GDP growth in 2023 is 4.2%, and it is expected to grow by 2.5% in 2024	Plan to increase oil production capacity to 5.7 million barrels/ day before 2030; reached an agreement with Russia to use their own currencies instead of the US dollar in commercial transactions, and the two sides can use the currency and banking platforms between the two countries for settlement

(continued)

Table 2 (continued)

Country	Politics	Economy	Policy
Iraq	The political situation is relatively stable, the current government was established in 2022, and it has the support of most political parties; affected by the spillover of the Israeli-Palestinian conflict, US military bases in Iraq are frequently attacked, and Iraq's domestic security risks are rising	The economic structure is single, heavily dependent on oil, and the development level of manufacturing and agriculture is not high due to factors such as incomplete industrial chains and lack of water resources. GDP growth in 2023 is –2.7%, and it is expected to grow by 2.9% in 2024	Actively attract foreign investment to develop the oil and gas industry, expand upstream and downstream oil and gas production capacity, and reduce dependence on Iranian natural gas imports; prohibit the withdrawal of US dollar cash and domestic transactions in US dollars from January 1, 2024; will establish a carbon emission management company to strengthen the supervision of carbon emissions and environmental protection in energy projects
United Arab Emirates	The political situation is stable, and the Emirate of Abu Dhabi appointed the Crown Prince in 2023. Actively develop relations with the United States, Russia, the European Union, and China. Relations with Iran have improved, and ambassadors have been reappointed to Iran. Diplomatic representative offices have been restored with Qatar	The oil and gas industry continues to develop, the non-oil economy grows strongly, the construction industry, wholesale and retail industry, transportation and warehousing, financial insurance services, and high-tech industries stand out, and it is the regional logistics, trade, and financial center. GDP growth in 2023 is 3.7%, and it is expected to grow by 3.8% in 2024	ADNOC plans to invest 150 billion US dollars, increase oil production capacity from the current 4 million barrels/day to 5 million barrels/day before 2027, and achieve self-sufficiency in natural gas consumption before 2030; ADNOC integrates natural gas processing business and liquefied natural gas operation, maintenance, and marketing business to establish ADNOC Gas; the UAE releases the national hydrogen energy strategy, sets hydrogen production targets, ADNOC announces to achieve methane zero-emission target in 2030, and net-zero emission target in 2045; the UAE starts to levy a 9% corporate tax from June 1

(continued)

Table 2 (continued)

Country	Politics	Economy	Policy
Oman	The domestic political situation has achieved a smooth transition. Sultan Qaboos, who ruled Oman for 49 years, passed away in 2020, and Haitham successfully took over as the new Sultan	The economy has recovered, with both the oil and gas industry and non-oil and gas industries continuing to grow, and the fiscal and debt situation has improved. GDP growth is estimated to be 2.7% in 2023 and is expected to increase by 2.9% in 2024	Plans to utilize its abundant solar and wind energy resources and advantageous geographical location to vigorously develop the hydrogen energy industry, with cumulative investment in hydrogen projects expected to reach $140 billion by 2050
Qatar	The domestic political situation is stable, with the Al Thani family hereditary Emir; in terms of diplomacy, it values relations with the West, restores diplomatic relations with Gulf countries, and actively mediates the Israeli-Palestinian conflict	Actively attracting foreign investment, the natural gas industry has developed. GDP growth is 2.5% in 2023 and is expected to increase by 3.1% in 2024	Invest heavily in the development of natural gas, making it a top priority for economic development, promoting the expansion of the North Field's production capacity, and selling minority stakes in the project to foreign oil companies such as BP, Total, ConocoPhillips, Sinopec, and PetroChina; in terms of taxation, the government has postponed the collection of VAT; the government has also launched the "2030 National Vision" plan, the core of which is to vigorously develop economic diversification and make Qatar a sustainable development, highly competitive, and high standard of living country by 2030

The number of regional asset transactions is limited. In 2023, regional asset transactions slowed down, mainly concentrated in Saudi Arabia and the UAE, with green and renewable energy transactions as the main. Major transactions include: Saudi Aramco acquired the global product business of Valvoline, a US-based Victory Oil Chemical Company; acquired shares of MidOceanEnergy, a US liquefied natural gas company, for $500 million; completed the acquisition of a 10% stake in China's Rongsheng Petrochemical Co., Ltd. for $3.4 billion; purchased 100% of the shares of Esmax, a Chilean downstream fuel and lubricant retailer, from Southern Cross Group, a private equity firm focused on Latin America, marking Saudi Aramco's first downstream retail investment in South America. In the UAE, ADNOC completed the acquisition of a 30% stake in the Absheron natural gas and condensate field in Azerbaijan; acquired all shares of OCI Global, a Dutch fuel producer, in Fertiglobe to support the company's chemical strategy and ammonia energy development strategy; acquired a 10.1% stake in Storegga, a UK-based company focused on the development of global carbon capture and storage (CCS) projects, marking its first international equity investment in carbon management. In addition, the Abu Dhabi National Energy Company (TAQA) completed the acquisition of a 5% stake in ADNOC Gas, and the UAE's Masdar Company completed the acquisition of a 50% stake in solar and battery storage projects in California, USA, owned by EDF Renewables, a subsidiary of the French Electricity Group (EDF).

Main Risk Points of Investment in Key Resource Countries in the Region

Regional security risks are rising. In 2023, a new round of Israeli-Palestinian conflict escalated and spilled over into countries such as Lebanon, Yemen, Iraq, and Syria. The United States was constantly entangled with the Houthi forces, tensions between Iran and its neighboring countries intensified, Israel and Hezbollah continued to exchange fire, and US military bases in Iraq and Syria were frequently attacked by missiles and drones. The regional humanitarian disaster and refugee crisis worsened, extremist terrorist forces gained room to cause trouble, the control of the central governments of Iraq and Lebanon declined, and the security situation in the Middle East became more severe.

The risk of sanctions against specific countries remains high. In 2023, the United States continued to impose sanctions on Iran, mainly involving individuals and entities supporting Iran's ballistic missile and drone programs. In November 2023, the U.S. House of Representatives passed a bill called "Stop Sheltering Iranian Oil," aimed at promoting the punishment of entities involved in Iranian oil trade and curbing the behavior of helping Iran sell oil under the sanctions framework. With the escalation of the Israeli-Palestinian conflict, the United States, as a staunch ally of Israel, may tighten sanctions on Iran again. The risk of sanctions against Iran remains high, and entities and individuals doing business with Iran need to be cautious about the risk of sanctions.

The business environment risks in some countries are worth noting. The central government of Iraq and the Kurdish government have deep contradictions, and there have been long-term disagreements on issues such as the distribution of oil export revenues. In 2023, the suspension of oil exports from northern Iraq to the Turkish port of Ceyhan led to a decline in Iraq's oil production; in addition, to curb the abuse of its foreign exchange reserves in the financial sector and evade US sanctions on Iran, Iraq will ban the withdrawal of US dollar cash and domestic transactions in US dollars from 2024, increasing business risks. On the UAE side, strict localization policies are implemented, with the government stipulating that the localization rate of employees increases by 2% each year, reaching 10% by the end of 2026, otherwise, economic penalties will be faced. In 2023, the government fined companies that failed to achieve the 2022 employee localization target a total of 400 million dirhams (approximately 109 million US dollars).

The Situation of International Oil Companies in Regional and Major Country Energy Cooperation

Oil and gas foreign cooperation is mainly concentrated in Iraq and Qatar. In 2023, international oil companies were quite active in cooperation in the Middle East, mainly in countries such as Iraq and Qatar. **In Iraq,** ExxonMobil withdrew from the West Qurna-1 project, and PetroChina became the lead contractor for the oilfield; TotalEnergies and Qatar Energy began to implement comprehensive energy projects in Iraq, including oil, natural gas, renewable energy, and seawater desalination, with a total investment of 27 billion US dollars; Lukoil extended the contract for the West Qurna-2 project in Iraq by 10 years to 2045; the Iraqi government signed the fifth round of oil and gas bidding exploration and development contracts with China Intercontinental Oil and Gas, United Energy Group, and UAE Crescent Petroleum. **In Qatar,** Sinopec respectively signed a 1.25% equity agreement for the North Field Expansion (NFE) project with Qatar Energy, and an integrated cooperation agreement for the North Field Expansion Phase 2 (NFS) project (including a 27-year LNG long-term purchase and sale agreement and a 1.875% equity in the NFS project); PetroChina signed a cooperation agreement with Qatar Energy, and Qatar Energy transferred a 1.25% stake in the North Field Expansion (NFE) project to PetroChina; Qatar Energy signed a 5-year crude oil supply agreement with Shell Eastern Trading, a subsidiary of Shell, supplying up to 18 million barrels of crude oil to Shell each year from January 2024.

Several foreign companies have reached long-term LNG trade agreements with GCC countries. In 2023, the position of GCC countries in the global LNG market continued to rise, and several companies from Asia and Europe signed long-term LNG trade agreements with the UAE, Qatar, and Oman. **On the Qatar side,** Sinopec, PetroChina, TotalEnergies, Eni, and Shell respectively signed 27-year LNG long-term trade agreements with Qatar, with trade volumes of 3 million tons/year, 4 million tons/year, 3.5 million tons/year, 1 million tons/year, and 3.5 million tons/year respectively. **On the UAE side,** TotalEnergies, Japan Petroleum Exploration Company (Japex), PetroChina International Co., Ltd. (PCI), and China New Energy Group respectively reached LNG trade agreements with ADNOC Gas, a subsidiary of the Abu Dhabi National Oil Company, with contract amounts (trade volumes) of 1 billion to 1.2 billion US dollars, 450 million to 550 million US dollars, 450 million to 550 million US dollars, and 1 million tons/year respectively. **On the Oman side,** Oman LNG and Germany's Securing Energy for Europe) has signed a four-year LNG supply agreement with a trade volume of 400,000 tons/year.

Major international oil companies are expanding their green low-carbon project cooperation in the Middle East. In Saudi Arabia, Norway's Aker company has signed a cooperation agreement with Saudi Aramco's carbon capture company. Aker will focus on reducing or eliminating carbon emissions in Saudi Arabia's industry and energy through CCUS projects. **In Oman,** Hydrom, a company focused on hydrogen energy development in the country, has signed six investment agreements with developers from Belgium, the Netherlands, the UK, Japan, Singapore, Germany, India, Kuwait, and the UAE, planning to produce hydrogen energy worth more than $51 billion.

Outlook for 2024

Geopolitical risks are becoming long-term. The spillover from the new round of Israeli-Palestinian conflict will become long-term, evolving into a wider regional war with the full participation of Hezbollah in Lebanon, Shiite militias in Syria and Iraq, and Houthi forces in Yemen, forming a confrontation with the US and Israel. The prospects for Israeli-Arab relations are bleak, and the normalization process may be shelved for a while. As the US supports Israel and pays little attention to the demands of Arab countries, the rift between Arab countries and the US will continue to widen. Iran will hold parliamentary elections in 2024, and the intensified sanctions by Western countries will make the economic and social pressures Iran faces long-term. Iran will further strengthen its relations with China and Russia while maintaining a state of easing relations with the Gulf Arab countries.

Regional economic development faces uncertainty. As the external shock brought by the Ukraine crisis gradually fades, coupled with the implementation of diversified trade and investment strategies by regional countries, the regional economy is expected to gradually recover, with an estimated economic growth of 3.4% in 2024. However, the spread of the Israeli-Palestinian conflict and the continued reduction in production by "OPEC+" may suppress the prospects for regional economic growth, and the overall economic development faces uncertainty.

Steady development of the oil and gas industry. In 2024, upstream investment in the oil and gas industry in the Middle East will continue to grow. Among them, the total upstream investment in oil and gas by Saudi Arabia and the UAE may reach $10 billion, mainly used to increase domestic oil and gas production capacity. **In terms of oil and gas production,** influenced by the plans of regional countries to increase oil production and new capacity construction, regional oil production is expected to increase, but "OPEC+" production cuts may bring uncertainty to oil production growth; regional natural gas production is expected to increase by 40 billion cubic meters, with the largest increase coming from Iran's South Pars Phase 11 and Saudi Arabia's associated gas in major oil fields. **In terms of oil and gas bidding,** in 2024, regional bidding will mainly focus on Iraq. Iraq will hold the "Fifth + " round and the sixth round of oil and gas project bidding activities, including 16 failed bids from the fifth round in the "Fifth + " round, and the sixth round includes 11 natural gas exploration blocks. The contract terms of this bid will be based on the fifth round of bidding contracts, but adjustments have been made to the financial and tax terms of the contract. **In terms of asset mergers and acquisitions,** the national oil companies of the Gulf Cooperation Council countries will be the main force in the M&A market. ADNOC is exploring opportunities to develop LNG projects in the international market in cooperation with Chinese companies; in addition, ADNOC is also considering participating in the natural gas development project in the Illizi Ghadames Basin in Libya. Kuwait Foreign Petroleum Exploration Company (KUFPEC) is discussing acquiring more oil and gas assets in Southeast Asian countries, and the details are expected to be announced in the first quarter of 2024. Qatar Energy will sign an LNG purchase and sale agreement for the North Field expansion project with ExxonMobil and PetroChina; a consortium formed by Mitsui & Co. and JERA may acquire a stake in the North Field expansion project.

Regional resource countries will actively implement COP28 related agreements. With the UAE hosting the 28th United Nations Climate Change Conference in 2024, regional countries will actively implement COP28 related agreements, and clean energy investment will accelerate: **First, some national oil companies will increase decarbonization investment**. For example, Bahrain's Bapco Energies signed the "Oil and Gas Decarbonization Charter" during the COP28 conference, but the company had not previously committed to achieving net-zero emissions. It is expected that Bapco Energies will increase its investment in decarbonization in 2024. **Second, regional natural gas flaring will decrease**. Iraq plans to eliminate natural gas flaring by 2030. With the operation of natural gas treatment facilities in the Hafaya and Nasiriya oil fields, it is expected that regional natural gas flaring will decrease by 5% in 2024. **Third, resource countries in the region will participate in more Carbon Capture and Storage (CCS) projects**. In October 2023, ADNOC announced that the company plans to increase its CCS production capacity from the current 4.1 million tons/year to 10 million tons/year by 2030, and will participate in more domestic and foreign CCS projects.

(This section was written by: Wang Ying, Lang Fengqiao. Reviewed by: Fu Lei)

Africa

Development of the Oil and Gas Industry in the Region and Major Countries

Oil and Gas Production and Trends

In 2023, Africa's crude oil production was stable and rising, with Libya's significant increase in production making up for the production gap in other traditional resource countries. Crude oil production was 335 million tons, an increase of 0.9% compared to 2022. Nigeria and Algeria are currently facing the problem of insufficient industry investment and the bottleneck of production decline due to "OPEC+" restrictions. Nigeria's crude oil production was 68.03 million tons, a decrease of 1.4% compared to the previous year; Algeria's crude oil production was 61.37 million tons, a decrease of 3.5% compared to the previous year. Libya's domestic situation is stabilizing, and the national oil company has launched a development plan for 2023–2027, aiming to revitalize the country's oil and gas industry, while implementing measures to improve production standards and workers' wages. These three factors ensure crude oil production, with crude oil production at 65.79 million tons, an increase of 18% compared to the previous year.

Natural gas production slightly increased, driven by European consumption demand. In 2023, natural gas production was 263.8 billion cubic meters, an increase of 3% compared to the previous year. Specifically, due to insufficient investment and gas field reduction, Nigeria's natural gas production was 41.6 billion cubic meters, a decrease of 3% compared to the previous year; Algeria's natural gas production was 106.3 billion cubic meters, an increase of 3% compared to the previous year; Mozambique's Coral FLNG project entered stable production, with natural gas production at 10.6 billion cubic meters, an increase of 58.2%. Since the Russo-Ukrainian conflict, European countries such as Italy have been continuously increasing their imports of Algerian natural gas in order to break away from their dependence on Russian natural gas. Eni plans to increase by 9 billion cubic meters per year on the current basis in 2024 (Figs. 5 and 6).

Oil and Gas Consumption, Imports and Exports, and Trends

Libya's export volume surged to fill the regional crude oil trade gap. In 2023, Africa's crude oil exports were nearly 260 million tons, a decrease of 0.5% year-on-year. Nigeria and Angola are the main crude oil exporting countries in Africa, but their production decline limits their crude oil export capacity. In 2023, Nigeria exported 65.38 million tons of crude oil, a decrease of 4.8% year-on-year; Angola exported 55.15 million tons of crude oil, a decrease of 0.5% year-on-year. Libya's

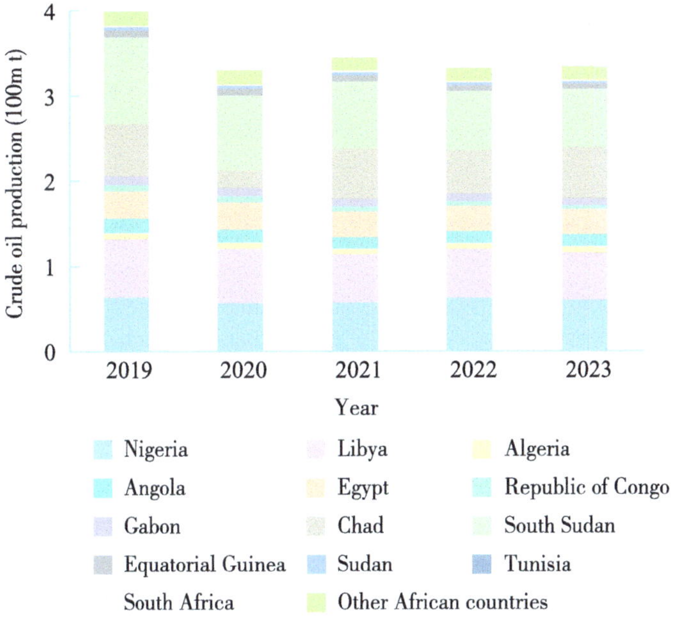

Fig. 5 Crude oil production in Africa 2019–2023. *Data source* PetroChina Economic Research Institute

crude oil production increased significantly, with annual exports of 52.48 million tons, an increase of nearly 20% (Fig. 7).

Natural gas exports slightly increased, and domestic consumption rose, limiting export potential. In 2023, Africa's natural gas exports were 123.6 billion cubic meters, an increase of 1% year-on-year. Algeria is Africa's largest natural gas exporter, with an export volume of 60.3 billion cubic meters, an increase of 7% year-on-year. Nigeria's domestic natural gas consumption demand is increasing year by year, and exports are declining. The annual export is 26.1 billion cubic meters, a decrease of 8% year-on-year (Fig. 8).

Investment Environment in the Region and Major Countries

Comparison of Investment Environments in Important Oil and Gas Resource Countries in the Region

The resurgence of coups, ongoing conflicts, and the development of the oil and gas industry is hindered. In 2023, military coups occurred in Niger and Gabon, and large-scale armed conflicts broke out in Sudan. Since 2020, the frequency and success

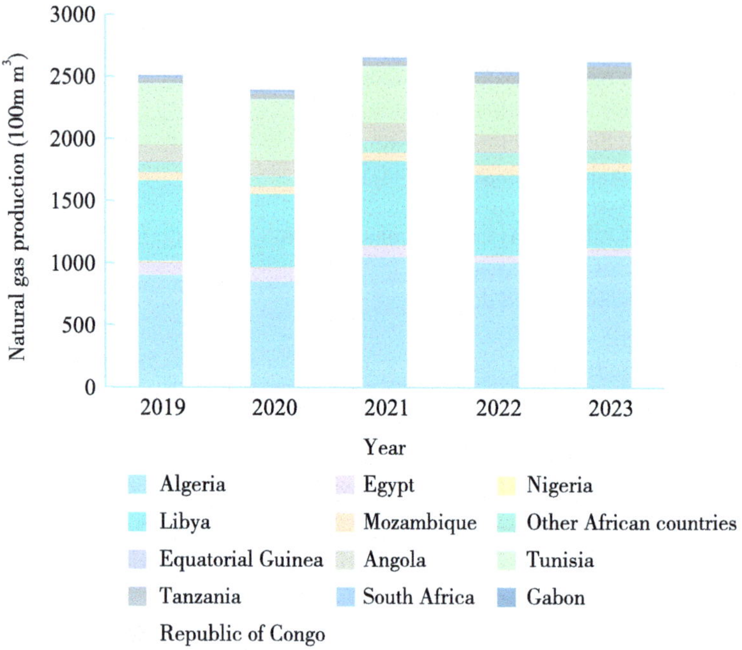

Fig. 6 Natural gas production in Africa 2019–2023. *Data source* PetroChina Economic Research Institute

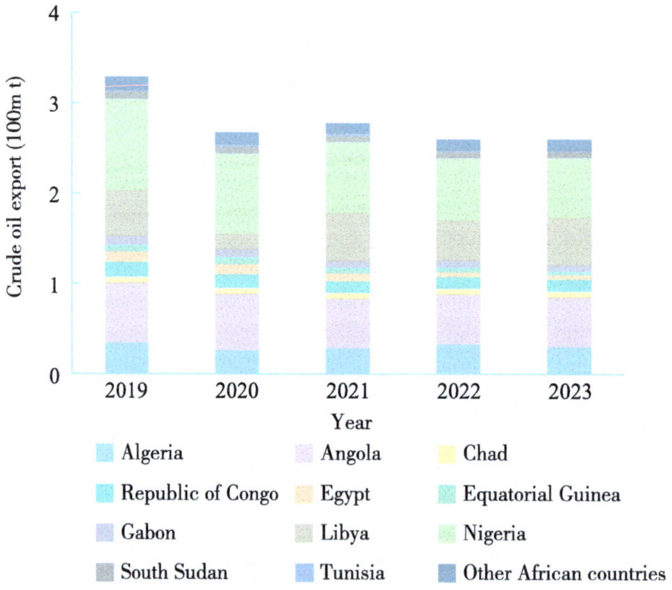

Fig. 7 Crude oil exports in Africa 2019–2023. *Data source* PetroChina Economic Research Institute

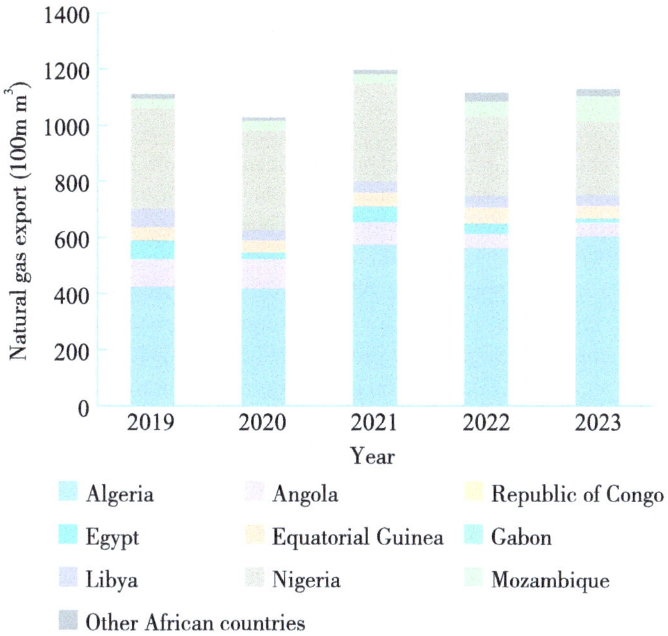

Fig. 8 Natural gas exports in Africa 2019–2023. *Data source* PetroChina Economic Research Institute

rate of coups in Africa have significantly increased, and political fluctuations have had a profound impact on the oil and gas industry. Because Benin supports ECOWAS's military action against Niger, the time for the Niger-Benin pipeline to be put into operation is uncertain; the armed conflict in Sudan is becoming protracted, and the situation is continuously deteriorating; South Sudan hopes that Russia will support the construction of a new oil export pipeline to ensure oil exports.

High inflation hits Africa's financial system, slowing economic recovery. In 2023, the economic growth rate of Sub-Saharan Africa was 3.3%, a decrease of 0.7% from the previous year. The coup in Niger led to financial and trade sanctions by the Economic Community of West African States (ECOWAS), causing foreign companies to significantly reduce external budget and project financing support. The GDP growth rate slowed to 4.1%, a nearly 8% decrease from the previous year. Armed conflict in Sudan resulted in significant damage to its economic foundation, causing economic development to regress by decades. The GDP growth rate was -18.2%, the lowest point since South Sudan's independence. Many countries faced severe inflationary pressures, with the currencies of Nigeria, Angola, South Sudan, and Egypt, four oil and gas producing countries, depreciating among the top 10% globally. Among them, the Nigerian Naira depreciated by 76%, the Angolan Kwanza by 64%, the South Sudanese Pound by 55%, and the Egyptian Pound by 25%. The debt levels of countries in the region continued to rise, with the median debt level in Africa rising from 51.5% in 2019 to about 60% in 2023. To control debt levels, Nigeria abolished approximately $10 billion in fuel subsidies annually, and Angola also announced that it would gradually abolish nearly $4 billion in energy subsidies per year.

Resource-rich countries actively adjust cooperation policies to attract foreign investment. In 2023, to increase oil and gas production, Libya introduced a number of policies to improve the domestic policy environment. The governments of eastern and western Libya established special committees to distribute oil revenues, and the transitional government approved the NOC's five-year development plan, aimed at simplifying the investment process for international oil companies and attracting foreign investment to develop energy projects. Nigeria abolished the fixed exchange rate to curb the significant depreciation of the Naira, hoping to improve the stability and liquidity of the local currency and reduce investors' concerns about exchange rate risk. Mozambique temporarily postponed the introduction of localization clauses to promote the progress of oil and gas projects.

Political instability allows terrorist organizations to take advantage. Frequent coups have made the Sahel region a hub for terrorism. In 2023, the number of deaths due to terrorism in Somalia, Burkina Faso, Mali, and Niger in Africa exceeded the total in South Asia, the Middle East, and North Africa (MENA), accounting for 43% of the global total. After the coup in Niger, its military bases were frequently attacked by the extremist organizations "Support for Islam and Muslims (JNIM)" and "Islamic State in the Greater Sahara (ISGS)". The security situation in Mozambique became complicated. Although terrorist organizations were dispersed under the joint crackdown of the Rwandan anti-terrorism forces and the African Union, they still launched sporadic armed attacks in many places. In addition, the Rwandan army, which is at odds with the Mozambican military, has been stationed on the Afungi Peninsula for a long time under the pretext of assisting in the fight against terrorism, increasing regional security uncertainty (Table 3).

Regional Foreign Cooperation Bidding and Asset Merger and Acquisition Situation

Traditional resource countries are increasingly willing to cooperate in oil and gas. In 2023, the Egyptian Ministry of Petroleum originally planned to launch 3 rounds of oil and gas block bidding, and has now achieved 2 rounds, which are the bidding for brownfield (industrial waste land) onshore oil and gas blocks, and the bidding for blocks along the Mediterranean coast. The Libyan National Oil Corporation announced during CERA WEEK that it plans to launch oil and gas block bidding in 2024, which is the first bidding since Gaddafi was overthrown in 2011. The Algerian Petroleum Development Agency had planned to launch bids for 10–12 blocks in the first quarter of 2023, but was forced to postpone due to the undetermined bidding blocks. Angola is promoting the bidding of undeveloped blocks in the Kwanza Basin

Table 3 Investment environment in major African countries

Country	Politics	Economy	Policy
Libya	The contradictions between the eastern and western governments of Libya are sharp, and the two sides have always been unable to reach a political compromise. Russia and the Libyan National Army are cooperating in defense, and Russia will establish a naval base in eastern Libya	A significant increase in crude oil exports has helped the economy recover quickly. The GDP growth rate in 2023 was 12.5%, and it is expected to be 7.5% in 2024	The National Oil Corporation aims to attract foreign investment to develop the energy industry through a five-year development plan, with the goal of increasing crude oil production to 2 million barrels/day within 5 years
Niger	A military coup broke out in Niger, and the military promised to hold elections after the three-year transitional government period. ECOWAS and the West African Economic and Monetary Union closed the border with Niger; Niger formed a military alliance with Mali and Burkina Faso	ECOWAS took a number of financial and economic sanctions against the coup in Niger, limiting Niger's oil exports. Niger's GDP growth rate fell to 4.1% in 2023, and it is expected to recover to 11.1% in 2024	After the military took power, it is committed to maintaining the development of the country's oil industry and will not adjust oil and gas policies in the short term

(continued)

Table 3 (continued)

Country	Politics	Economy	Policy
Sudan	Disagreements arose within the Sudanese military due to restructuring, intensifying conflicts. Full-scale armed conflict broke out between the armed forces and the rapid support forces, mediation failed, and the conflict is still ongoing	The protraction of armed conflict in Sudan has further exacerbated economic recession. The GDP growth rate in 2023 was − 18.1%, and it is expected to be 0.3% in 2024	Since the independence of South Sudan, the Sudanese government has actively attracted foreign investment with favorable fiscal terms. However, due to the fluctuating domestic political and security situation, investors are adopting a cautious wait-and-see attitude
Mozambique	The presidential election will be held in 2024. The current ruling party, the Mozambique Liberation Front, has strong governance capabilities and political legitimacy, and the political situation is stable	The growth of natural gas exports has driven the economic recovery of Mozambique. The GDP growth rate in 2023 was 8%, and it is expected to be 4.9% in 2024	Mozambique's oil and gas industry localization law has not yet been introduced; the international market is full of challenges, and policy stability helps the stable and orderly development of the natural gas development industry

and the Lower Congo Basin, which is another attempt to develop marginal oil fields and increase crude oil production.

International oil companies adjust their investment portfolios and focus on core assets. In 2023, several international oil companies began to transfer their non-core assets within their domains. In 2023, bp sold its Senegal Yakaar-Teranga FLNG assets to fill the budget overrun caused by the delay of the Greater Tortue Ahmeyim (GTA) LNG project it operates in the country; TotalEnergies transferred 40% of its rights in Angola's 20/11 block to Petronas of Malaysia to focus on the development of the Venus deepwater oil field in Namibia. The trend of international oil companies exiting onshore assets in Nigeria is becoming increasingly prominent. Eni and Equinor have both decided to transfer all their businesses, thus completely withdrawing from the Nigerian oil and gas field.

Main Risk Points of Investment in Key Resource Countries in the Region

Factional and power disputes are frequent, and the political risks of resource countries are prominent. The political and security situation in the Sahel region continues to fluctuate and poses spillover risks, negatively affecting the political situation in neighboring countries, and the possibility of interruption or delay of oil and gas projects in resource countries is increasing.

Terrorist attacks are erupting in many places, and security risks are rising. Multiple terrorist organizations such as Boko Haram and ISWAP are entrenched in northeastern Nigeria, and the security risk remains high and continues to infiltrate into Niger, Chad and other countries; the coup in Niger has led to a resurgence of terrorism, multiple forces have launched armed attacks, and security risks are rising; although the overall strength of extremist organizations in Mozambique has declined and the risk of terrorist attacks has decreased, the relationship between the Mozambican military and the Rwandan army is tense, making the regional security situation more complex.

The currency has depreciated significantly, and exchange rate risk continues to rise. The currencies of most Sub-Saharan African countries have weakened to varying degrees against the US dollar, the fall in oil prices, the Fed's interest rate hike, and the soaring prices of commodity imports have exacerbated inflationary pressures in the African region.

The Situation of International Oil Companies' Energy Cooperation in the Region and Major Countries

European energy companies are increasing their cooperation in North Africa's new energy. Against the backdrop of the Russo-Ukrainian conflict, European countries have begun to seek alternative energy sources to reduce their dependence on Russian natural gas. By 2023, the abundant solar and wind energy resources in North Africa have attracted increasing investment from European energy companies in new energy cooperation. The European Bank for Reconstruction and Development (EBRD) is working with Morocco on energy transformation and sustainable development cooperation to accelerate the decarbonization of Morocco's power sector; Denmark's Maersk has signed a cooperation framework agreement worth $300 million with Egypt to produce green fuels and their derivatives in the Suez Canal Economic Zone; Germany is negotiating with Algeria to make it a major supplier of green hydrogen. The investment of European energy companies in North Africa's new energy will promote economic development in North Africa and promote the application of renewable energy in Europe, which is sustainable.

International oil companies are increasing their exploration and development investment in key resource countries. Egypt has announced cooperation plans with BP, Shell, Eni and other companies, and 30 exploration wells will be drilled offshore in the next two years. BP stated that it will invest $35 billion in natural gas projects in Egypt over the next three years. TotalEnergies announced several oil and gas development plans in Algeria, covering oilfield production increase, natural gas cooperation, and supplying LNG to Europe. After Shell discovered the large deepwater oilfield Joker in Namibia, the proven recoverable reserves of deepwater crude oil in Namibia reached 11 billion barrels of oil equivalent, with considerable resource potential. TotalEnergies and Shell respectively announced that they will invest 50% and 25% of their exploration budgets in Namibia's deepwater oilfields.

Outlook for 2024

In 2024, the political situation in Africa will continue to fluctuate, with the conflict in Sudan developing over the long term, undercurrents still flowing in countries such as Niger and Gabon that have experienced coups, and elections being held by transitional governments in Chad and Mali, posing risks of political instability. The IMF predicts that Africa's economic growth rate will increase by 0.7–4.0% in 2024, inflation levels in many countries are expected to fall, but fluctuations in oil prices affect the economic recovery of resource countries, and some countries have high debt risks. In the future, the security situation in Africa will continue to be tense, with terrorist organizations becoming active after the coup in Niger, sporadic terrorist attacks continuing in northern Mozambique, armed conflicts in Sudan leading to refugees flowing into neighboring countries, adding uncertainty to the security of countries such as Chad and South Sudan.

(This section was written by: Zhao Jie, Zhang Yanyun, Luo Jiyu. Reviewed by: Fu Lei)

Latin America

Development of the Oil and Gas Industry in the Region and Major Countries

Trends in Oil and Gas Production

Significant achievements in oil and gas exploration, with production maintaining steady growth. In 2023, of the top ten oil and gas discoveries worldwide, Latin America accounted for four, mainly located in Guyana (2), Suriname (1), and

Colombia (1), with discovered oil and gas 2P recoverable reserves of 150 million tons and 278 billion cubic meters, respectively, accounting for 35% and 3% of the total new discoveries worldwide that year. Among them, the Lancetfish 1 discovery made by ExxonMobil and its partners offshore Guyana was the largest oil discovery of the year worldwide, with oil and gas 2P recoverable reserves of 70 million tons and 170 billion cubic meters respectively; the Roystonea 1 discovery offshore Suriname and the Wei 1 discovery offshore Guyana ranked fourth and fifth in the world's largest oil discoveries of the year. In recent years, major oil and gas discoveries in Latin America have all been offshore, demonstrating the huge potential of offshore oil and gas exploration.

In 2023, with strong production growth in Brazil and production recovery in Mexico, Colombia, and Venezuela, Latin America's crude oil production continued to grow, increasing by 7.4% year-on-year to 456 million tons. Among them, Brazil's crude oil production was 181 million tons, up 11.8% year-on-year, mainly due to the production growth of traditional subsalt fields such as Tupi, the advancement of the Marlim field production increase plan, and the start-up of new projects such as Petrobras' Búzios 5 and Mero 2 subsalt fields, Equinor's BM-C-22 project, and TotalEnergies' Lapa South project. Mexico remained the second-largest oil producer in the region, with crude oil production increasing to 104 million tons in 2023, up 6.9% year-on-year, due to the increase in condensate production offsetting the sharp decline in mature field production. Argentina's crude oil production increased to 35.65 million tons, up 8.7%, due to increased investment by national oil company YPF and smaller regional producers such as VistaEnergy. Venezuela's crude oil production increased to 41.83 million tons, up 8.7%, second only to Brazil, due to the US easing sanctions on its oil and gas industry at the end of 2022, allowing Chevron to resume oil production in Venezuela. Colombia increased its upstream investment, reversing the trend of declining production, with production increasing to 41.53 million tons, up 4.5%. Ecuador and Peru saw production declines due to ongoing political and economic turmoil affecting investment confidence (Fig. 9).

In 2023, the natural gas production in Latin America was 178.5 billion cubic meters, an increase of 0.9% year-on-year, with the increase mainly coming from Brazil, Argentina, and Trinidad and Tobago. Thanks to the development of deep-sea sub-salt gas fields, Brazil's natural gas production in 2023 was 21.2 billion cubic meters, an increase of 8.2%. In July 2023, the Nestor Kirchner natural gas pipeline in Argentina was put into operation, connecting the gas-rich Neuquén province with the capital Buenos Aires, promoting the development of natural gas. To compensate for the decline in production from mature gas fields, Trinidad and Tobago actively promoted the development of offshore gas fields, especially the development of cross-border offshore gas fields shared with Venezuela. In January 2023, the US government issued a two-year permit at the request of Trinidad and Tobago, allowing it to do business with the Venezuelan National Oil Company (PDVSA). In September, Trinidad and Tobago signed an agreement with Venezuela to jointly develop the cross-border Dragon gas field and build a natural gas pipeline (Fig. 10).

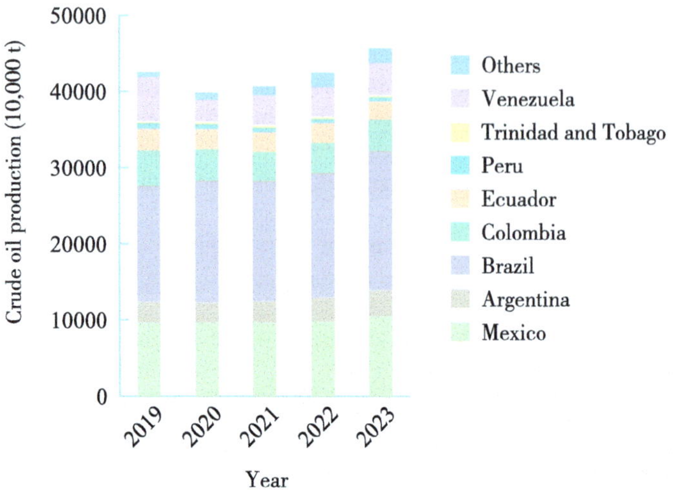

Fig. 9 Crude oil production in Latin America from 2019 to 2023

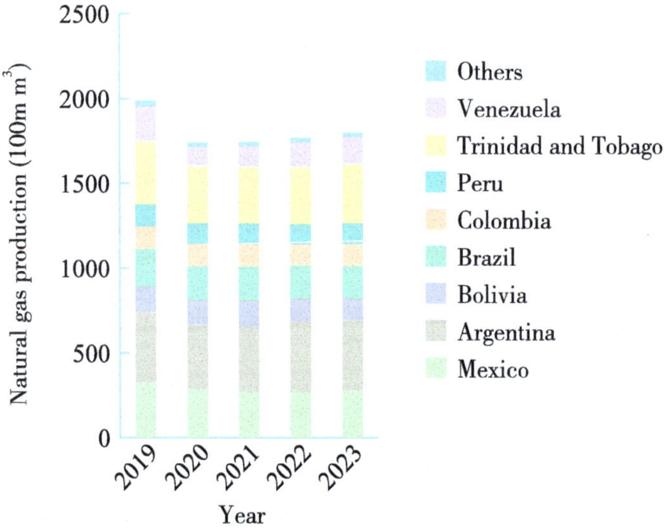

Fig. 10 Natural gas production in Latin America from 2019 to 2023

Trends in Oil and Gas Consumption and Import and Export Changes

The growth rate of refined oil consumption slows down. In 2023, the average consumption of refined oil in Latin America was 6.3 million barrels/day, still lower than the level of 6.4 million barrels/day before the COVID-19 pandemic in 2019. The growth rate of refined oil consumption slowed from 7.4% in 2022 to 3.5%.

Natural gas consumption has declined slightly. Due to improved weather conditions, natural gas demand slightly decreased from 2022 to 2023. Increased rainfall in Brazil led to an increase in electricity generation from hydroelectric power stations, reducing the electricity industry's demand for natural gas.

Investment Environment in the Region and Major Countries

Frequent elections in Latin American countries, political changes and economic crises, changes in market policies, and social unrest are intertwined, making the investment environment complex and chaotic. In 2023, the left still dominated Latin American politics, but the competition between the left and right wings intensified, and the ruling position was unstable. Regional unity, diplomatic autonomy, and integration strengthened, the economy continued to be depressed, the social situation did not improve significantly, and violent conflicts in some countries intensified.

Analysis of the Investment Environment in Important Oil and Gas Resource Countries

The regional political form shows the characteristics of polarization, fragmentation, and complex governance. The political pendulum effect of competition and alternating development between the left and right wings is still present in Latin America. The left wing in Latin America usually represents the interests and demands of the middle and lower classes, advocating for changes in the existing political and social order, redistribution of social power and benefits; advocating for a certain degree of intervention in the economy; tending to prioritize fairness over efficiency; and having varying degrees of anti-American or "de-Americanization" tendencies. The right wing usually represents the interests of the middle and upper social classes, advocates for maintaining vested interests, agrees with moderate social reforms, opposes drastic social changes; tends to laissez-faire market economic policies and elite governance political concepts; advocates for efficiency over fairness; and has obvious pro-American tendencies or "Americanization" preferences.

In 2023, Paraguay, Guatemala, and Argentina held presidential elections as scheduled, and Ecuador held early general elections due to the impeachment of the former president. The old right-wing ruling party in Paraguay continued to govern; the left-wing party won in Guatemala; the far-right candidate Miley won in Argentina; the center-right candidate Novoa was elected in Ecuador. Although the left wing still holds the majority in Latin America, the election results show a split and opposition in the political stance of the people. The victorious left-wing forces are often in a relatively weak position in the parliament and are restrained by right-wing forces. Whether it is a left-wing or right-wing ruling country in Latin America, they all face difficulties and challenges in governance, and their ruling positions are not stable.

Venezuela's government and opposition reach an agreement. On October 17, 2023, the Venezuelan government and the opposition Unified Program Party held talks again in Barbados, and the two sides reached a consensus on several topics such as democratic guarantees and signed a political agreement. The agreement stipulates that the Venezuelan presidential election will be held in the second half of 2024, and international observers including the European Union and the United Nations will supervise the voting. At present, the opposition has elected a far-right candidate, and the ruling party is confident of winning. The competition between the two sides is fierce, and there is still uncertainty about the political process in Venezuela and the direction of the 2024 general election.

Political turmoil in Ecuador, increasing demands from indigenous people. On May 17, 2023, President Lasso, facing impeachment over corruption issues, announced the dissolution of the parliament and the early holding of presidential and parliamentary elections. On October 15, the 35-year-old center-right party candidate Novoa won the presidential election, becoming the youngest president in Ecuador's history, with a term until May 2025.

Argentina's left wing loses, far-right wins. On November 19, 2023, Argentina's far-right presidential candidate Javier Milei unexpectedly won with nearly 56% of the vote, a huge advantage of about 12 percentage points ahead of his opponent, defeating the center-left Peronist party candidate and current Minister of Economy Sergio Massa. This marked a dramatic change in Argentine politics. Milei's election shows that right-wing forces are still very powerful on the Latin American political stage, and it also indirectly indicates the economic difficulties currently faced by Argentine society and the general disappointment with the establishment.

The United States' influence in Latin America has weakened. The United States regards Latin America as a core interest, with strategic goals to consolidate its "backyard", suppress the left, and squeeze out external forces. The United States uses ideology, military, and financial means to fully control Latin America, ensuring Latin America as its source of raw materials and goods market. On November 3, 2023, the first Americas Economic Prosperity Partnership (APEP) Leaders Summit was held in Washington, with 11 American countries participating. The summit topics included supply chains, immigration, and sustainable development. Currently, Latin American countries are generally in varying degrees of economic and social development difficulties, looking forward to external assistance. The Biden administration's attempt to control Latin America and squeeze out China by promoting APEP may not be fulfilled. Latin America's "leaders" Brazil and Argentina have not yet joined APEP, and the Mexican president was also absent from the summit. Against the backdrop of the decline in US global influence, Latin America's position in US foreign strategy has declined, and the US's control over Latin America has relatively weakened. More and more Latin American countries are seeking to unite for independent development, advocating diversified foreign relations, and getting rid of dependence on the United States.

China-Latin America relations are flourishing. High-level visits are frequent. In January 2023, President Xi Jinping was invited to deliver a video speech at the 7th Summit of the Community of Latin American and Caribbean States (CELAC). In March, China and Honduras established diplomatic relations. Presidents from Latin American countries including Brazil's Lula, Honduras's Castro, Venezuela's Maduro, Argentina's Fernandez, Chile's Boric, Colombia's Petro, and Uruguay's Lacalle visited China. The relationship between China and Venezuela was upgraded to an all-weather strategic partnership, and the relationship between China and Uruguay was upgraded to a comprehensive strategic partnership. China has signed free trade agreements with Chile, Costa Rica, Ecuador, and Peru, and 21 Latin American countries have joined the "Belt and Road" initiative. The internationalization of the RMB is solidly advancing in Latin America. In November 2022, the RMB clearing service in Argentina officially started. In February 2023, China and Brazil signed a memorandum of cooperation to establish an RMB clearing arrangement. In April, the Argentine government announced the use of RMB to settle trade in goods imported from China. In June, the Central Bank of Argentina announced that the RMB would be included in the country's bank savings accounts and current accounts. In July, Bolivia became a South American country that uses the RMB for import and export trade settlement.

The trend of regional integration is strengthening. With the majority of Latin American countries being governed by the left, regional relations have shown a significant warming trend, and regional unity and cooperation have once again become the core topics of interaction between countries in the region. The 7th Summit of the Community of Latin American and Caribbean States (CELAC) held in Argentina in January 2023 passed the "Buenos Aires Declaration", emphasizing the need to accelerate the process of Latin American integration. The South American Heads of State Summit held in May signed the "Brasilia Consensus", clarifying the necessity and urgency of deepening South American integration.

Economic growth is weak, inflation remains high. Due to the complex impact of the weak international economic situation and the gloomy regional economic prospects, countries in Latin America continue to face a situation of low economic growth. In 2023, the economic growth rate in Latin America fell to 2.2%, a year-on-year decrease of 1.5%. Among the major economies in the region, the economic growth rates of Brazil, Mexico, Colombia, and Peru were 2.5%, 2.9%, 1.2%, and 1.3% respectively, while Argentina and Chile experienced negative growth of 3.0% and 0.3% respectively. Latin American countries generally face problems such as high inflation, high interest rates, high debt ratios, high unemployment rates, and high poverty rates. The average inflation rate in the region in 2023 was 4.9%, a decrease from the previous year, but it is still not optimistic. There are 26 countries in Latin America with inflation rates lower than at the end of 2022, with the largest declines in Chile, Costa Rica, Guatemala, Guyana, Honduras, and Trinidad and Tobago, with declines exceeding 4 percentage points, but the inflation situation in Argentina and Venezuela is still very severe, with inflation rates as high as 117% and 404% respectively (Table 4).

Table 4 Investment environment of important resource countries in Latin America in 2023

Country	Politics	Economy	Policy
Brazil	The moderate left-wing ruling coalition led by Lula faces a disadvantageous situation of "small morning and big evening", and the "palace dispute" is fierce. The government will use foreign relations and the reshaping of Brazil's great power status as a starting point	The economy is in a state of low growth, with a GDP growth rate of 2.5% and an inflation rate of 3.2%	A series of measures have been taken to optimize the investment environment and attract foreign capital and technologically advanced companies to participate in Brazil's oil and gas resource exploration and development. But at the same time, it maintains the government's control over the energy industry and prohibits the privatization of energy industries such as refineries, old oil fields, and power companies
Ecuador	The political situation is complex, drug trafficking and organized crime bring serious security problems, and the new president faces challenges in governance	GDP growth rate is 2.3%, inflation rate is 1.7%	The government encourages foreign companies to participate in upstream oil and gas investment, but the continuous opposition from indigenous and environmental organizations will have a negative impact on oil and gas production
Peru	The political situation is unstable, and the risk of violent incidents caused by political conflicts rises before the 2024 presidential election	Economic growth slows down, GDP growth rate is 1.3%, inflation rate is 6.4%	The government may nationalize expiring oil and gas projects
Venezuela	The Maduro government and the opposition have reached a new round of reconciliation, easing the political situation. The United States temporarily lifted some energy and trade sanctions against Venezuela	The economy is growing in a recovery manner, with a GDP growth rate of 4.2%, and the inflation rate has eased	The government intends to attract investment and increase oil production by improving contract terms

Situation of Foreign Cooperation Bidding and Asset Mergers and Acquisitions

In 2023, five rounds of oil and gas block bidding were held in the Latin American region. From November 28, 2022, to May 31, 2023, Suriname held bids for six offshore oil and gas blocks. From December 9, 2022, to July 15, 2023, Guyana held bids for 14 offshore blocks, eventually receiving bids for eight blocks. From August 17 to December 13, 2023, the Brazilian National Petroleum Agency (ANP) held the fourth round of permanent open license block bidding and the second round of permanent open production sharing contract block bidding. The fourth round of license block bidding offered blocks in nine basins, with 192 winning blocks, setting a record for the permanent open bidding model. In this round of bidding, 15 companies won, among which, the Brazilian National Oil Company (operator, holding 50%) jointly won three blocks in the offshore Pelotas Basin with Shell (holding 30%) and CNOOC (holding 20%). The Brazilian National Oil Company (holding 70%) also jointly won 26 blocks in the basin with Shell (holding 30%), and Chevron won 15 blocks in the basin. Australia's Karoon Company won two blocks in the Santos Basin, and CNOOC and the Norwegian National Oil Company each won one block in the basin. Brazil's Elysian, PetroRecôncavo, and other local companies won onshore blocks in various coastal basins. In the second round of production sharing contract block bidding, bp won the Tupinambá block in the Santos Basin. On October 3, Trinidad and Tobago held bids for 13 offshore blocks, with the bidding period ending on April 2, 2024. On November 7, Suriname launched a new round of offshore block bidding, including 11 blocks in nearshore and shallow water areas, with the bidding period ending on May 31, 2024. In 2023, a total of 85 block contracts were awarded in the Latin American region. Of these, Brazil awarded 48 block contracts. Forty-four onshore blocks were all won by local independent oil companies; bp Brazil Company won one deepwater block in the Santos Basin; the Brazilian National Oil Company won one deepwater block in the Santos Basin and two deepwater blocks in the Campos Basin. Guyana awarded seven block contracts, all located in the Guyana Basin, of which two were in deep water, four were on the continental shelf, and the contract models were production sharing contracts and mineral tax system, respectively awarded to consortia composed of ExxonMobil, Hess (acquired by Chevron in October 2023), and China National Offshore Oil Corporation, Total Energies, Qatar Energy and Petronas, Delcorp, Watad Energy and Arabian Drillers, Liberty Oil and Cybele Energy, International Group Investment and Montego Energy, and Guyana's Sispro Company. Argentina awarded seven block contracts, respectively won by Argentina's National Oil Company, Pan American Energy, and Phoenix Global Resources. In addition, Bolivia, Colombia, Ecuador, and Peru onshore and Suriname, Trinidad and Tobago, and Uruguay offshore awarded a total of 23 blocks, mainly won by the national oil companies of their own countries as well as bp, Shell, Total, etc. 3. Key Risk Points for Investment in Key Resource Countries in the Region **Venezuela still faces the risk of U.S. sanctions.** For more than a decade, the United States has been imposing sanctions on Venezuela's oil industry, government agencies, and certain individuals to curb the anti-American actions of the former

President Chavez and the current President Maduro. Since 2019, the United States has expanded economic sanctions against Venezuela, including banning the import of Venezuelan crude oil, banning the export of diluents to Venezuela, and freezing the assets of the Venezuelan Oil Company in the United States. In November 2022, the United States relaxed sanctions against Venezuela, allowing Chevron's joint venture with PDVSA to increase oil production. On October 18, 2023, in response to the political agreement reached between the Venezuelan government and the opposition, the United States temporarily lifted sanctions on Venezuela's oil and gas sector, gold sector, allowing some secondary trading of Venezuelan sovereign bonds and PDVSA debt and equity. At the same time, the U.S. government reserves the right to revoke the above authorizations and requires the Maduro government to fulfill its promise to hold democratic elections recognized by the West by the end of 2024. **Political turmoil leads to a severe social security situation.** Since 2023, due to the spread of drug trafficking activities, intense election competition, and other factors, violent incidents have frequently occurred in Ecuador, with riots occurring in many prisons, and many candidates have been assassinated or attacked, causing the country to enter a state of emergency twice for 60 days. Since the political crisis in Peru in December 2022, social order has been chaotic, with various demonstrations occurring repeatedly, and some areas have even evolved into violent conflicts resulting in casualties. **Social environmental risks are rising.** Under the continuous push of indigenous communities and environmental organizations for a decade, Ecuador's Supreme Court and the National Election Commission agreed in May and June 2023 to "permanently close the Ishpingo Tiputini Tambococha Block (ITT Block, also known as Block 43) held a national referendum. On August 20, Ecuador held a national referendum on whether to close large oil fields in the Amazon rainforest region during the presidential and parliamentary elections. The results showed that nearly 60% of voters supported the closure of the oil fields. This move is not only the first in Ecuador's history, but also the first in the world, indicating that the social and environmental compliance risks in Latin American countries where indigenous people gather, such as Ecuador, are rising.

The Situation of International Oil Companies in Regional and Major National Energy Cooperation

The United States relaxes sanctions on Venezuela, and international oil companies accelerate oil and gas cooperation in Venezuela. The United States has significantly relaxed sanctions on Venezuela, providing a rare window of opportunity for oil companies. Many global oil companies are flocking to Venezuela to enhance their oil and gas cooperation with Venezuela. Shell, Spain's Repsol, Hungary's MOL, Sweden's Maha Energy, Trinidad and Tobago's National Gas Company, and Bolivia's National Gas Company have sent representatives to Venezuela, hoping to change contract terms, obtain operating rights for oil and gas fields, or resolve Venezuela's unpaid debts. In early November 2023, Venezuela signed a supplementary agreement with French oil exploration company Etablissements Maurel & Prom (M&P), which provides more operational and financial control to the joint venture, referring to the cooperation between Venezuela and Chevron. On December 18, Spain's Repsol also signed a similar supplementary agreement with PDVSA, and it is reported that Italy's Eni is also negotiating a similar contract with Venezuela.

Deepwater development in countries such as Brazil and Guyana attracts foreign oil companies. As the difficulty of conventional oil and gas exploration on land and in shallow water increases, international oil companies are implementing strategic shifts, and deep water has become an important area for the discovery of oil and gas reserves. Guyana has become a hot spot for global oil exploration. In addition to multinational oil companies such as CNOOC, ExxonMobil, Hess, Tullow Oil, and Repsol that have entered the Guyanese market, Qatar Petroleum has also announced that it will establish a joint venture with French oil giant Total to enter the Guyanese shallow water oil and gas development field. Encouraged by ExxonMobil's significant exploration achievements in Guyana's offshore area, international oil companies such as Total Energies, Shell, and Chevron have also shown great interest in the oil and gas exploration potential of Suriname's offshore area. The Brazilian government has taken a series of measures to optimize the investment environment and attract foreign capital and advanced companies with technical advantages to participate in the exploration and development of Brazil's oil and gas resources.

Outlook for 2024

The game between left and right intensifies, and the turnover of power accelerates. The victory of the far right in Argentina further strongly impacts the momentum of the region's left turn, and the left-wing government faces challenges. In Chile, due to economic contraction, public dissatisfaction with the Boric left-wing government is accumulating, and opinion polls show that Boric's support rate has fallen to 29%. In contrast, the momentum of the right-wing conservative Republican Party is growing, and the party won a majority of seats in the constitutional committee election in May. In June 2023, the party's founder, Cast, led in the opinion polls for the 2025 presidential election. Mexico and Venezuela will both hold presidential elections in 2024. The left and right parties in Mexico have started a competition in advance. Former Venezuelan parliamentarian Maria Corina Machado has been appointed as the opposition presidential candidate, intending to compete with Maduro.

Oil and gas production will continue to grow. In Brazil, with the Búzios 5 and Mero With the commissioning of the second phase, the output of the salt-under oil field will gradually expand. Coupled with the output of traditional salt-under layer oil fields including the Tupi oil field and the advancement of the Marlim oil field's production increase plan, the prospects for oil production increase are promising. In

Guyana, the Liza-1 and Liza-2 projects operated by the ExxonMobil consortium are steadily advancing, and it is expected that the Payara and Yellowtail projects will be put into operation in 2024 and 2025 respectively. Together with the Uaru project approved in May 2023 and exploration activities in other offshore blocks, the potential for future oil production increase is great. In Mexico, the output will show a slight growth momentum under the promotion of new oil fields such as the Trion deepwater project. Increased upstream investment in Argentina will contribute to the increase in oil production. In Venezuela, due to the United States' relaxation of sanctions on its oil and gas industry, oil production is expected to gradually recover, but due to long-term lack of investment and infrastructure issues, the short-term oil growth is limited. In Colombia, due to new oil and gas discoveries, the trend of declining oil production will be reversed. The increase in natural gas production will mainly come from Argentina and Trinidad and Tobago. The continuous development of midstream infrastructure projects in Argentina will promote natural gas production, and the three discoveries of the Brazilian BM-C-33 project, Pão de Açúcar, Gávea, and Seat, will promote the growth of natural gas production. Ecopetrol, the Colombian oil company, will increase its investment in strategic natural gas assets, which will also promote the development of natural gas projects. Bolivia has always been one of the largest natural gas producers in the region, but due to lack of investment and weakened export demand from its main traditional trading partner Argentina, natural gas production is facing a decline.

Oil and gas consumption continues to grow at a slow pace. Due to factors such as weak international market demand, reduced foreign trade vitality, difficulty in obtaining external funds, and insufficient internal consumption capacity, it is expected that the Latin American economy will continue to grow at a slow pace of 1.5% in 2024. In terms of refined oil, the lack of infrastructure in the region will affect the popularization of electric vehicles. It is expected that the transportation departments of major Latin American countries will continue to rely on traditional energy, but as the regional economy maintains slow growth, the demand for refined oil will continue to grow slowly, and it is optimistically estimated that it will return to pre-COVID-19 levels by 2025. In terms of natural gas, due to the growth of electricity demand in Mexico and Argentina, natural gas consumption may increase, but as more and more countries implement carbon reduction policies and accelerate the application of non-hydropower renewable energy, regional natural gas consumption growth will slow down, and it is expected to grow by 3.0% year-on-year in 2024.

(Authors of this section: Sun Yimin, Zhong Wenxin, Jin Huandong, Yan Baihui. Reviewer: Fu Lei)

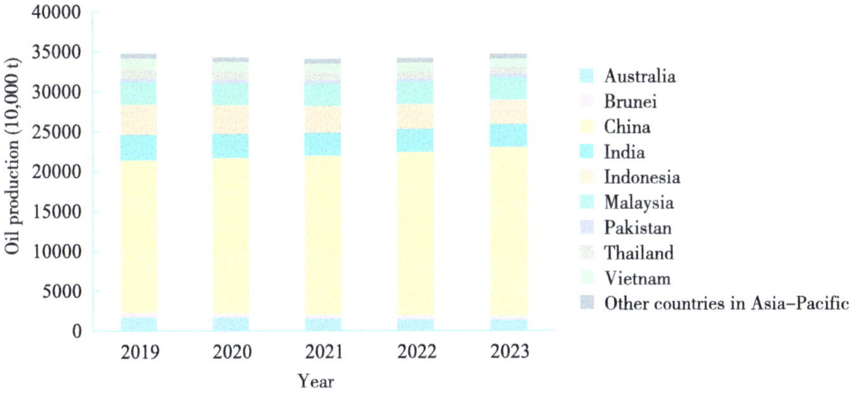

Fig. 11 Crude oil production in the Asia–Pacific region from 2019 to 2023

Asia–Pacific Region

Development of the Oil and Gas Industry in Major Countries in the Region

Oil and Gas Production and Trends

Upstream investment continues to grow, driving both oil and gas production. In 2023, oil production was 347 million tons, an increase of 5.08 million tons, an increase of 1.5%. The increase in production mainly came from China, with an estimated increase of 6.5 million tons. Affected by the reduction in production of old oil fields, oil production in Australia and Indonesia declined (Fig. 11). Compared with the previous year, the two countries decreased by 600,000 tons and 500,000 tons respectively. Due to the lack of large-scale development projects put into operation, it is expected that the region's oil production will decline in 2024. The region's natural gas production was 690.7 billion cubic meters, an increase of 10 billion cubic meters, an increase of 1.0%. Both China and India's natural gas production increased, increasing by 12 billion cubic meters and 1.2 billion cubic meters respectively. Due to the decrease in production of old oil and gas fields, Thailand's natural gas production decreased by 1.4 billion cubic meters (Fig. 12).

Oil and Gas Consumption and Trends

Economic recovery continues, crude oil processing and refined oil consumption increase. According to the "Asia–Pacific Economic Outlook Report" released by the International Monetary Fund, it is predicted that the annual economic growth

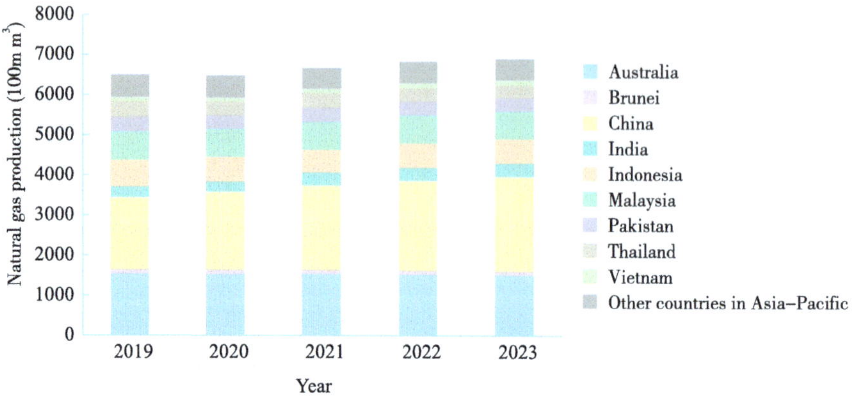

Fig. 12 Natural gas production in the Asia–Pacific region from 2019 to 2023

rate of the Asia–Pacific region in 2023 is expected to reach 4.6%, higher than the growth expectation of 3.9% made in 2022. The annual crude oil processing volume is 30.755 million barrels/day, an increase of 1.275 million barrels/day from the previous year, an increase of 4.3%. The growth mainly comes from China, India, and South Korea, with China increasing by 1.21 million barrels/day, India increasing by 115,000 barrels/day, and South Korea increasing by 29,000 barrels/day. Due to the closure of refineries, Australia's crude oil processing volume decreased by 24,000 barrels/day. The consumption of refined oil in the region, mainly gasoline and diesel, was 20.18 million barrels/day, an increase of 1.28 million barrels/day from the previous year, an increase of 6.6%. Among them, Vietnam's refined oil consumption was 408,000 barrels/day, an increase of 17.2%; India's refined oil consumption was 2.844 million barrels/day, an increase of 5.8%; Australia's refined oil consumption was 985,000 barrels/day, an increase of 6.2%; Indonesia's refined oil consumption was 1.29 million barrels/day, an increase of 3.8%.

The economy is growing rapidly, driving up natural gas consumption. In 2023, the consumption of natural gas was 9230 billion cubic meters, an increase of 159 billion cubic meters from the previous year, an increase of 1.75%. Among them, the consumption of natural gas in India and China increased by 7.4% and 6.6% respectively. Affected by the start-up of nuclear power and coal-fired units, the consumption of natural gas in Japan and South Korea decreased, falling by 9.0% and 6.3% respectively compared to the previous year.

Oil and Gas Import and Export and Changing Trends

Both crude oil imports and exports increased. The region's crude oil imports were 25.692 million barrels/day, an increase of 1.275 million barrels/day from the previous year, an increase of 5.2%. The increase mainly came from China and

India, which increased by 1.093 million barrels/day and 168,000 barrels/day respectively. It is expected that by 2024, the region's crude oil imports will maintain their current level. The region's crude oil exports were 854,000 barrels/day, an increase of 55,000 barrels/day from the previous year, an increase of 6.8%. The main exporting country, Indonesia, increased by 20,000–37,000 barrels/day, and Australia increased by 13,000 barrels/day.

Both refined oil imports and exports increased. The region's refined oil imports were 11.721 million barrels/day, an increase of 693,000 barrels/day from the previous year, an increase of 6.3%. The increase in imports mainly came from Australia, Singapore, Vietnam, and Thailand, with refined oil imports increasing by 81,000 barrels/day, 73,000 barrels/day, 60,000 barrels/day, and 20,000 barrels/day respectively. It is expected that by 2024, the region's refined oil imports will continue to grow. Asia–Pacific refined oil exports were 6.438 million barrels/day, an increase of 98,000 barrels/day from the previous year, an increase of 1.55%. The increase in exports mainly came from China and Singapore, with refined oil exports increasing by 101,000 barrels/day and 46,000 barrels/day respectively.

Both natural gas imports and exports increased. The total natural gas imports of the countries in the region were 4473 billion cubic meters, an increase of 269 billion cubic meters from the previous year, an increase of 6.4%. As China's natural gas consumption increased, its import volume also increased. It is estimated that the total imports for 2023 will be 1674 billion cubic meters, an increase of 147 billion cubic meters from the previous year, an increase of nearly 10%. Among them, pipeline gas imports were 680 billion cubic meters, and LNG imports were 994 billion cubic meters. Due to a decrease in domestic production, Thailand's LNG imports increased by 3.3 million tons, an increase of 39.8%. India's LNG imports increased by 2.3 million tons, an increase of 11.3%. However, traditional LNG importing countries such as South Korea and Japan saw a significant decrease in imports, decreasing by 6 million tons and 2 million tons respectively. The region's liquefied natural gas exports were 137 million tons (1893 billion cubic meters), slightly higher than the previous year. Among them, Indonesia's export volume was 16.4 million tons, an increase of 1.2 million tons from the same period of the previous year, an increase of 8.0%. However, the export volumes of Australia, Malaysia, and Brunei all slightly declined. The region's pipeline gas exports continued to shrink, with a total export of 185 billion cubic meters, a decrease of 20 billion cubic meters from the previous year. Among them, Myanmar exported 110 billion cubic meters, a decrease of 21 billion cubic meters; Indonesia exported 72 billion cubic meters, which was the same as the previous year.

Investment Environment in the Region and Major Countries

Political Security Situation and Outlook

The United States is rallying countries in the region to contain China, and the geopolitical risk in the Asia–Pacific region continues to rise. As the competition between China and the United States continues to intensify, the United States, in conjunction with Australia, Japan, and South Korea, is rallying ASEAN, establishing an Indo-Pacific cooperation framework, and other series of actions, intending to create and establish a NATO in the Asia–Pacific region, provoking a comprehensive containment and deterrence against China. In August 2023, the United States, Japan, and South Korea held a leaders' summit at Camp David in Maryland, USA, and jointly issued a joint statement called the "Spirit of Camp David", confirming the "Camp David Principles" of maintaining consistency in policy positions. This meeting formed a "mini-NATO" in the Northeast Asia region led by the United States, meaning a further strengthening of military cooperation between the three countries. In addition, the United States, in conjunction with countries in the region, held several military exercises near China's sea, with obvious intentions against China, causing the geopolitical risk in the Asia–Pacific region to continue to rise.

The Australian Prime Minister's visit to China may promote the development of bilateral cooperation. In November 2023, Australian Prime Minister Albanese made his first visit to China after taking office, and it was also the first visit to China by an Australian Prime Minister since 2016, which is of great significance for restoring the momentum of bilateral relations. During his visit to China, Albanese attended the China International Import Expo held in Shanghai, with about 250 Australian companies participating in this Expo. China and Australia have broad prospects for cooperation in traditional energy and emerging energy fields, especially in dealing with climate change, new energy, and green infrastructure. Both countries have huge potential for cooperation in technology, capital, talent, and market.

Macroeconomic Situation and Outlook

With strong willingness to cooperate, the regional economy shows a clear growth momentum. The Chinese economy continues to recover and improve, providing new opportunities for regional and even global economic recovery. China is implementing the "Regional Comprehensive Economic Partnership Agreement" with high quality, actively promoting the construction of the China-ASEAN Free Trade Area 3.0, continuously advancing to join the "Comprehensive and Progressive Agreement for Trans-Pacific Partnership" and the "Digital Economy Partnership Agreement", continuously deepening trade and investment cooperation with regional economies, and the Asia–Pacific region will still be the most dynamic locomotive of the global economy. On the 10th anniversary of the "Belt and Road" initiative, China successfully held the third "Belt and Road" International Cooperation Summit Forum, India held the Shanghai Cooperation Organization Leaders Summit and the 18th G20 Leaders Summit, Indonesia held the 43rd ASEAN Summit, and Asia once again ushered in a "highlight moment" to promote global governance. The IMF estimates that the economic growth rate of the Asia–Pacific region is expected to be 4.6% in 2023, higher than 3.9% in 2022.

Policy Environment Analysis of Important Oil and Gas Resource Countries in the Region

Australia, Indonesia, Vietnam and other countries have introduced incentives to increase oil and gas development efforts. Queensland, Australia has launched a new natural gas exploration and development financial support plan, intending to provide a total of no more than 21 million Australian dollars (about 13.5 million US dollars) of financial support for developers of selected exploration and development projects in the new frontier areas of natural gas and coal seam gas in the northern Bowen and Galilee basins of Queensland from 2024 to the first half of 2025. This move aims to support energy security on Australia's east coast and attract more oil and gas investment to Queensland. Indonesia has formulated effective fiscal and tax policies, improved one-stop investment services, and promoted tax incentives to encourage investors to increase production and exploration in Indonesia to achieve the goal of increasing reserves and production. The newly revised "Petroleum Law" of Vietnam was approved by the National Assembly and will take effect on July 1, 2023, providing incentive measures for frontier exploration and marginal oil field development in Vietnam. The income tax of oil fields enjoying investment incentive policies is reduced to 32%, and the income tax of oil fields enjoying special investment incentive policies is reduced to 25%. The new law stipulates that the Prime Minister does not have to approve all oil applications, and some approvals will be undertaken by the Vietnamese Ministry of Industry and Trade.

Malaysia, Indonesia and other countries have introduced incentives and new regulations to seek cooperation in CCUS projects. The Malaysian government has introduced a series of tax incentives to promote the development of the CCS industry, including: companies engaged in CCS services will be eligible for a 10-year income tax preference, 100% investment can offset 100% or 70% taxable income, exemption from all import duties and sales taxes on CCS technology equipment from January 1, 2023 to December 31, 2027, and assessment of the carbon pricing mechanism of the Emissions Trading System (ETS). At the end of March, Indonesia introduced new regulations, providing specific guidelines for the scope and requirements of CCS/CCUS activities, supplementing provisions on carbon pricing and carbon trading, and providing a more complete legal framework for the development of CCS/CCUS.

Australia and Indonesia's increased natural gas export controls will affect the supply in the Asia–Pacific market. In response to the ongoing impact of the sudden international energy supply tension since the beginning of 2022, the Australian government has proposed a number of interventions for the domestic energy market. One is to extend the Domestic Gas Security Mechanism (ADGSM) established by the previous government to 2030, requiring LNG suppliers on the east coast to restrict some LNG exports that have not yet signed long-term trade agreements when necessary to strengthen domestic market supply guarantees. The second is the completion of the draft design of the "Mandatory Natural Gas Behavior Code" ("Natural Gas Code"). The "Natural Gas Code" emphasizes two aspects: one is to maintain the domestic natural gas wholesale price cap at 12 Australian dollars/gigajoule in the short term to ensure additional domestic supply obligations and protect east coast natural gas consumers from the impact of international price fluctuations; the second is to ensure that producers invest in new natural gas supply projects to avoid domestic natural gas shortages and maintain reasonable natural gas supply prices in the medium term. The Indonesian government said that in order to meet domestic natural gas consumption needs, it plans to introduce a policy to restrict LNG exports, but existing LNG export contracts will not be affected by the new policy.

Australia is accelerating the promotion of new regulations on carbon reduction and carbon trading, posing new compliance requirements for energy companies. On March 30, 2023, the Australian Parliament formally passed the "Safeguard Mechanism (Credit) Amendment Act". The Act stipulates that large industrial facilities already in operation should control their carbon emissions within the carbon emission baseline; before 2030, this carbon emission baseline will decrease by 4.9% each year, and after 2030, it will decrease by 3.285% each year. For newly developed LNG projects, it is required to adopt industry best practices for carbon emission reduction from the start date, and in unclear situations, it should not be lower than the industry average emission reduction requirements. For carbon emissions exceeding the baseline, Australian carbon credits can be purchased or offset in other ways; for carbon emissions below the baseline, they can be retained or transferred as carbon credits. The new Australian regulations aim to promote substantial emission reductions, contribute to achieving net-zero emissions by 2050, and assist in the implementation of its national energy transition strategy. Given the extensive carbon reduction commitments made by the international community and the carbon reduction-related legislative background in Australia since 2011, energy companies in Australia, including international oil companies and local companies, have proactively prepared in advance to adapt to this trend. In the short term, it may increase the additional production costs of projects already in operation and raise the economic threshold and company management level requirements for the development of new coal and natural gas projects; in the long term, it helps to enhance the price competitiveness of Australian LNG products with carbon–neutral labels in the international market, especially in OECD countries, and transfer carbon reduction costs.

Regional Foreign Cooperation Bidding and Asset Acquisition Situation

Bidding rounds are numerous, the area is wide, and the distribution of countries is diverse. In 2023, the region completed 32 rounds of exploration block bidding, authorized 104 exploration blocks, of which 32 were marine blocks; the authorized exploration block area was about 348,000 km^2, of which the marine block authorized area was 227,000 km^2, accounting for 65.2%. The authorized blocks in Asia–Pacific mainly include: 22 onshore blocks in India, 14 onshore blocks in China, 13 onshore blocks in Australia, 10 onshore blocks in Indonesia, 6 onshore blocks in Pakistan, 6 onshore blocks in Papua New Guinea, 14 offshore blocks in Malaysia, 6 offshore blocks in India, 6 offshore blocks in Indonesia, 3 offshore blocks in Thailand, etc. Currently, there are 15 rounds of exploration blocks in Asia–Pacific that are being bid and will end at the end of the year, with a total block area of 669,000 km^2, mainly distributed in India, Mongolia, Indonesia, Pakistan, Australia, China, Malaysia, and other countries. Asset acquisition business is light, Australia is relatively active. In 2023, the region's oil and gas asset acquisition transaction volume was less than 15 billion US dollars, of which Australia's upstream and liquefied natural gas transaction volume exceeded 7 billion US dollars, setting a new historical high. Australian M&A activities are mostly concentrated in the natural gas field, including: EIG and Brookfield's acquisition of Origin Energy, bp's acquisition of Shell's Browse giant natural gas project interests, mining giant Hancock Energy's acquisition of Warrego Energy, Taiwan's CPC's purchase of Santos Dorado project shares, Japan's LNG company (Sojitz and Sumitomo Corporation joint venture) acquisition of Woodside's Scarborough natural gas project 10% stake, Total Energies and Japan's Inpex acquisition of Thailand's upstream company PTTEP's interests in the Australian Cash/Maple oil field.

Main Risk Points of Investment in Key Resource Countries in the Region

In 2023, with the continuous recovery of the Asia–Pacific economy and the continuous advancement of energy transformation, the oil and gas industry faces certain challenges. The security situation in some countries in the region is still severe, with high security risks in Myanmar. Due to environmental assessment issues and labor disputes, there are continuous strikes in Australia, and social risks are rising. The political, security, and social situation in Myanmar continues to be complex. Ethnic issues in Myanmar are prominent, and related political disagreements and armed struggles occur from time to time. Since October 27, 2023, with the outbreak of war between the Kachin Independence Army-led local armed forces (ethnic armed forces) and the Myanmar government forces in northern Myanmar, armed conflicts have frequently occurred nationwide, plunging it back into a turbulent situation. Affected by this, the Myanmar government forces are facing unprecedented challenges, involving not only ethnic armed forces and other anti-military armed forces, but also telecom fraud groups in northern Myanmar, and the security situation in some areas is complex and severe. The risk of escalating sanctions in the energy sector in Myanmar is increasing. Since 2023, the West, led by the United States, has continuously expanded economic sanctions against Myanmar, often involving the energy sector. At the end of January, the United States for the first time listed the management of Myanmar Oil and Gas Enterprise (MOGE) and Myanmar's Minister of Energy, U Moe Myint Oo, as targets for sanctions; in April, it discussed sanctions against MOGE with allied countries; in late June, it imposed sanctions on Myanmar's Ministry of Defense, Myanmar Foreign Trade Bank, and Myanmar Investment and Commercial Bank; at the end of October, it announced that from December 15, it would prohibit U.S. citizens from directly or indirectly providing specific financial services to MOGE, including loans, account management, insurance, and investment. The European Union, Canada, Australia, and the United Kingdom have also imposed new sanctions on the Myanmar military, including the Myanmar Minister of Energy and Myanmar entities supplying aviation fuel to the Myanmar military. In addition, many Myanmar citizens, social organizations, and international non-governmental organizations have called for the international community to sanction the Myanmar military regime, with particular attention to oil and gas companies like MOGE that provide important sources of income for the military regime. If the political and security situation in Myanmar continues to be unstable, it is not ruled out that the West will continue to increase sanctions on Myanmar's energy sector.

Australian workers' strikes and environmental disputes have impacted the business environment. In November 2023, the Federal Court of Australia issued a temporary injunction requiring Santos to not lay the Barossa natural gas pipeline before a specified date, on the grounds that it had not sufficiently consulted with local indigenous groups about the project's environmental impact. The project's capital expenditure is $3.6 billion, and delays or shelving of the project could lead to a few percentage points of investment growth. In addition, workers at Chevron's Gorgon and Wheatstone LNG projects have gone on strike several times during the year due to disputes over treatment and working conditions, posing serious challenges to project production and operation.

International Oil Companies' Energy Cooperation Situation in Regions and Major Countries

Due to the advancement of energy transition, international oil companies have reduced offshore exploration activities. Major oil and gas companies operating in Australia, including Woodside, Chevron, Shell, and ExxonMobil, have abandoned several offshore exploration blocks. Apart from exploring near their production facilities, these companies are basically no longer deploying new exploration operations. The scenario of drilling 15–20 exploration wells offshore Australia each year may be hard to replicate.

Western oil companies continue to divest regional upstream assets. After Shell sold its SC38 natural gas assets in the Philippines, its shares in the Baram Delta EOR and SK 307 non-operating product sharing contracts outside the coast of Sarawak, Malaysia, and its 26.25% stake in the Queensland Curtis LNG (QCLNG) public facilities in Australia, it is now selling a 35% stake in the Badak LNG project in Indonesia; ExxonMobil continues to advance its upstream asset divestment plan, with assets mainly located offshore Malaysia; OMV of Austria is selling upstream assets, including a 50% stake in SapuraOMV Upstream Sdn in Malaysia and a 100% stake in OMV New Zealand Limited; Repsol of Spain is negotiating with the Indonesian government to abandon the Andaman III deepwater exploration block. The block is located in the North Sumatra Basin, covering an area of 4683 km^2, with operator Repsol owning a 51% interest and partner Petronas of Malaysia owning a 49% interest; Harbour Energy of the UK has decided to exit upstream operations in Vietnam, selling its 53.125% interest in the offshore oil and gas fields Chim Sao and Dua in the Nam Con Son Basin's 12W block to Big Energy of Vietnam for $84 million.

Some international oil companies are focusing on investing in the region's natural gas sector. Eni has further strengthened its position in the upstream natural gas sector in Indonesia, not only acquiring Chevron's interests in natural gas assets such as the Ganal block (Chevron holds 62%), Rapak block (Chevron holds 62%), and Makassar Straits block (Chevron holds 72%), but also increasing exploration efforts in Indonesian waters and discovering a large amount of natural gas. On October 3, Eni discovered a large amount of natural gas in the Geng A large amount of natural gas resources were discovered in the North-1 exploration well. According to preliminary estimates, the newly discovered geological reserves of natural gas are about 141.5 billion cubic meters, and the condensate oil reserves reach 400 million barrels. Shell is advancing oil and gas exploration and development activities in Malaysia that are primarily gas-based. In June 2023, Shell formulated an oil and gas exploration and development plan in Malaysia, including geophysical exploration of the production sharing contract blocks in the eastern waters of Malaysia; on August 22, the Timi gas field operated in the eastern waters of Malaysia was put into production. The gas field is about 200 km away from the coast, and the recoverable reserves of oil and gas total 177 million barrels of oil equivalent, of which the natural gas reserves are 28.3 billion cubic meters, accounting for 94.3%, and the peak production of natural gas will reach 8 million cubic meters per day.

International companies are actively laying out regional CCS/CCUS businesses. Many countries in the region are launching incentive measures and new regulations to seek cooperation in CCUS projects, and international oil companies are actively laying out in response. Under the push of operators Woodside and major shareholders bp and other partners, the Browse project in which PetroChina has a stake has determined to include the CCS project in the basic development plan and has completed the conceptual design and carbon back-injection block application. In March, ExxonMobil, Chevron and other international oil companies signed a preliminary agreement with Pertamina, the Indonesian national oil company, to develop a CCS center.

Chevron has the world's largest commercial CCS project in the Gorgon LNG project in Western Australia, and plans to invest in two low-carbon pilot projects in Western Australia, one of which is a soil carbon sequestration pilot project led by Carbon Sync, and the other is a research project in cooperation with Deakin University to explore potential carbon dioxide sequestration opportunities in the coastal wetlands of the region. Shell, SK Group, Samsung, Lotte and Petronas are jointly developing the Shepherd CCS project in Korea and Malaysia, which will capture carbon dioxide from Korean industrial facilities and transport it to Malaysia for sequestration. The Abu Dhabi National Oil Company (ADNOC) and Santos of Australia have signed a strategic cooperation agreement focusing on cooperation in the field of carbon capture and storage.

Outlook for 2024

Geopolitical risks remain high. The United States is constantly peddling the so-called "Indo-Pacific strategy" in the Asia–Pacific region, piecing together various closed and exclusive "small circles" to incite division and confrontation. Against the backdrop of the U.S. election year, it is possible that the U.S. will increase its efforts to stir up trouble in the Asia–Pacific region for domestic political needs, and the situation in the Taiwan Strait, the South China Sea, and the Korean Peninsula may be under pressure as a result. In addition, many countries in Asia–Pacific will have important elections, including the election of the president of the Liberal Democratic Party, the ruling party in Japan, the Korean parliamentary elections, the Lok Sabha (lower house of parliament) elections in India, the presidential elections in Indonesia, and the National Assembly (lower house of parliament) elections in Pakistan. These elections will have a significant impact on the political situation in the relevant countries, which in turn will have a spillover effect on the regional situation.

The economy will maintain relatively rapid growth. It is expected that the Asia–Pacific region will still maintain overall peace and relatively rapid economic growth in 2024. The International Monetary Fund predicted in October 2023 that the economic growth rate of the Asia–Pacific region will reach 4.2% in 2024, higher than the global economic growth rate of 2.9%. As the strong dollar era is nearing its end, the financial environment is improving. In 2024, the Monetary Authority of Singapore may adopt a more relaxed policy; affected by rising inflation, the central banks of Thailand and Bangladesh may remain on hold; the Reserve Bank of Australia will maintain a hawkish stance; the Bank of Japan is expected to end its negative interest rate policy; China will adopt a prudent monetary policy, maintain reasonable and sufficient liquidity, and match the scale of social financing and money supply with the expected targets of economic growth and price levels.

Oil and gas consumption and imports continue to grow. The main natural gas producing countries in the region, such as Indonesia, Malaysia, and Brunei, are under great pressure to supply natural gas due to the fact that the main producing gas fields have passed their peak production period and it is difficult to maintain growth in production. With the relatively rapid economic growth in the countries in the region and the strong demand for clean fuels, it is expected that the import volume of natural gas in the region will continue to grow in 2024.

(Authors of this section: Yu Gongming, Wang Yijun, Reviewer: Fu Lei)

North America

Development of the Oil and Gas Industry in the Region and Major Countries

Oil and Gas Production and Trends

Crude oil and natural gas production have increased significantly. In 2023, the regional oil production is estimated to be 1.129 billion tons, an increase of 5.6% year-on-year; natural gas production is estimated to be 13.42 trillion cubic meters, an increase of 3.8% year-on-year, reaching a historical high. Among them, the U.S. crude oil production is 847 million tons, an increase of 8.3% year-on-year; natural gas production is 11.525 trillion cubic meters, an increase of 3.3%. Canada's crude oil production is 282 million tons, a decrease of 1.7% year-on-year; natural gas production is 1.889 trillion cubic meters, an increase of 6.8% (Figs. 13 and 14). The main growth area of U.S. crude oil production is the Permian Basin, where oil production in the basin has led to a significant increase in associated gas production and has driven the growth of domestic natural gas production. The slight decrease in Canada's crude oil production is mainly due to the continuous forest fires in the west affecting the production of major oil and gas producing areas.

Trends in Oil and Gas Consumption, Imports and Exports

The consumption of refined oil and natural gas both increased slightly. In 2023, the regional crude oil processing volume is 911 million tons, which is flat compared to the previous year and still lower than the 2019 level (944 million tons). The consumption of refined oil and natural gas both maintained a slight increase. Among them, the consumption of refined oil is 716 million tons, an increase of 1.3% year-on-year; the consumption of natural gas is 10.2 trillion cubic meters, a decrease of 1.4% year-on-year. The growth in refined oil consumption mainly comes from gasoline and aviation kerosene, while diesel consumption continues to decline. Due to the reduction of extremely cold weather in the warm winter, the use of gas by

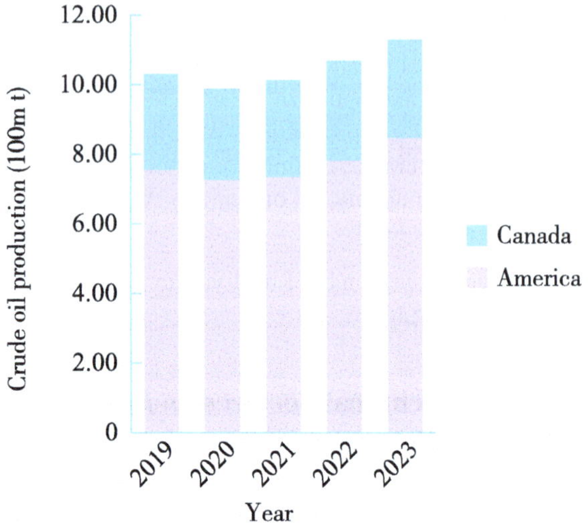

Fig. 13 North American crude oil production 2019–2023. *Data source* EIA, CER

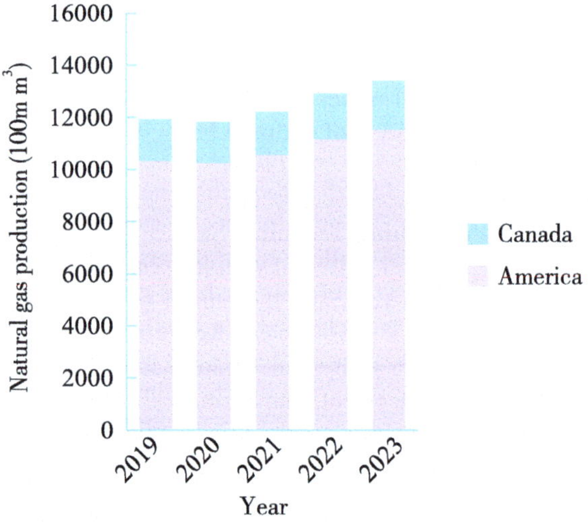

Fig. 14 North American natural gas production 2019–2023. *Data source* EIA, CER

U.S. residents and businesses has decreased compared to the previous year, but the commissioning of two large natural gas power plants has driven the growth of gas used for power generation (Figs. 15 and 16).

Crude oil and natural gas imports have dropped significantly, exports have increased significantly, and the proportion of intra-regional trade has decreased. In 2023, North America's crude oil imports are 329 million tons, a decrease of 8.6% year-on-year; exports are 522 million tons, an increase of 7.4% (Fig. 17). The growth of U.S. crude oil production has increased the supply to domestic refineries, and crude oil imports have dropped significantly. Canada's imports have continued to grow due

Fig. 15 North American refined oil consumption 2019–2023. *Data source* EIA, CER

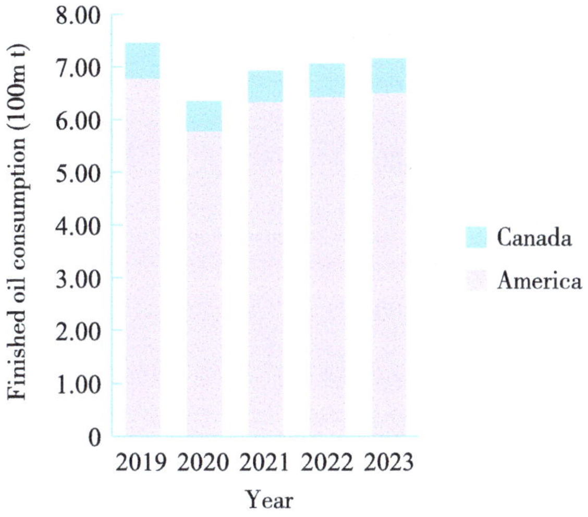

Fig. 16 North American natural gas consumption 2019–2023. *Data source* EIA, CER

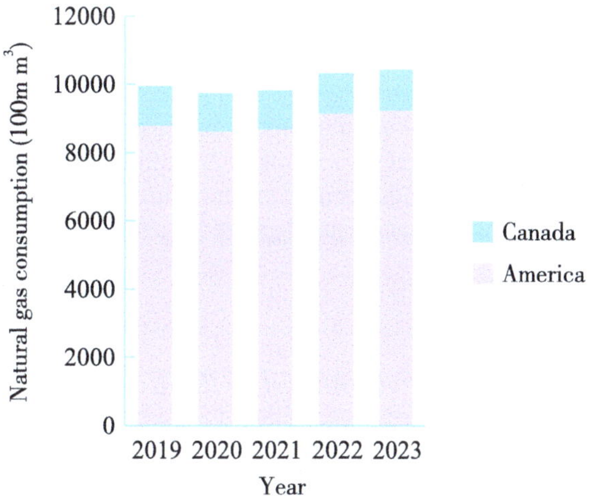

to the decline in crude oil production. As U.S. oil and gas production continues to grow rapidly, oil and gas exports continue to increase significantly, leading to a shift in the global oil and gas trade pattern. As the U.S. continues to rapidly increase its exports of crude oil and natural gas outside the region, the proportion of oil and gas trade between the U.S. and Canada has decreased, and Canada's dependence on exports to the U.S. is greater than the U.S.'s dependence on imports from Canada. If the Trans Mountain pipeline expansion project can be put into operation as scheduled in 2024, Canada's crude oil exports will become more diversified.

The U.S. has become a net exporter of both crude oil and natural gas. In 2023, U.S. exports of crude oil and natural gas continue to reach new highs, with crude

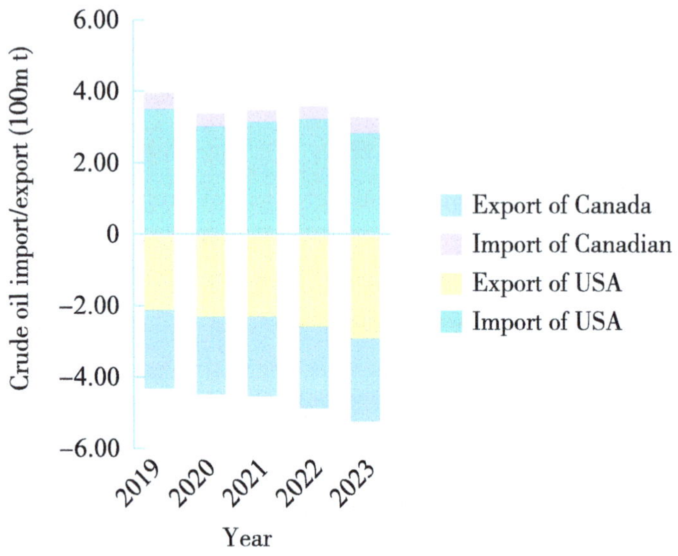

Fig. 17 North American crude oil imports and exports 2019–2023. *Data source* EIA, CER

oil exports of 289 million tons, an increase of 11.6% year-on-year, exceeding crude oil imports; natural gas exports of 2.1 trillion cubic meters, an increase of 7.6%, and imports of 820 billion cubic meters (Fig. 18). The U.S. has become a net exporter of both crude oil and natural gas and is a major supplier of oil and natural gas in the world.

Investment Environment in the Region and Major Countries

Comparison of Investment Environment in Important Oil and Gas Resource Countries in the Region

The political situation is generally stable, but party struggles are intensifying. In 2023, the election of the Speaker of the U.S. House of Representatives was chaotic, and the Speaker, McCarthy, who was elected after a historic 15 rounds of voting, was impeached less than 9 months after taking office. Subsequently, after several rounds of elections, radical Republican Mike Johnson was elected Speaker of the House, the Republican Party will turn further to the right, and the contradictions with the Democratic Party will become more prominent, which will hinder the smooth implementation of the Biden administration's policies.

The U.S. economy is growing moderately, but the risk of recession still exists, and the Canadian economy has experienced a technical recession. From March

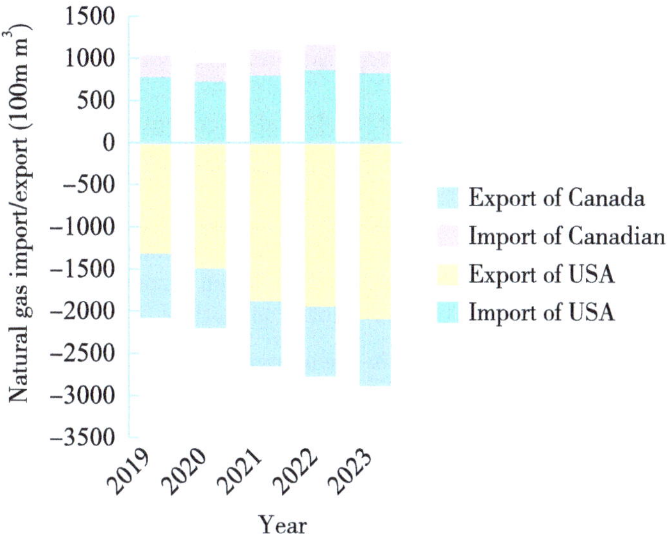

Fig. 18 North American natural gas imports and exports 2019–2023. *Data source* EIA, CER

2022 to July 2023, the Federal Reserve raised interest rates 11 times, accumulating a total increase of 525 basis points; the Bank of Canada raised interest rates 10 times, with the benchmark interest rate at its highest level since April 2001. Supported by the Chip Act and the Infrastructure Act, the United States significantly increased its fiscal expenditure, with fiscal expenditure in fiscal year 2023 reaching \$6.13 trillion, driving growth in both industrial construction investment and personal consumption in the United States. The World Bank estimates that the U.S. GDP growth rate in 2023 will be 2.5%. However, the U.S. consumer confidence index has been falling for four consecutive months, and the international credit rating agency Moody's has downgraded the outlook for the U.S. sovereign credit rating from "stable" to "negative", indicating a risk of recession in the U.S. economy. The impact of Canada's tight monetary policy on economic growth is more evident, with GDP growth in the first half of the year only 1.6%; according to data released by Statistics Canada, the actual GDP growth rate in the third quarter was − 0.3%, indicating that the economy has entered a technical recession. The IMF estimates that Canada's GDP growth rate in 2023 will be 1.3%, and does not rule out the possibility of negative growth.

Policies encouraging the development of low-carbon and new energy sectors have been introduced one after another, but the progress is slow. Under the support of the "Bipartisan Infrastructure Act" and the "Inflation Reduction Act", in 2023, various departments of the U.S. government have successively launched a number of plans and measures to promote carbon reduction and the development of new energy. The Department of Energy launched the "Grid Resilience and Innovation Partnership (GRIP) Program", with a total planned investment of \$10.5 billion, to enhance the resilience and reliability of the power grid, covering projects that fund microgrids and integrated energy storage technologies. The Department of Energy also announced several funding support plans for new energy and energy storage technology research and development projects. The Environmental Protection Agency (EPA), the Department of the Interior (DOI), and other departments have launched support plans for the development of clean energy. On June 5, the first national clean hydrogen strategy roadmap was released in the United States, outlining a sustainable development strategy for hydrogen production, transportation, storage, and application in the United States. On November 14, in conjunction with the release of the Fifth National Climate Assessment (NCA5), President Biden announced an investment of over $6 billion for community climate change construction.

Policies limiting carbon emissions in the oil and gas industry have been introduced, but implementation challenges exist. On July 24, the Canadian government published the "Canadian Inefficient Fossil Fuel Subsidy Government Self-Review Assessment Framework" and the "Canadian Inefficient Fossil Fuel Subsidy Government Guide", which will gradually phase out subsidies for fossil fuels to promote clean technology and emissions reduction in all sectors. On August 19, the Canadian federal government released a draft of the Clean Electricity Regulations, introducing a plan to achieve net-zero emissions in the power grid by 2035, but it faced opposition from oil and gas producing provinces, mainly Alberta, and the implementation of the bill faces significant challenges (Table 5).

Main Risk Points for Investment in Key Resource Countries in the Region

In 2023, there are constant frictions between China and the United States, and between China and Canada. Although the leaders of China and the United States have reached some consensus, the main tone of competition between China and the United States will not change. Optimistically, the relationship between China and Canada will also maintain the status quo under this situation. Chinese companies may still face potential political risks when investing in North America. Under the continuous pressure to address climate change, the United States and Canada will continue to increase restrictions on carbon emissions from the oil and gas industry, and investment in the oil and gas industry in both countries still faces certain policy and legal risks.

The Situation of International Oil Companies in Regional and Major National Energy Cooperation

The core oil and gas production areas in the United States are becoming more concentrated and large-scale. The two giant oil and gas mergers reached in 2023

Table 5 Comparison of investment environments in North American countries

Country	Politics	Economy	Policy
United States	The political situation in the United States remains stable, but the House of Representatives election is chaotic, partisan competition intensifies, and this struggle will become more intense in the election year of 2024	The U.S. economy, supported by growth in fiscal spending, performs better than expected, maintaining moderate growth, but the risk of recession still exists. Economic growth may continue to slow in 2024	The "Grid Resilience and Innovation Partnership (GRIP) Program" has a total investment of $10.5 billion and is implemented in 58 projects across 44 states in the United States. The first phase allocates $3.46 billion, which will be used to deploy over 35 GW of additional renewable energy capacity and build 400 independent microgrids. The second phase will allocate $3.9 billion for grid modernization construction The Environmental Protection Agency (EPA) plans to provide $2 billion in funding through its Environmental and Climate Justice Community Change Grants program to support community-led clean energy projects The Department of the Interior (DOI) has allocated $100 million for upgrading water infrastructure On June 5, 2023, the first national clean hydrogen strategy roadmap was released in the United States, with an investment of $7 billion to establish seven regional hydrogen production centers in the United States and assess the potential contribution of hydrogen to the decarbonization goals of various sectors over the next 30 years On November 14, 2023, Biden announced an investment of over $6 billion for community climate change construction, including strengthening the aging grid infrastructure in the United States, reducing flood risks in communities, supporting conservation work, and promoting environmental justice
Canada	The political situation in Canada is stable, but competition between parties continues. The current Prime Minister's approval rating is declining, which will lead to more intense party struggles before the next election	Canada's interest rate policy to curb inflation has had a significant impact on economic development. The economy grew slowly in the first half of the year and has entered a technical recession in the second half	The "Canadian Inefficient Fossil Fuel Subsidy Government Self-Review Assessment Framework" and the "Canadian Inefficient Fossil Fuel Subsidy Government Guide" will gradually eliminate fossil fuel subsidies, with federal support for oil, natural gas, and coal production expected to be reduced by $1 billion each year. The federal government has launched a net-zero emissions plan for the power grid by 2035, which Alberta Premier Danielle Smith has explicitly stated she will not comply with, believing that the federal government's plan to reduce emissions from electricity, oil, and natural gas production infringes on provincial constitutional rights

have made the pattern of the core oil and gas production area in the United States—the Permian Basin more concentrated and large-scale. The acquirers are the two largest oil giants in the United States, ExxonMobil and Chevron, and the acquired are Pioneer Natural Resources Company and Hess Corporation, respectively. Chevron and Pioneer Natural Resources Company are the first and third largest shale oil producers in the Permian Basin, and Hess Corporation is the fourth largest oil and gas producer in the United States. After the merger, the oil and gas assets in this area are more concentrated in the hands of large oil companies. Through mergers and acquisitions, adjacent oil and gas assets will form a contiguous scale, which will also drive the cost of barrel operations to further decrease, and the scale advantages of large oil companies will become more apparent. For example, after ExxonMobil acquired Pioneer Natural Resources Company, the proportion of oil and gas production in unconventional areas will account for about 40% of the company's total production, and drive the company's barrel operation costs to decrease by at least 5%.

Outlook for 2024

Party struggles will become more intense. 2024 is the year of the U.S. presidential election, and the political struggle between the two parties will become more chaotic. The current leading candidates are still Biden and Trump, the former's approval rating continues to decline and faces impeachment investigation, the latter faces legal challenges, and the possibility of other candidates cannot be ruled out. The political situation and government operations will become more uncertain. Canadian Prime Minister Trudeau's approval rating is declining, and domestic party competition will become more intense before the new round of elections in 2025. In the U.S. election year, the policy towards China inevitably becomes a focus, Sino-U.S. relations will face frequent fluctuations, but there will be no major turn overall.

The U.S. economy is heading for a soft landing, and the Canadian economy is at risk of recession. In 2024, the impact of the U.S.'s fiscal expansion policy will decrease, but the reflow of manufacturing will continue, and the U.S. economy is heading for a soft landing. The growth of the Canadian economy is not optimistic. After the inflation rate is effectively controlled, the central bank plans to implement a rate cut policy, but due to concerns about a sharp rebound in house prices, the timing and intensity of the rate cut policy are difficult to determine, and it cannot be ruled out that the Canadian economy will fall into recession.

Oil and gas production will continue to grow, but the growth rate will be limited. In 2024, the oil and gas production in North America will continue to grow, with the growth in crude oil production contributed by the United States and Canada, and the growth in natural gas production mainly coming from the United States. Under the current international oil price level and international market demand, U.S. crude oil and natural gas producers will continue to strive to increase production.

The expansion project of the Canadian Trans Mountain pipeline is expected to be completed and put into operation in 2024, increasing Canada's crude oil export capacity and significantly increasing oil sands production. However, major U.S. oil companies will still focus on improving shareholder returns and cash flow, and will remain restrained in increasing investment and production without encouraging policies from the government. The growth rate of U.S. oil and gas production may be less than that in 2023.

Oil and gas exports will continue to grow. With the advancement of energy transition, U.S. refined oil consumption will begin to decline, domestic refinery operating rates will fall, some refining capacity will continue to shut down, and domestic crude oil demand will decrease; the contrast between high international natural gas prices and persistently low North American natural gas prices will encourage natural gas producers to export to international markets to earn more profits, and U.S. crude oil and natural gas exports will continue to grow. However, the continuous drought in Panama in 2023 led to a significant reduction in the capacity of the Panama Canal, and the Panama Canal Authority imposed restrictions on passing ships, requiring ships to pay an advance passage fee, which led to a sharp increase in the cost of oil and gas transportation from the United States to Asia, and U.S. oil and gas exports will be affected in the short term. Due to the significant increase in export pipeline capacity, Canadian crude oil exports will see a significant increase, with the main target market being Asia.

The trend of mergers among large U.S. companies will continue. In recent years, resource integration centered on the Permian Basin resource area in the United States has been ongoing. After ExxonMobil and Chevron completed two giant acquisitions, the merger wave of large independent oil companies with assets in this region as their core business will continue. At the beginning of 2024, APA Corporation and Callon Petroleum reached an agreement, with APA acquiring Callon in an all-stock transaction valued at approximately $4.5 billion. Callon's assets complement APA's assets in the Permian Basin, supporting APA's scale expansion in the region. Top producers in the Permian Basin, such as Devon Energy and natural gas producer Chesapeake Energy, are also interested in acquisitions, and mergers between large independent oil companies may occur in 2024.

(Written by: Jin Huandong Reviewed by: Fu Lei)

Europe

Development of the Oil and Gas Industry in the Region and Major Countries

Oil and Gas Production and Trends

In 2023, crude oil production increased slightly, while natural gas production decreased slightly. Regional crude oil production is estimated to be 162 million tons, an increase of 3.65% year-on-year; natural gas production is estimated to be 1881.21 billion cubic meters, a decrease of 1.47% year-on-year. Oil and gas production is mainly concentrated in Norway and the United Kingdom. Norway's oil and gas production were 103.99 million tons and 114.6 billion cubic meters, respectively, with oil production increasing by 5.78% year-on-year and natural gas production decreasing by 4.10%; UK oil and gas production were 38.85 million tons and 36.5 billion cubic meters, respectively, decreasing by 2.06 and 2.00% year-on-year. The development of the Sakarya gas field in the Black Sea of Turkey has driven natural gas production to increase significantly from 400 million cubic meters in 2022 to 40 billion cubic meters, becoming the main source of increase in natural gas production in Europe (Figs. 19 and 20).

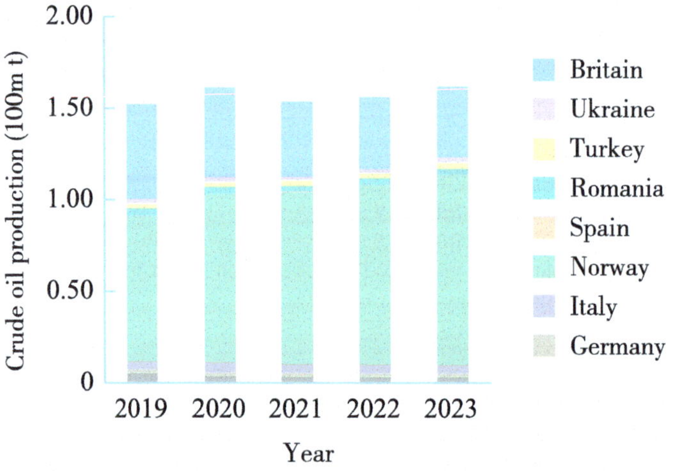

Fig. 19 Crude oil production in Europe 2019–2023. *Data source* Fitch Solutions

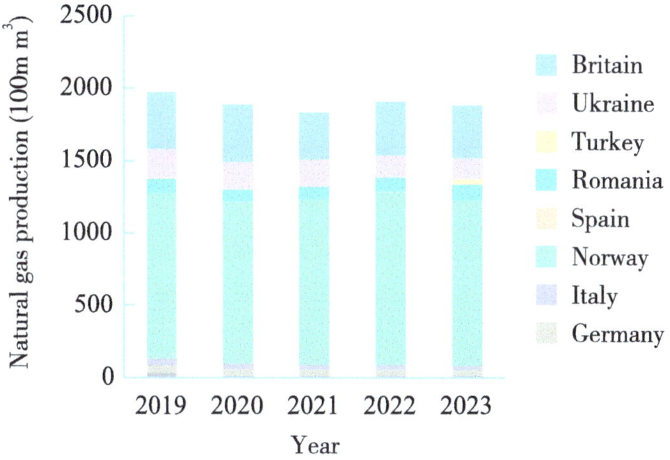

Fig. 20 Natural gas production in Europe 2019–2023. *Data source* Fitch Solutions

Oil and Gas Consumption, Imports and Exports, and Trends

In 2023, both oil and natural gas consumption declined, and the production of refined oil also showed a downward trend. The regional natural gas consumption was 3498 billion cubic meters, a decrease of 5.99% from the previous year. Among them, due to the decline in natural gas demand in the industrial, household, and power sectors, Italy's natural gas consumption is estimated to be at its lowest level since 1997, a decrease of 8.12% from the previous year; France and Spain's natural gas consumption decreased significantly, estimated to decrease by 8.85% and 15.41% from the previous year, respectively. The high natural gas prices have dragged down industrial activities in Germany and the UK, with estimated annual decreases in natural gas demand of 8.15% and 5.56%, respectively. According to current natural gas consumption data in various countries, France and Italy have achieved the EU's target of reducing natural gas demand by 15%. Regional refined oil consumption was 325 million tons, a decrease of 0.21% from the previous year. The decline in regional refined oil demand, coupled with a decrease in refining profits, as well as increased competition in the refining industry from Russia, Asia, and the Middle East, put pressure on the refining industry, and regional refined oil production decreased by 2.91%.

In 2023, both the volume of oil imports and the net volume of natural gas imports declined, with the structure of natural gas imports undergoing a change. The region's crude oil imports were 337 million tons, a decrease of 4.53% from the previous year. It is expected that due to the decline in refinery processing demand, crude oil imports will further decline in 2024 (Fig. 21). The region's natural gas imports increased slightly, but the net import volume decreased from 1565.51 billion cubic meters in the previous year to 1459.09 billion cubic meters (Fig. 22).

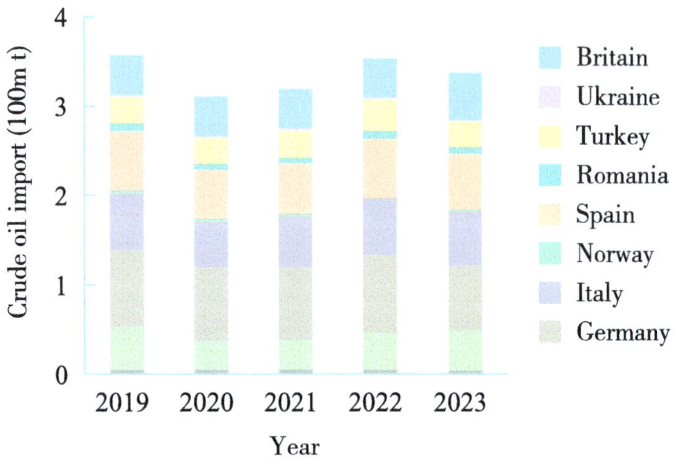

Fig. 21 Crude oil imports in Europe from 2019 to 2023. *Data source* GTT

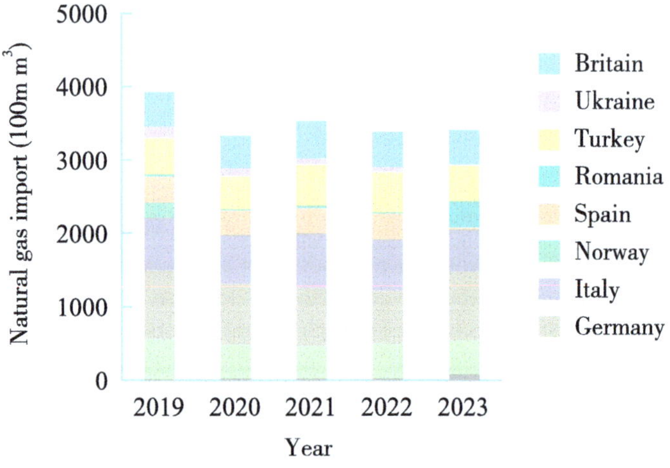

Fig. 22 Natural gas imports in Europe from 2019 to 2023. *Data source* Joint oil data initiative

In 2023, dependence on Russian oil and gas decreased, and the sources of oil and gas imports diversified. Since the beginning of 2023, when the EU fully banned Russian oil imports, Germany's crude oil import sources have diversified. EU statistical bureau data shows that in the first half of the year, the proportion of imports from Norway was 18.4%, from the United States was 15.7%, and from the UK was 13.6%. Due to the significant reduction in natural gas exports from Russia to the EU, Norway has replaced Russia as the EU's largest pipeline natural gas supplier. European countries have increased LNG imports, mainly from the United States, to offset the decline in pipeline natural gas imports. Germany has signed long-term supply contracts for LNG imports with the United States and Oman, and Italy's LNG imports are expected to reach a record 155 billion cubic meters. In addition, the EU launched a natural gas procurement plan before the arrival of winter, and the natural gas storage levels of various countries are close to full, providing a buffer for the impact of winter natural gas supply.

Investment Environment in the Region and Major Countries

Comparison of Investment Environment in Important Oil and Gas Resource Countries in the Region

Contractionary monetary policy to curb inflation has increased the risk of economic recession. The European Central Bank has raised interest rates 10 times in a row since July 2022, with a total increase of 450 basis points, and the deposit mechanism interest rate has reached 4%, the highest level since the establishment of the euro area. The Bank of England has raised interest rates 14 times in a row since December 2021, with the key interest rate rising to 5.25%, the highest level in 15 years. Signs of recession have appeared in the European economy, with the euro area's GDP falling by 0.1% quarter-on-quarter in the third quarter, and the UK's economy contracting in the third quarter, with GDP revised down by 0.1% quarter-on-quarter. As the inflation rate has not yet fallen back to the expected target, interest rates in European countries will remain at a high level in the short term. An IMF report shows that the European Central Bank should continue to maintain a relatively high interest rate of around 4% in 2024 to bring the price level back to the target level, and the continued contractionary monetary policy will expose the European region to the risk of economic recession.

The EU's policy towards China will continue to maintain a dual strategy. The EU emphasizes the importance of cooperation with China, while also imposing restrictions on China in certain areas. Since the beginning of 2023, the common interests between China and the EU have prompted both sides to actively engage in exchanges and cooperation. On December 7, the 24th China-EU Leaders' Meeting was held in Beijing. The two sides had in-depth exchanges on strategic issues in China-EU relations, dialogue and cooperation in various fields, key concerns of both sides, global governance, and international and regional hotspots. European Council President Michel and European Commission President von der Leyen stated that the EU and China are important economic and trade partners, with increasingly broad areas of cooperation. Managing and developing China-EU relations directly affects the interests of the people of both sides and world prosperity and security. However, the EU has successively introduced the "Chip Act", "Key Raw Materials Act", "Foreign Subsidies Regulation", and "Economic Security Strategy" and other regulations to strengthen regulatory review. The implementation of "de-risking" and diversification policies has a certain targeting towards China, and it has launched anti-subsidy investigations into China's electric vehicle and steel manufacturing industries in strategic coordination with the United States.

The wave of strikes has exacerbated the economic predicament. In 2023, several European countries experienced strikes, involving multiple industries, and even multiple rounds of continuous strikes. France erupted in multiple rounds of strikes involving multiple industries due to pension reform, and the "May Day Parade" was considered "one of the largest social movements in history". Germany experienced repeated strikes by postal, transportation, and medical institution workers, and the public transportation union organized the largest nationwide strike in nearly 30 years. Since March 2023, about 400,000 people in Spain have participated in protest activities, reaching the largest scale in history. The strikes in various countries have caused serious economic and social losses.

The oil and gas-related policies of some European countries lack stability. In 2022, the UK introduced an energy profit tax, levying an additional 25% tax on oil and gas, with plans to end it in 2025. However, in the budget autumn statement of the same year in November, it was announced that the energy profit tax would be increased to 35% from January 1, 2023, and the deadline would be extended to the end of March 2028. In June 2023, to alleviate the concerns of oil and gas companies about the profit tax, a price floor for the profit tax was set, and no 35% profit tax needs to be paid when the oil and gas prices are below this level. At the same time, there is a significant disagreement between the Conservative Party and the Labour Party on oil and gas exploration and development, and the results of the UK general election will affect the prospects of UK oil and gas investment (Table 6).

Regional Foreign Cooperation Bidding and Asset Acquisition Situation

Limited growth space for upstream investment. Among the major resource countries in Europe, only Norway issued new licenses in 2023. Denmark's decision to end all oil and gas exploration and development in the North Sea by 2050 will suppress upstream investment. As a mature oil and gas production basin, the exploration licenses issued by the UK in 2022 did not reverse the declining trend of exploration activities. In January 2023, Norway issued 47 offshore oil and gas exploration licenses to 25 companies, distributed in the North Sea (29), the Norwegian Sea (16), and the Barents Sea (2). Among them, Equinor received 26 licenses, Aker BP received 17 licenses, VarEnergi received 9 licenses, and Wintershall Dea received 11 licenses. Europe has good potential for the development of unconventional natural gas such as coalbed methane and shale gas, but most countries' governments have banned hydraulic fracturing for unconventional resource development, and the investment space for unconventional resource development is also limited.

Significant increase in investment in new and upgraded regasification facilities. In September 2023, Europe's under-construction natural gas regasification output

Table 6 Investment environment of major oil and gas resource countries in Europe

Country	Politics	Economy	Policy
Norway	The Labour Party lost in the local elections held on September 11, 2023, and the Conservative Party became the party with the most votes in this local election	The inflation rate in October 2023 was 3.34%, and inflation slowed down. Interest rates were raised by 25 basis points in September and again by 25 basis points in December. The manufacturing PMI was 47.94, below the boom-bust line, and the IMF predicts that the real GDP growth rates for 2023 and 2024 will be 2.3% and 1.5% respectively	In 2021 and 2022, the center-left government gave up exploration incentives and included more border areas to ensure support for the annual budget proposal. Political opposition to border exploration is growing, especially in the Barents Sea, where the Socialist Left Party is seeking to permanently remove the area from future exploration
UK	The UK government plans to hold a general election in the autumn of 2024, which may affect future oil and gas policies	The inflation rate in September 2023 was 6.70%, still at a relatively high level. The Bank of England paused interest rate hikes in November, and the interest rate hike cycle is nearing its end, but interest rates will remain at a relatively high level in the short term. The initial value of GDP quarter-on-quarter in the third quarter was 0%, and the economy stagnated. The IMF predicts that the real GDP growth rates for 2023 and 2024 will be 0.5% and 0.6% respectively	From January 2023, the energy profit tax will be increased to 35%, making the overall tax rate of the UK oil and gas industry reach 75%. In June 2023, a price floor for the profit tax was set. When the oil and gas prices are below this level, no 35% profit tax needs to be paid. The proportion of oil and gas producers applying for investment subsidies has been reduced from 80 to 29%

reached 54 million tons/year, a significant increase from 22.7 million tons/year in February 2022. S&P Global predicts that from 2021 to 2027, Europe will have an additional 90 million tons/year of regasification capacity put into operation, a significant increase from the forecast value in February 2022. Germany is the main country for new regasification capacity, and the government supports the acceleration of regasification facility construction projects with multiple incentives such as allowing projects to start construction before obtaining all permits. Italy has a relatively diverse source of natural gas imports, has a weak desire to build large onshore receiving stations, and mainly improves receiving station capacity through optimizing operations or small-scale infrastructure upgrades.

M&A transactions are relatively active compared to global trends. In 2023, the number and value of global mergers and acquisitions showed a downward trend, remaining below the historical average, but the number and value of transactions in Europe were relatively high. In the first quarter, Europe had two large-scale transactions, namely Equinor's acquisition of Suncor Energy's remaining investment portfolio in the UK for $600 million plus contingent payments, and the sale of a 28% stake in the Statfjord field in the North Sea of Norway for $220 million plus contingent payments. In the second quarter, there were 13 transactions in Norway with high transaction values, including Var Energy's acquisition of Neptune Energy Group Ltd.'s Norwegian business for $2.3 billion, Kistos PLC's agreement to acquire Mime Petroleum of Norway to enter the Norwegian continental shelf market, Attica Exploration AS's acquisition of Concedo, and Japan Petroleum Exploration's announcement of a joint venture with Longboat Energy PLC's Norwegian subsidiary, among others.

Key Risks of Investment in Regional Resource Countries

Intensified risk of economic recession in Europe. In 2023, the European economy was sluggish, with the EU's year-on-year economic growth rates for the first three quarters being 1.1%, 0.4%, and 0.1% respectively, and the Eurozone's year-on-year economic growth rates being 1.2%, 0.5%, and 0.1% respectively, all showing a significant slowdown. The final PMI for the Eurozone in October was 46.5, the lowest level since November 2020, and below the boom-bust line for five consecutive months. Among them, the final PMI for the service sector fell from 48.7 in September to 47.8, shrinking for the third consecutive month and hitting a 32-month low; the final PMI for the manufacturing sector fell from 43.4 in September to 43.1, below the boom-bust line for the 16th consecutive month. Both the service and manufacturing sectors continued to contract, increasing the likelihood of a recession in the Eurozone.

Adjustments to the tax system may increase the cost of oil and gas investment. In 2023, Norway abolished the tax incentives for oil and gas investment during the COVID-19 pandemic, and the tax on oil and gas investment returned to its previous higher level. The UK will increase the energy profit tax to 35% from January 2023, making the overall tax rate for the oil and gas industry 75%, one of the highest in the world. Some oil and gas companies have indicated that they will reduce their investments in the North Sea of the UK. For example, Total plans to reduce its investment in the North Sea of the UK by 25% in 2023. In addition, from October 1, 2023, the EU's Carbon Border Adjustment Mechanism (carbon tariff) will begin trial operation, which may be extended to the oil and gas industry in the future, increasing costs and squeezing profits in the oil and gas industry.

Environmental organizations influence oil and gas investment through various legal means. Some environmental organizations analyze the annual reports and other public information of oil and gas companies, and publish research reports questioning the green commitments of oil and gas companies. For example, Greenpeace released an analysis report, arguing that the investment in green energy by 12 oil companies including Shell, Total, and bp is only a small part of their investment in fossil fuels, accusing these companies of "greenwashing". They prevent investment and development in oil and gas fields through legal action. For example, in September 2023, Greenpeace took legal action against the government's approval of projects on the grounds that the Jackdaw oil and gas project and the Cambo oil field development do not comply with the UK's climate commitments.

The Situation of International Oil Companies in Regional and Major National Energy Cooperation

Divestment and divestiture of non-core oil and gas assets. bp plans to divest $2 billion to $3 billion in 2023; in early 2023, Repsol sold 25% of its oil and gas exploration department to the American fund EIG for about $5 billion; Lukoil plans to sell its gas stations in Romania.

Promoting the clean transformation of assets. Total is transforming the Grandpuits refinery into a bioplastics factory, while adding a desulfurization unit and a hydrogen unit at the Donges refinery. In the future, international oil companies may actively promote the transformation of refineries, or turn to biofuels, to benefit from the growing space for sustainable fuel development.

Expanding layout in the field of renewable energy. bp plans to allocate more than 40% of its capital expenditure to energy transition by 2024, has acquired a 40% stake in the UK's Viking carbon capture and storage project, has invested about 50 million euros in the Pangbourne Laboratory Center for testing electric vehicle batteries, and is cooperating with Iberdrola to produce green hydrogen in the UK, Spain, and Portugal. Shell also values the diversification of its assets in the UK from the oil and gas industry.

The business of oil and gas service companies is gradually transitioning. This can be divided into four transition directions: one is to divest most of the oil and gas related business, and the main business turns to low-carbon activities, that is, a complete transformation into a clean energy service company; the second is to still focus on oil and gas business, and take low-carbon measures under the mandatory requirements of customers or national regulations; the third is to increase low-carbon projects, develop low-carbon products, but give equal weight to low-carbon business and oil and gas business; the fourth is to provide decarbonization services according to customer needs.

Outlook for 2024

The Ukraine crisis will have a lasting and profound impact on Europe. It is predicted that by 2024, European inflation will not be able to fall back to the target range, and the economic outlook is still not optimistic. Industrial countries such as Germany may passively experience a new round of deindustrialization, with energy prices soaring and increasing manufacturing costs. A report from the European Commission shows that about a quarter of German small and medium-sized enterprises are considering relocation. The prospects for Europe's strategic autonomy are dim, and the interests of countries with the United States in the fields of energy, food and other bulk commodities are more closely tied, and achieving strategic autonomy will face more obstacles.

The energy crisis cannot be alleviated in the short term. The stability of the EU's natural gas supply is still fragile. Firstly, countries such as Spain and France still depend on Russia for LNG imports. With the prolongation of the Ukraine crisis and the US sanctions on Russian LNG, there is a risk of supply in the future; secondly, the escalating conflict between Palestine and Israel may pose a threat to Europe's LNG imports from the Middle East. In November 2023, the European Commission passed the "State Aid Temporary Crisis and Transition Framework" amendment, extending the state subsidy policy to June 2024 to cope with the impact of the Ukraine crisis and rising energy prices. It will take longer for Europe to achieve energy self-sufficiency, and it will continue to face the risk of an energy crisis in the short term.

Continuously promoting the acceleration of energy transformation. The EU has continuously introduced policies and regulations to promote energy transformation. In 2019, the European Commission issued the "European Green New Deal", incorporating the EU's goal of achieving carbon neutrality by 2050 into the legal framework, and ensuring the achievement of the carbon neutrality goal through legislation. After the Ukraine crisis, the EU proposed the "RePowerEU" energy plan in 2022 to accelerate the energy transformation in Europe. In October 2023, the new Renewable Energy Directive was approved by the European Council, which significantly increased the proportion of renewable energy in total energy consumption from the previous 32% to 42.5% by 2030, and added an indicative supplement of 2.5% to achieve a target of 45%. After the implementation of the new Renewable Energy Directive, renewable energy such as wind and solar energy will develop rapidly, and the demand for green hydrogen will also increase significantly.

Pay more attention to domestic oil and gas production. As an important natural gas supplier to the EU, Norway approved 19 new oil and gas projects in 2023, and the future oil and gas production has room for increase. The goal of the Sakrya gas field development project in Turkey is to increase Turkey's natural gas production to 18 billion cubic meters by 2031, and the future natural gas production is on the rise. Italy's Selva gas field began production in 2023, and the Argo-Cassiopea gas field is expected to start production in 2024, and the production in 2024 may increase significantly. Countries such as Germany have proposed to reconsider the use of hydraulic fracturing to increase domestic natural gas supply, but due to environmental pressures and other reasons, the possibility of lifting the ban on hydraulic fracturing in the short term is small.

Continue to actively carry out energy diplomacy. Leaders of countries such as Germany, the United Kingdom, and Italy have made multiple visits to important resource countries in the Middle East and Africa to carry out energy diplomacy. For example, in January 2023, the new Italian Prime Minister Giorgia Meloni officially visited Algeria and Libya, and in October, German Chancellor Scholz visited Nigeria and Ghana. It is expected that in 2024, European countries will continue to actively carry out energy diplomacy, strengthen energy trade relations with important resource countries, and promote upstream oil and gas investment and oil and gas trade.

(This section was written by: Duan Yixuan Reviewer: Fu Lei)

International Oil and Gas Cooperation

Overview of Overseas Investment Cooperation of Chinese Oil Companies

Current Situation and Progress of Overseas Oil and Gas Cooperation of Chinese Oil Companies

Steadily Advancing High-Quality Exploration and Development, Overseas Oil and Gas Equity Production is Stable and Increasing

In 2023, Chinese companies focused on key areas for increasing production, accelerated the pace of production, and overseas oil and gas equity production was stable and increasing, reaching 188 million tons, an increase of 1.9% over the previous year. Among them, crude oil production was 149 million tons and natural gas production was 480 billion cubic meters. The increase in production mainly came from the production of new projects in countries such as Brazil, Guyana, and Australia. Looking at the situation of each company, CNOOC's overseas equity production was 30 million tons, an increase of 11% over the previous year, mainly benefiting from the production of Guyana's Liza Phase II, Payara project, and Brazil's Buzios 5, Brazil's Mero 2 projects and other projects. PetroChina's stable production was 102 million tons, which was stable compared with the same period last year, mainly benefiting from the production of a batch of landmark projects such as Mozambique, Niger Phase II, and the high-level operation of main projects in countries such as Abu Dhabi, Iraq, Kazakhstan, Turkmenistan, and Chad. Sinopec's annual production was 35 million tons, basically the same as the previous year, mainly benefiting from the increase in production of Egyptian projects to make up for the decline in production of other projects (Fig. 1). With the efficient development of key projects in countries such as Iraq, the production of private enterprises such as United Energy is gradually increasing, with an annual increase of 1.5%.

X. Qian et al., *Global Oil and Gas Industry Development Report: 2023*,
https://doi.org/10.1007/978-981-95-2375-7_10

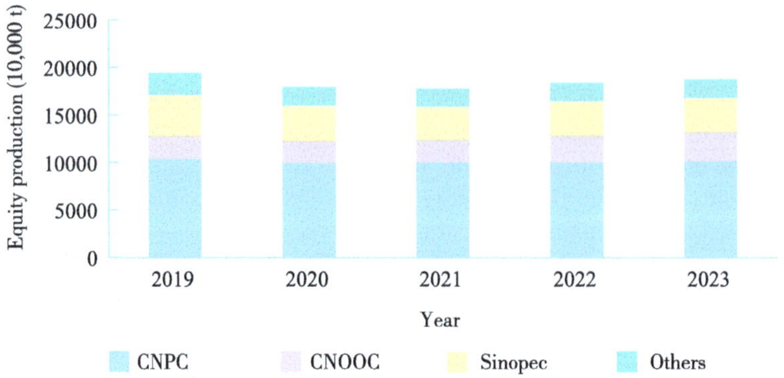

Fig. 1 2019–2023 China's petroleum enterprises overseas oil and gas equity production. *Data source* Company reports

Accelerate the Advancement of Midstream and Downstream Businesses, and Effectively Coordinate the Development of the Entire Industry Chain

Chinese enterprises pay more attention to the coordinated development of the entire industry chain of overseas businesses. In 2023, while stably operating upstream businesses, Chinese enterprises accelerated the advancement of LNG, refining, and pipeline businesses. The Coral Floating LNG project in Area 4 of Mozambique by PetroChina started trial production, and the second phase of the Niger project overcame various difficulties and was put into operation on schedule. In the midstream and downstream fields, Sinopec invested in the largest polyethylene project in Kazakhstan, the second phase of the Hengyi Petrochemical Brunei project received preliminary approval from the government, and after the East Africa Crude Oil Pipeline project was confirmed to be financed by several Chinese banks, various construction agreements were signed one after another, and the progress was accelerated.

Continue to Promote Asset Optimization, and the Asset Structure Continues to Improve

Chinese enterprises continue to accelerate the disposal of inefficient and high-cost assets, while strengthening the injection of high-quality assets. In 2023, Chinese enterprises completed nearly 20 asset transactions (Table 1), sold projects such as Australia's Arrow and Indonesia's Madura, and the total quantity and amount of assets sold increased.

Actively strive for the extension of high-quality assets and the acquisition of high-quality new projects. PetroChina has replaced ExxonMobil as the main operator of the West Qurna 1 oil field in Iraq; the PK project in Kazakhstan and the Akjubin oil contract have been extended. Sinopec has amended the terms of its project contract with Algeria through negotiations, and the new contract for the Zarzaitine oil field

Table 1 Major overseas oil and gas asset transactions of Chinese petroleum enterprises in 2023

Enterprise	Asset name	Transaction type
Sinopec	Qatar North Field Expansion	New project
Sinopec	Qatar North Field Expansion phase II	New project
Sinopec	Algeria Zarzaitine Oilfield	Old project Optimization cooperation terms
Sinopec	Kazakhstan Polypropylene Project	New project
PetroChina	Qatar North Field Expansion	New project
PetroChina	Ecuador Block 94	New project
PetroChina	Iraq West Qurna 1	Old project succession becomes operator
PetroChina	Kazakhstan PK Project No. 39 Oil Contract	Contract extension
PetroChina	Kazakhstan Aktobe No. 76 Oil Contract	Contract extension
Continental oil and gas	Iraq Southern Oil and Gas Field	New project
United energy	Iraq Southern Oil and Gas Field	New project
PetroChina	Tajikistan Kebogda Project	Divestiture
PetroChina	Australia Bowen Project	Divestiture
PetroChina	Indonesia Madura Project	Divestiture

Data source Rystad Energy, company announcements, etc.

has officially come into effect. Sinopec and PetroChina have each invested in the expansion project of the North Field in Qatar, and Sinopec has also signed the Phase II agreement for the capacity expansion project of the North Field in Qatar. Intercontinental Oil and Gas and United Energy Company have each won bids for the exploration and development project of the southern oil and gas field in Iraq. The new cooperation agreement provides strong support for further optimizing the structure of overseas assets.

Characteristics of Overseas Oil and Gas Cooperation of Chinese Petroleum Enterprises

Highlighting the Countries and Regions Along the "Belt and Road", International Oil and Gas Cooperation Has Achieved New Breakthroughs

Chinese enterprises actively use the national energy diplomacy platform, focus on the "Belt and Road" route, and achieve new results in deepening energy cooperation.

In 2023, Chinese enterprises earnestly implemented the achievements of the "China-Arab States Summit" and the "China-Gulf Cooperation Council Summit", leveraged cooperation with key resource countries along the "Belt and Road", and successfully extended international cooperation to key resource countries such as Qatar. A series of new cooperation agreements have been signed with Qatar and other countries, expanding cooperation between China and Arab countries, and China and the Gulf in oil and gas development and trade, and international energy cooperation has reached a new level. Among them, half of the new projects are concentrated in Arab countries such as Qatar and Iraq. In addition, PetroChina has reached the first cross-border RMB settlement transaction for LNG trade with GCC countries, promoting the internationalization of the RMB. With the "China-Central Asia Summit" and the third "Belt and Road" International Cooperation Summit Forum as windows, PetroChina, Sinopec, CNOOC, China Energy Construction and other central enterprises have reached important energy cooperation agreements with Central Asian countries and Brazil, and signed multiple cooperation agreements and memorandums of understanding in the fields of oil and gas exploration and development, downstream refining, oilfield services, and oil and gas trade, further expanding cooperation in traditional energy fields such as oil and natural gas, and strengthening cooperation in renewable energy fields such as hydropower, solar energy, and wind energy. Sinopec, CNOOC, and PetroChina have each determined a number of intentions and plans to deepen cooperation with Niger, Tanzania, Mozambique, and other countries. Rongsheng Petrochemical and Saudi Aramco plan to deepen cooperation through equity swaps, and are discussing Rongsheng Petrochemical's acquisition of 50% of Saudi Aramco's Jubail Refining Company and Saudi Aramco's acquisition of no more than 50% of Ningbo Zhongjin Petrochemical Co., Ltd., a subsidiary of Rongsheng Petrochemical. The synergy and quality of international cooperation continue to rise.

Emphasizing Risk Control and Business Synergy, the Quality of International Energy Cooperation Continues to Improve

In 2023, Chinese enterprises followed the practice of recent years, strictly controlled investment risks, controlled the pace of investment, and strived to improve investment returns. For projects with huge investment, long investment cycles, and high investment risks, the progress of the project is delayed; for investments with low relevance, they are reduced or suspended. In line with the principle of "not going to dangerous places, not going to chaotic places", new investments in high-risk countries are strictly controlled, and the pace of investment in existing projects is slowed down.

Persist in promoting oil and gas investment and trade, and drive the cooperation of the entire industry chain with points and lines. Use trade connections to find upstream investment opportunities, and while strengthening upstream investment, drive cooperation in engineering construction and equipment manufacturing.

Sinopec and PetroChina have extended the industrial chain to upstream cooperation based on Sino-Qatari natural gas trade. Sinopec has established an overseas investment holding company to manage the professional management of overseas refining business investment, construction, and operation. Chinese enterprises have introduced new energy power generation technology in multiple oil and gas cooperation projects to promote the green transformation of oil fields. The cooperation field expands from the upstream of oil and gas to the entire industry chain, from oil and gas to new energy, the synergy and integration are further enhanced, and the cooperation mode is more diversified from bilateral to multilateral.

It is expected that in 2024, Chinese enterprises will seize the strategic opportunity to deepen cooperation with "Belt and Road" countries, continue to develop new projects, adjust and optimize the asset structure, and maintain a steady increase in overseas oil and gas production. With the implementation of proposed and under-construction natural gas, refining, engineering technology services and other cooperation projects, international oil and gas cooperation across the entire industry chain will reach a new level.

(This section is written by: Luo Jiyu Reviewer: Fu Lei)

Overseas Technical Services and Engineering Construction of Chinese Petroleum Enterprises

Current Situation of Overseas Technical Services and Engineering Construction of Chinese Petroleum Enterprises

Eliminating the Impact of the New Crown Epidemic, the Execution of Overseas Projects of Chinese Enterprises Has Fully Returned to Normal

In 2023, **PetroChina Technical Service** Overseas markets hit a recent high, with the overall utilization rate of overseas engineering and technical teams increasing by 7.2% year-on-year, and the annual cumulative utilization rate of drilling rigs increasing by 4.4% year-on-year. Key projects such as UAE drilling and well repair, and Saudi Arabia's physical exploration and collection have achieved "contract signing, commencement, and revenue generation in the same year". **PetroChina Engineering** Key projects are progressing well, the Niger Agadem oilfield phase II integrated project is waiting for the owner's instruction to start production, and the Iraq Hafaya natural gas processing plant (GPP) project is in trial operation. **Sinopec Oilfield Services** The early layout of the project is progressing smoothly, the annual drilling plan for the EBANO oilfield in Mexico has been started in advance, and the long-cycle surface business projects in Uganda and other places are progressing well. **Sinopec Refining and Chemical Engineering** (hereinafter referred to as **SEG**) Multiple overseas projects have achieved significant results, with a total of 19 projects in Saudi Arabia with a total contract amount of 1.9 billion US dollars, and the localization employment ratio exceeds 85%; ExxonMobil Singapore integration, Thailand clean energy, Algeria LNG storage tank and MTBE and other projects are progressing smoothly. **CNOOC Services** Overseas business is fully promoted, "Zhenhai 6", "COSLSeeker", "COSLHunter", "Offshore Oil 936" platforms have completed adaptive modifications and went to the Middle East for operations, "Deep Blue Exploration" platform successfully completed drilling operations in Southeast Asia projects, and three semi-submersible platforms in Europe were fully operational in the first half of the year. **Offshore Oil Engineering** Projects in North America, the Middle East, Africa and other places are effectively controlled. The Saudi Arabia Marjan project, the assembly of the upper part of the P79 FPSO project in Bahrain, the prefabrication and installation of pipelines, etc., are fully promoted. The PAD14 project in Canada was completed ahead of schedule for the first phase of mechanical completion and delivery. After the completion of the domestic module construction work for the Longhu Northwest project, it arrived at the site to prepare for installation. The Uganda Kingfisher project delivered the temporary camp in the east in the first half of the year and began to level and backfill the CPF site and manufacture the equipment tank foundation. The BASF module construction project and the newly awarded Saudi Aramco LTA CRPO 122 project started smoothly.

Seize the Opportunity of the International Market Recovery, and the Development Momentum of the Overseas Market is Strong

In the first half of 2023, **SEG** The new contract amount in the overseas market has doubled, reaching 1.112 billion US dollars, a year-on-year increase of 107.1%; **CNOOC Services** The international business has grown significantly, with a new contract amount of 3.95 billion yuan, a year-on-year increase of 64.6%; **Jerry Shares** Overseas income reached 2.52 billion yuan, a year-on-year increase of 77.6%, accounting for 46.4%. In the first three quarters, **PetroChina Engineering** The new contract amount for overseas has doubled, signing 29.123 billion yuan, a year-on-year increase of 169.28%; **Sinopec Oilfield Services** The new contract amount in the overseas market is 14.04 billion yuan, a year-on-year increase of 1.9%, and the overseas market revenue increased by 29% year-on-year; **Offshore Oil Engineering** The new contract amount for overseas has doubled, signing 7.635 billion yuan, a year-on-year increase of 276%; **Anton Oilfield Services** The new contract amount in the overseas market is 2.23 billion yuan, a year-on-year increase of 58%. In 2023, **PetroChina Technical Services** The new contract amount in the overseas market is 4.05 billion US dollars, and the operating income increased by 10.3% year-on-year. Among them, the new contract amount in the overseas market of Great Wall Drilling is 1.67 billion US dollars, a year-on-year increase of 19.5%, and the revenue increased by 16.6% year-on-year.

Actively Create New Paths, New Markets, and New Areas of Cooperation Growth Points

In January 2023, **PetroChina Engineering** Signed the UAE offshore pre-service project for the first time, and entered the downstream refining construction field in Iraq for the first time. In June, **BGP Inc.** Won the bid for the B72 block of about 3000 km of 2D seismic data processing project of Oman Western Petroleum Company (oxy) for the first time, once again expanding the map of the Middle East high-end seismic data processing market. In May, **CNOOC Services** The Saudi Aramco LTA CRPO 122 project undertaken held a groundbreaking ceremony, taking a "key step" in the transition from international engineering subcontractor to international engineering general contractor. In May, **Offshore Oil Engineering** The completion ceremony of the peripheral engineering of the LNG project in Canada was held, realizing the world's first case of delivering all modules of the peripheral engineering of the LNG modular factory. In addition, Chinese companies have also made significant progress in the detailed design and construction of large-scale FPSOs internationally. In May, **Dalian Shipbuilding Industry** Group Co., Ltd. and Japan's MODEC jointly developed and designed the world's first M350 type FPSO—Bacalhau was officially delivered; **COSCO Shipping Heavy Industry** Co., Ltd. and Yantai CIMC Raffles Offshore Science and Technology Group Co., Ltd. are carrying out the construction of large FPSO for Petrobras, which has enhanced the technical strength and voice of Chinese companies in the field of deep-sea oil and gas development.

Cultivate High-End Business and Strive to Create a "High Ground" for Overseas Market Competition

In 2023, **PetroChina Technical Service Continuously optimizing market layout and business structure, actively cultivating high-end markets, expanding the market share of technical services, forming 7 markets with revenue scale of over 1 billion yuan, including Iraq, UAE, Saudi Arabia, etc., signing 8 contracts of over 100 million US dollars, including Great Wall Drilling Mexico, Algeria, UAE, Iraq 4 super 100 million US dollars drilling and well repair contracts, Oriental Geophysical Saudi Arabia large-scale land and sea seismic exploration contract, Sichuan Drilling Turkmenistan Revival Gas Field contract, etc. Sinopec International Petroleum Engineering Company (hereinafter referred to as Guogong Company) actively explores high-end markets such as drilling package, fracturing, directional wells, and offshore drilling, constructs a relatively reasonable business level of middle and high-end proportions, and strives to transform from labor output to capital, technology, and management output. Offshore Oil Engineering The core modules of the North American Shell LNG project built at the Qingdao site were all completed and delivered in March, continuing to consolidate the industry position of the "first echelon" of international high-end modular LNG construction. PetroChina Tech Development, Baoshi Machinery jointly won the bid for the Abu Dhabi Oil Company's 10 sets of 5000 m hybrid**

power desert rapid transport drilling rig project, further consolidating the international high-end market. Dongfang Honghua series of high-end drilling rigs as the most advanced and quality-assured drilling rigs of ADNOC so far have been repeatedly rated as the annual drilling rig, recognized by ADNOC and ADNOC Drilling, one of which, a 3000 HP artificial island drilling rig, set a world record for a 15,540 m displacement well in May (Table 2).

Going Out and Inviting in, the International Business Circle of Friends is Further Expanded

In 2023, **PetroChina Technical Services** conducted high-level customer visits and market marketing with Iraq, the UAE, Chad and other countries, jointly held a global technology cooperation roadmap seminar with Schlumberger, conducted multi-level high-frequency work exchanges with Baker Hughes, and jointly drew a new blueprint for cutting-edge technology cooperation in engineering. In the Middle East, it organized market marketing activities with the theme of "Integrated Solutions for Efficient Exploration and Benefit Development of Oil and Gas Fields", fully demonstrating the overall technical strength of PetroChina Technical Services. Oriental Geophysical and PetroChina Logging successfully held international seminars in the second half of the year, receiving good market feedback and high customer recognition. **SEG** maintained high-frequency interaction with oil companies and financial institutions in Algeria, Japan, Saudi Arabia, the UAE, Kuwait, Malaysia, Thailand, Kazakhstan, Cambodia and other countries in the first half of the year, conducted high-level visits with ExxonMobil, Spain's United Technologies Corporation (TR), Italy's Sapeim, JGC Corporation of Japan, MaireTecnimont of Italy, and engineering companies in Korea (Samsung, Hyundai Engineering and SKEcoplant), and signed a strategic cooperation agreement with TR to exchange views on traditional energy opportunities, new energy development, and expanding new project cooperation. In March, **Bomesc** (BOMESC) signed a memorandum of understanding for cooperation with Samsung Heavy Industries Geoje Shipyard in Korea. In September, **Jerry Equipment** signed a strategic cooperation agreement with Sweden's Alfa Laval Group in Yantai. The two sides will work together to explore the marine industry and establish a strategic cooperation relationship of complementary advantages and resource sharing in marine engineering, high-end equipment manufacturing, new energy and other fields.

Leveraging International Platforms to Increase Brand Awareness and International Influence

In August 2023, the "Top 250 International Contractors" list by Engineering News-Record (ENR) was released, **China Chemical Engineering** With an annual revenue of 26.18 billion USD, the American company Bechtel continues to top the list of contractors in the global oil and gas service sector, followed by the second-ranked

Table 2 New engineering technical services, engineering construction, and equipment business signed by Chinese companies overseas in 2023

Company	Project name
PetroChina	Drilling Service Contract of Mexico National Oil Company
	Large-scale Land and Sea Seismic Exploration Contract in Saudi Arabia
	Drilling Service Contract of Algeria National Oil Company
	UAE Well Repair Rig Service Contract
	Turkmenistan Revival Gas Field Inspection, Evaluation and Maintenance Service Contract
	Indonesia Drilling Service Contract
	Well Repair Service Contract of Qatar Energy Company
	Georgia Drilling Service Contract
	Iraq Hafaya Oilfield Billion Yuan Fracturing Project
	Malaysia Bioenergy Engineering
	East Africa Crude Oil Export Pipeline Project (EACOP Project) General Contract
	Algeria Propane Dehydrogenation/Polypropylene Unit
	UAE Hail and Ghasha Development Project EPC Pre-service Project
	Iraq Basra Refinery Upgrade and Renovation Construction Project
	Abu Dhabi National Oil Company Onshore Company Bab Oil Ninth Column Project and Bu Hasa Oilfield Central Processing Station Sixth Column Project
	US Shale Gas Market 7051 Meters 80 Type Perforating Gun and Supporting Equipment Order
Sinopec	Saudi Aramco AMIRAL Project EPC Contract
	Saudi Aramco Drilling and Well Repair Project, Three-dimensional Seismic Acquisition Project
	Kuwait National Oil Company Drilling and Well Repair Project
	Ecuador SACHA Turnkey Project
	Mexico National Oil Company IXACHI 3D Seismic Acquisition and Processing Project
	Saudi Arabia NEOM New City Wind Power Installation Project Contract
CNOOC	Saudi Aramco LTA CRPO 122 Project
	Southeast Asia 2023–2025 Land Abandonment Services Contract
	Southeast Asia Abandonment Integration Project
	Americas Drilling Completion Integration Services Project
	Americas Solid Control Equipment Technology Large Service Contract
	Southeast Asia Large-scale Cementing and Drilling Completion Fluid Joint Service Project
	Drilling Platform Service Contracts with Fixed Terms and Option Periods up to Five Years with Two Norwegian International Oil Companies

(continued)

Table 2 (continued)

Company	Project name
	Land Integration Technology Service Contract Signed with a Large Oil Company in Mexico
Private Enterprises	Honghua Oil Services Signed a Contract for Drilling Engineering Services for 8 + 3 Wells in Iraq's Zubair Oilfield
	Bomesc and Mitsui Ocean (MODEC) Signed a Contract for the Construction of the Topside Modules of the Floating Production Storage and Offloading Vessel (FPSO)
	Jerry Won a Contract for a Complete Set of Turbo Fracturing Vehicles from a Well-known Oil Service Company in North America
	Jerry Natural Gas Provides Fuel Gas Boosting Services for the Largest Single 800 Megawatt Gas Turbine Combined Cycle Power Station Abroad
	Bomesc and the Subsidiary OFS of Mitsui Ocean Development Company (MODEC) of Japan Signed a Contract for the Construction of the Topside Modules of the Floating Production Storage and Offloading Vessel for ExxonMobil
	Anton Oil Services Renews Contract for Integrated Oilfield Management Project in Large Oilfield in Southern Iraq

Data source Semi-annual and third quarter reports of Chinese enterprises in 2023

BECHTEL. **China Petroleum Engineering** ranks fourth on the list of "Top Ten International Oil and Gas Engineering Companies". In March, the mainstream Egyptian energy industry media "Future of Petroleum" published a report titled "The First Land Drilling Rig of Egypt's Honghua Drilling Manufacturing Company EPHH", introducing the first land drilling rig project cooperatively produced by China and Egypt, which was undertaken by Egypt's Honghua Drilling Manufacturing Company. In October, at the Abu Dhabi International Petroleum Exhibition (ADIPEC) held in the UAE, **China Petroleum** participated as a whole, highlighting the digitization of oil and gas production, energy transition, and the concept and practice of low-carbon sustainable development, demonstrating the comprehensive strength and integrated solutions of China Petroleum. **BGP** Oman project department operates and provides supporting collection technology solutions. The PDO three-dimensional node collection processing project, which won the ADIPEC 2022 Remote Operation Excellence Award, was showcased. **Jerry Shares** fully demonstrated its new achievements and breakthroughs in localization, low-carbon, and intelligent in the Middle East. **Oriental Honghua** brought high-end products such as desert fast-moving drilling rigs and 10,000-m drilling rigs to the event, deepening the international market's recognition of the strength of Chinese enterprises.

Characteristics of Overseas Technical Services and Engineering Construction Cooperation of Chinese Petroleum Enterprises

Continuously Deepening the "Belt and Road" Country Markets, Making Positive Contributions to Promoting Practical Cooperation in the Energy Field

Great Wall Drilling has seized the opportunity to participate in the development of the "Belt and Road". From 2005 to 2023, it completed 101 wells in Cuba, maintained a record of "zero pollution, zero accidents", trained more than 300 petroleum engineering technicians for the Cuban National Oil Company, and built 4 qualified drilling teams. SEG has signed over 230 new engineering project contracts in 26 "Belt and Road" countries, with a contract value of over 16 billion USD. The large-scale refining and chemical projects it undertook have driven the export of domestic complete sets of equipment and materials, realizing the "same ship sailing" of engineering construction and "Made in China". Some projects actively connect with the development strategies of the host countries, promoting the upgrading of the petrochemical industry and economic transformation of the host countries. The State Construction Company continues to expand cooperation with the upstream sector of the petroleum industry in "Belt and Road" resource countries. At present, there are 203 teams overseas in 28 countries executing 297 contracts, providing valuable opportunities for the company to open up markets and promoting "Made in China" to go overseas. At the same time, it has effectively promoted resource development in the host countries, forming a good situation of win–win development. Sinopec Oilfield Service Corporation has a market share of over 40% in Kuwait's drilling and well repair market, with fast equipment mobilization, efficient and safe construction, and overall improvement of the operation level in the oil area. China Petroleum Engineering's business has now expanded to 77 countries and regions in five major regions: the Middle East, Russia-Central Asia, Asia–Pacific, Africa, and the Americas. It has formed a market scale of over 100 million USD in contracts in 24 countries including Iraq, UAE, Algeria, Niger, Turkmenistan, Uzbekistan, Thailand, Malaysia, etc. The company has successively helped Sudan, Chad, and Niger to establish petroleum industrial systems, constructed a large number of classic projects such as Iraq's Halfaya oilfield, Turkmenistan's Amu Darya natural gas gathering and transportation, Uzbekistan's Karakuli gas field phase I, Chad's RONIER oilfield development, etc. It has successfully constructed a large number of classic projects such as Iraq's oilfield export pipeline, Russia's Far East pipeline, India's East-to-West gas transmission, Central Asia's natural gas pipeline, Tanzania's natural gas pipeline, etc. It has successfully executed many classic projects such as Kazakhstan's PKOP refinery, Algeria's refinery, Vietnam's Ningping coal head fertilizer general contract, etc., which have been highly recognized by project owners and even heads of state in the host countries.

Actively Exploring and Innovating International Cooperation Models, Enhancing Adaptability to Changes in the International Market

At present, some national oil companies are beginning to explore new project contracting models, such as the EPCM (Engineering, Procurement, and Construction Management) model, which places higher demands on the overall control and agile management capabilities of contractors. Chinese enterprises are also actively exploring diversification of their main business and business models, hoping to transform from "engineering contractors" and "equipment manufacturers" to "investors/financiers + engineering service providers", while extending investment and financing to new technology research and development, new technology cooperation and other fields, to promote the transformation and upgrading of SEG, and boost competitiveness and value creation. For example, China Petroleum Engineering is actively exploring and developing new business models such as "F (Financing) + EPC", "BOT (Build, Operate, Transfer)", and "Investment-Build-Operate Integration" to adapt to market demand changes, leverage professional advantages, and quickly tap into target markets. SEG is actively exploring a new model of Chinese engineering construction, "International Rules + Chinese Model", striving to enhance two capabilities: high-level front-end engineering and low-cost project execution, and strengthen three types of businesses: "technology + front-end engineering business, large-scale EPC business with variety advantages, and construction and installation business", with four measures: "engineering technology integration optimization, combination of Chinese and Western work systems, investment and financing driving projects, and increasing international cooperation" as the focus, to promote high-quality development of international business.

(This section was written by: Li Boyuan Reviewed by: Fu Lei)

Operating Status and Development Trends of Oil Companies

Operating Status of Oil Companies

In 2023, the operating performance of foreign oil companies declined slightly due to fluctuations in oil prices, but overall it remained at a relatively good historical level, with abundant cash flow and a positive performance in the capital market. The capital expenditure of major international oil companies continued to grow, and their willingness to invest and expand capital significantly increased. The operating strategies of the three major Chinese oil companies showed significant differences: PetroChina focused on improving quality and efficiency, transforming and developing, strictly controlling capital expenditure, and achieving a record high in operating efficiency; Sinopec maintained steady development, with both operating efficiency and capital expenditure increasing steadily; CNOOC continued to expand its scale, with a significant increase in capital expenditure.

The Oil and Gas Production of Foreign Oil Companies Has Slightly Decreased, While the Three Major Chinese Oil Companies Have Reached a Record High

In 2023, the overall oil and gas production of the 23 Chinese and foreign oil companies included in the statistics showed a downward trend, with an anticipated decrease of 3% year-on-year. Among the 20 foreign oil companies, 4 companies saw a decrease in oil and gas production, mainly due to a decrease in natural gas production. The oil and gas production of the three major Chinese oil companies continued to rise, reaching a record high.

X. Qian et al., *Global Oil and Gas Industry Development Report: 2023*,
https://doi.org/10.1007/978-981-95-2375-7_11

Major international oil companies saw a rebound in crude oil production, while natural gas output fell. The oil and gas production of major international oil companies slightly decreased. Except for Shell and bp, which saw a significant decline, the other three companies did not show significant changes, with an estimated annual decrease of 2% for the five companies. Among them, crude oil production is expected to increase by 3% year-on-year for the whole year, which is the first rebound since the continuous decline in crude oil production of each company since 2019, mainly due to the increased upstream investment by major international oil companies since 2022, and the production of each company has increased to varying degrees; it is estimated that the annual natural gas production will decrease by 8% year-on-year, with the decline mainly affected by the natural gas production of TotalEnergies, mainly due to the divestiture of 19% of its shares in Russia's Novatek. (Table 1; Figs. 1 and 2). **The production of crude oil and natural gas by national oil companies declined,** with an estimated annual decrease of 5%. The decline was mainly affected by Saudi Aramco, which significantly reduced its production in 2023, resulting in a significant decrease in output. **Independent oil companies continued the growth trend of crude oil and natural gas production from the previous year,** with an estimated annual increase in oil and gas production of 8%.

The oil and gas production of the three major Chinese oil companies has reached a historical high. The three major Chinese oil companies continue to strengthen oil and gas exploration and development and increase reserves and production capacity, with an estimated annual increase in total oil and gas production of over 6%, setting a record high for the same period. Among them, crude oil production continues to grow

Table 1 Changes in oil and gas production of oil companies from January to September 2023

Company	Growth of oil and gas production	Growth of crude oil production	Growth of natural gas production
ExxonMobil, BP, Shell, Chevron, TotalEnergies	− 1.78	3.12	− 7.84
Petrobras[a], Petronas[b], Equinor[c], Eni, Pemex[d], Saudi Aramco[e], Kuwait Petroleum Corporation, Iraq Petroleum Company	− 4.18	− 2.97	− 1.02
Suncor, Marathon Oil, Occidental Petroleum, ConocoPhillips, Devon Energy, Apache, EOG Resources	6.58	5.7	7.7
PetroChina, Sinopec, CNOOC	5.5	4.4	7.2

Unit %
[a]Petrobras; [b]Petronas; [c]Equinor; [d]Pemex; [e]Saudi Aramco

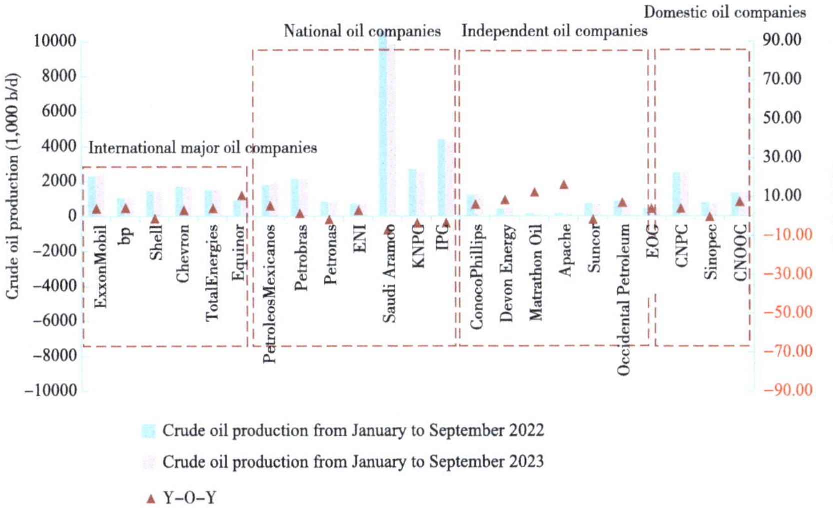

Fig. 1 Crude oil production and change rate of oil companies from January to September 2022 and January to September 2023. *Data source* Quarterly Reports of Various Companies, OPEC Bulletin

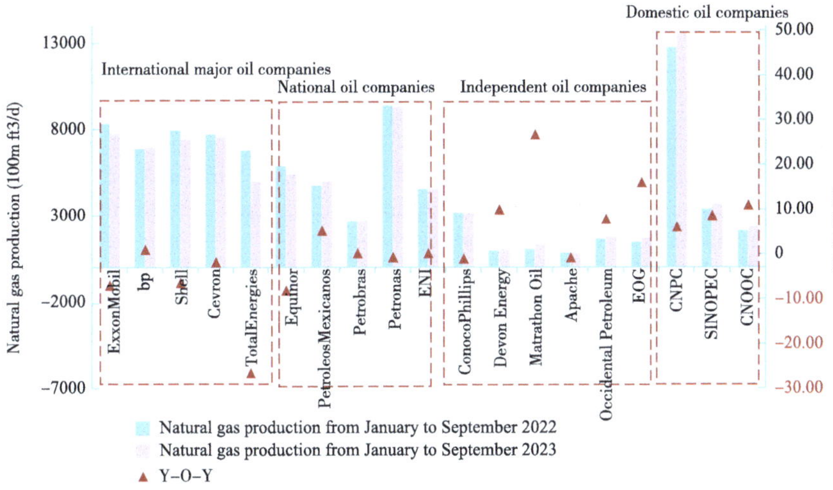

Fig. 2 Natural gas production and change rate of oil companies from January to September 2022 and 2023. *Data source* Quarterly Reports of Various Companies, OPEC Bulletin

based on the rebound in 2022, with an estimated annual increase of 5%, the highest since 2016; natural gas production continues to grow rapidly due to market demand and the requirements of companies for low-carbon transition, with an estimated annual growth rate of over 7%, and the proportion of natural gas production in oil and gas production continues to increase.

Foreign Oil Companies Maintain Stable Crude Oil Processing Volume, Slightly Increase Refined Oil Sales Volume, and the Three Major Chinese Oil Companies Significantly Increase Refined Oil Sales Volume

In 2023, the gross profit of refinery oil refining is at a high level, and the crude oil processing volume of Chinese and foreign oil companies is generally on the rise. However, due to the difference in regional refining gross profit, European and American companies show a differentiated situation, and the scale of Chinese refining and sales has significantly increased.

Foreign oil companies maintain stable crude oil processing volume and slightly increase refined oil sales volume. In 2023, the crude oil processing volume of major international oil companies remained stable. However, due to the refining gross profit in the United Statesbeing twice that of Europe, European downstream transformation is stronger and refinery divestiture occurs faster than in the U.S. As a result, three European companies saw a decline in crude oil processing volume, while American companies ExxonMobil and Chevron experienced slight increases. It is estimated that the total crude oil processing volume of the five major international oil companies will decrease by 0.5% year-on-year, and the sales volume of refined oil of major international oil companies will continue to grow due to global demand growth, with an estimated annual growth of 3% (Table 2).

The scale of refining sales of Chinese oil companies has significantly increased. In 2023, PetroChina and Sinopec adjusted their refining business strategies, increased the load of the equipment, expanded the total processing volume, and the crude oil processing volume turned from decrease to increase. In terms of refined oil sales, PetroChina and Sinopec benefited from the recovery of Chinese macroeconomics and oil market demand, continuously increased the development of efficient markets, and the sales volume of refined oil increased by about 15% year-on-year, reaching the highest level since 2020, recovering to 90% of the level before the COVID-19 pandemic.

The Operating Performance and Net Profit of Foreign Oil Companies Generally Declined, But They are still at a Relatively High Level. The Operating Performance of the Three Major Chinese Oil Companies is Differentiated

In 2023, the operating performance of 20 foreign oil companies generally declined, with significant declines in operating income and net profit, a significant decline from the historical high in 2022, but still at a relatively good level in the past decade; free cash flow declined but remained ample. The sales revenue of the three major Chinese

Table 2 Crude oil processing volume and refined oil sales volume of major international oil companies and three major Chinese oil companies in 2023

Company	Crude oil processing volume			Refined oil sales volume		
	Processing volume for the first three quarters of 2022 (Thousand Barrels/ Day)	Processing volume for the first three quarters of 2023 (Thousand Barrels/ Day)	Year-on-year growth (%)	Sales volume for the first three quarters of 2022 (Thousand Barrels/ Day)	Sales volume for the first three quarters of 2023 (Thousand Barrels/ Day)	Year-on-year growth (%)
ExxonMobil	4046.00	4129.00	2.05	5321.00	5496.00	3.29
bp	1547.00	1444.00	6.66	2958.00	3081.00	4.16
Shell	1391.00	1360.00	2.23	2490.00	2570.00	3.21
Chevron	1493.00	1563.00	4.69	2593.00	2731.00	5.32
TotalEnergies	1497.00	1456.00	2.74	3953.00	3886.00	1.69
Total International Oil Companies	9974.00	9952.00	0.22	10,586.00	10,862.00	2.59
PetroChina	3317.41	3865.19	16.51	3030.89	3437.91	13.43
Sinopec	4901.91	5274.03	7.59	4126.89	4914.97	19.10
TotalEnergies Chinese Oil Companies	8219.32	9139.22	11.19	7157.78	8352.88	16.70

Data source Quarterly reports of each company

oil companies all declined, the net profit performance was significantly differentiated, and the cash flow situation was generally improving.

The operating income, net profit, and free cash flow of major international oil companies have significantly decreased compared to the previous year. Major international oil companies have significantly lower realized oil prices than in 2022, and it is estimated that the annual operating income will decrease by 17% year-on-year. Excluding the negative impact of bp's profit from withdrawing from Russia last year, it is estimated that the average annual net profit will decrease by 46% year-on-year. In 2022, bp's profit was negative due to the impact of withdrawing from Russia, and there was a recovery growth in 2023, and the other four companies all showed a significant downward trend (Figs. 3 and 4). In the first three quarters of 2023, major international oil companies were affected by the decline in performance and a significant increase in capital expenditures. The free cash flow significantly decreased compared to the previous year, a year-on-year decrease of 43.26%, but it was still ample, mainly because the good operating conditions in 2022 and measures such as cost reduction and efficiency improvement accumulated sufficient cash flow (Fig. 5).

The operating income, net profit, and free cash flow of national oil companies and independent oil companies have significantly decreased compared to the previous year.

Average exchange rate of US dollar to RMB from January to September 2022: 1 US dollar = 6.6068 RMB; Average exchange rate of US dollar to RMB from January to September 2023: 1 US dollar = 7.0148 RMB.

The performance of the three major Chinese oil companies is differentiated, and the overall cash flow situation is good. In 2023, despite an increase in oil and gas production and sales volume among the three major Chinese oil companies, declining oil prices and the of the RMB depreciation have led to a decrease in sales revenue. It is estimated that the total annual operating income will drop by over 5% year-on-year. Affected by the business structure, the net profit of the three major oil companies will be significantly differentiated. It is estimated that CNOOC, which is mainly upstream, will be greatly affected by the decline in oil prices; PetroChina and Sinopec will take advantage of the synergistic advantages of upstream and downstream integration, and their net profit performance will be better, especially PetroChina, which has benefited from the continuous cost reduction and efficiency improvement in recent years, and the effective development of new energy and new materials as well as refining and chemical business. In the unfavorable environment of the chemical industry in 2023, the operating profit of downstream business increased by more than 30% year-on-year, and it is estimated that the annual net profit will increase by 10% year-on-year, creating the best performance in the same period in history. The cash flow situation of the three major Chinese oil companies is generally improving, and the free cash flow is higher than the same period in 2022.

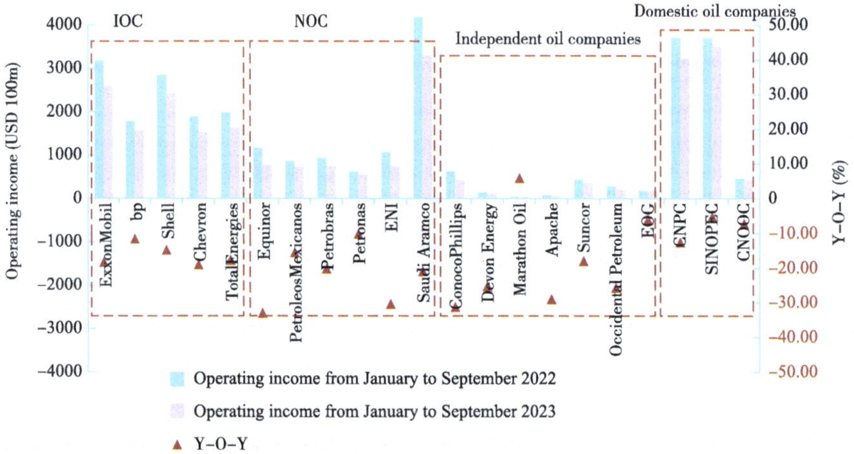

Fig. 3 Operating income of oil companies from January to September 2022 and January to September 2023. *Data source* Quarterly reports of each company

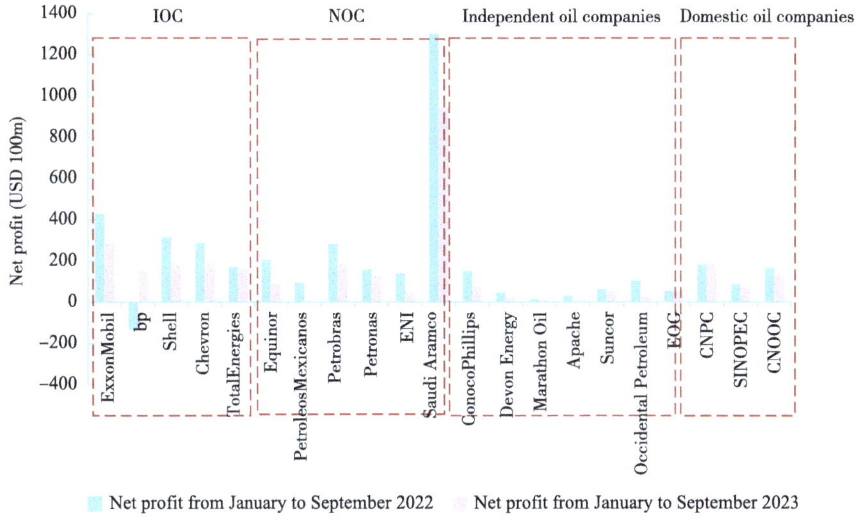

Fig. 4 Net profit of oil companies from January to September 2022 and January to September 2023. *Data source* Quarterly reports of each company

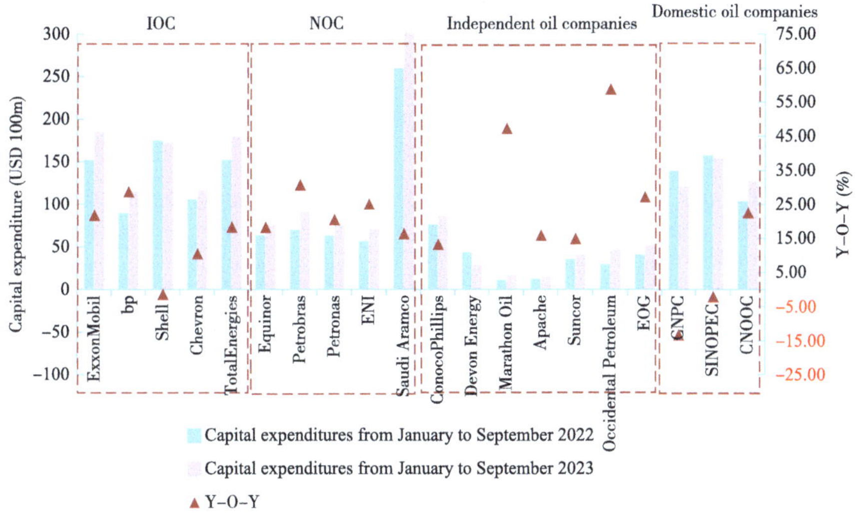

Fig. 5 Free cash flow of oil companies from January to September 2022 and 2023. *Data source* Quarterly Reports of Each Company

Looking at the segments, the operating profits of upstream businesses of Chinese and foreign oil companies have declined, and the downstream presents a differentiated situation. In the first three quarters of 2023, due to the realized oil price significantly lower than 2022, it is estimated that the operating profit of the upstream

Table 3 Operating profit of oil company segments from January to September 2023

Company	Upstream			Downstream		
	Operating profit for the first three quarters of 2022 (Billion USD)	Operating profit for the first three quarters of 2023 (Billion USD)	Year-on-year growth (%)	Operating profit for the first three quarters of 2022 (Billion USD)	Operating profit for the first three quarters of 2023 (Billion USD)	Year-on-year growth (%)
ExxonMobil	306.56	173.43	− 43	141.89	104.28	− 27
bp	183.79	96.28	− 48	69.79	43.95	− 37
Shell	317.62	116.67	− 63	59.41	61.31	3
Chevron	247.98	158.52	− 36	63.83	49.9	− 22
TotalEnergies	232.06	142.1	− 39	70.31	51.73	− 26
Equinor	572.79	250.87	− 56	28.18	28.18	0
Occidental Petroleum	75.46	34.39	− 54	15.88	9.89	− 38
Subtotal	1936.26	972.26	− 50	449.29	349.24	− 22
PetroChina	210.29	189.17	− 10.04	74.55	92.04	23.47
Sinopec	65.98	55.85	− 15.35	57.41	57.79	0.66

Data source Quarterly reports of each company
Note PetroChina's downstream includes refining, chemical, finished oil sales, and natural gas sales; Sinopec's downstream includes refining, chemical, and finished oil sales

business of major international oil companies will significantly decrease by 50% throughout the year, and the operating profit of the exploration and production business of the three major Chinese oil companies will also decline, with a decrease of more than 10%. In terms of downstream, major international oil companies are affected by the high return of refining gross profit and the continuous downturn in the chemical market, and it is estimated that the operating profit of downstream business will decrease by 22% year-on-year (Table 3). With the warming of the Chinese market and the increase in sales volume of refined oil, the operating profit of the downstream business of the three major Chinese oil companies has rebounded.

Most Global Oil Companies are Increasing Their Oil and Gas Investments, and Chinese Oil Companies are Showing a Certain Trend of Differentiation

In 2023, due to geopolitical conflicts and the fluctuation of international oil and gas prices, the pace of energy transition has slowed down. Countries are strengthening their focus on energy security. Chinese and foreign oil companies continue to increase their confidence in the development of the oil and gas industry and generally increase their investment.

Foreign oil companies are increasing their investments in the oil and gas sector, with capital expenditures increasing by 16% compared to the previous year. Among them, except for Shell, which has slightly cut its capital expenditure due to the replacement of the CEO and stricter discipline, other companies' capital expenditures have achieved varying degrees of growth, exceeding the average level of the previous year. It is estimated that the total capital expenditure of the five major international oil companies will increase threefold year-on-year (mainly driven by ExxonMobil and Chevron's mega-mergers and acquisitions in October, see below), significantly increasing on the basis of the investment rebound in 2022. The investment direction is mainly towards unconventional, deepwater, and LNG fields. The increase in upstream investment by major international oil companies reflects their converging understanding of the strategic position of the oil and gas business, that is, the oil and gas business is still the main source of profit/performance for current oil companies. **National oil companies continue to maintain a relatively high level of investment.** It is estimated that the total capital expenditure will increase by 20% year-on-year, still firmly optimistic about the oil and gas business, and increase investment (Fig. 6). **Independent oil companies are also generally increasing their capital expenditure, expected to increase by 15% for the whole year.** This is mainly because American independent oil companies have significantly increased their investments under the encouragement of Chinese policies to increase energy supply.

PetroChina's capital expenditure is for January–June data

The total capital expenditure of the three major Chinese oil companies has increased year-on-year. Due to the increase in the workload of projects under construction, CNOOC's development business expenditure has increased significantly, and it is estimated that the total capital expenditure for the year will be 120–130 billion yuan; Sinopec optimizes its capital structure and focuses on improving the efficiency of fund use, and it is estimated that the total capital expenditure for the year will be 165.8 billion yuan; PetroChina continues to strengthen investment control, according to the concept of rigorous investment, precise investment, and benefit investment, it is estimated that the total capital expenditure for the year will be 243.5 billion yuan, although it has decreased year-on-year, it is still much higher than Sinopec and CNOOC.

Looking at the segments, major international oil companies have significantly increased their investments in upstream businesses, reaching the highest level in nearly a decade. In 2023, the investment in traditional upstream businesses of the five major international oil companies is estimated to increase by 250% year-on-year. Among them, ExxonMobil and Chevron saw the largest increases, showing a growth trend of 6 to 7 times, with these two companies successively acquiring Pioneer Natural Resources and Hess Corporation in October. The total capital expenditure

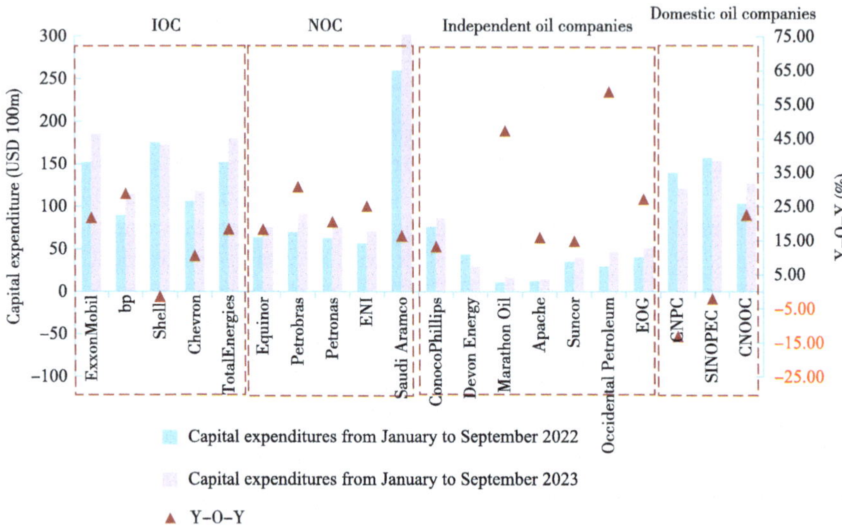

Fig. 6 Capital expenditure of oil companies from January to September 2023. *Data source* Quarterly reports of each company

for downstream businesses reached 18 billion US dollars, a year-on-year increase of 21.62%. Among them, bp's downstream capital expenditure increased significantly by 74% (Table 4), mainly due to the upgrade and renovation of the Cherry Point refinery in 2023, and the acquisition of truck fuel supplier TravelCenters of America, further expanding the terminal sales network.

Major Chinese and Foreign Oil Companies All Value Shareholder Returns, Foreign Companies' Stock Prices Fluctuate Slightly, Chinese Companies' Stock Prices Rise Significantly

In 2023, international major oil companies invested prudently, increased dividends, repurchased stocks, and adjusted their transformation strategies more steadily, with stock prices fluctuating slightly. As of December 31, the average stock price of the five major international oil companies fell by 1.62%, with Shell performing the best, rising by 5.78% (Fig. 7).

International major oil companies continue to use more funds for dividends, stock repurchases, and debt repayment. In 2023, the dividends and repurchases of international oil companies accounted for 53% of operating cash flow, and the capital market performance was good. Major international oil companies have announced

Table 4 Capital expenditure of oil company sectors from January to September 2023

Company name	Upstream			Downstream		
	Capital expenditure from January to September 2022 (Billion US dollars)	Capital expenditure from January to September 2023 (Billion US dollars)	Year-on-year growth (%)	Capital expenditure from January to September 2022 (Billion US dollars)	Capital expenditure from January to September 2023 (Billion US dollars)	Year-on-year growth (%)
ExxonMobil	115.87	139.91	20.75	34.71	39.24	13.05
BP	64.9	75.97	17.06	21.03	36.5	73.56
Shell	90.37	89.06	– 1.45	58.89	64.33	9.24
Chevron	85.87	117.69	37.06	18.41	21.53	16.95
Total Energy	136.93	161.09	17.64	14.82	18.23	23.01
Subtotal	493.94	583.72	18.18	147.86	179.83	21.62
PetroChina	110.22	113.51	2.99	28.88	7.39	– 74.42
Sinopec	78.10	72.29	– 7.44	75.07	78.89	5.08

Data source Quarterly reports of various companies

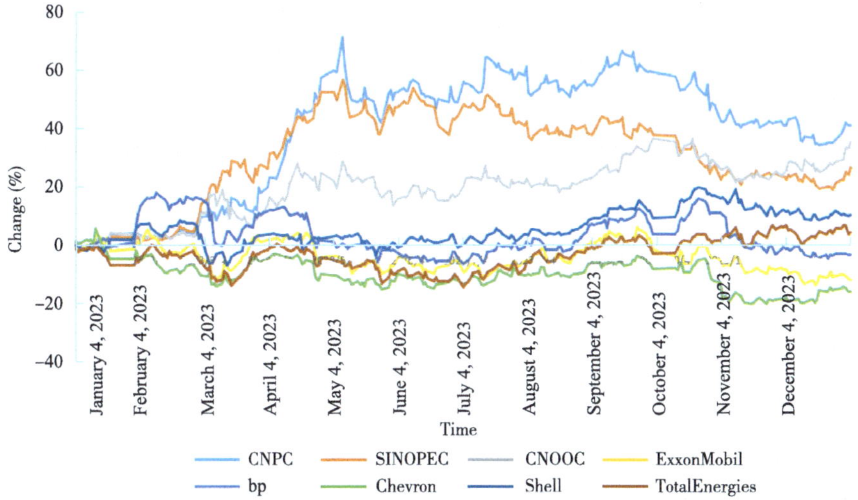

Fig. 7 Changes in the stock prices of Chinese and foreign oil companies in 2023. *Data source* investing.com

a new round of large-scale repurchase plans. TotalEnergies estimates that it repurchased a total of $9 billion in 2023. Shell conducted more repurchases in the fourth quarter and raised the original plan of $2.5 billion to $3.5 billion, estimating a total repurchase of $13.2 billion for the year. ExxonMobil estimates a total repurchase of $17.5 billion in 2023 and raised the dividend from $0.91 per share to $0.95. Chevron maintained a quarterly dividend level of $1.51 per share but will increase the 2024 dividend by 8% and increase the annual stock repurchase scale by 14% to $20 billion.

The capital market performance of the three major Chinese oil companies is stronger than the market, and they highly value shareholder returns. The stock prices of the three major oil companies have risen by an average of about 34% since the beginning of the year, among which, PetroChina performed the best, with an increase of 41%. The three major oil companies have also increased their efforts to return to shareholders. PetroChina distributed an interim dividend of 0.21 yuan per share, with a dividend amount of 38.43 billion yuan and a dividend rate of 45.1%. Both the dividend per share and the total dividend amount set a record for the same period in history. Sinopec distributed an interim dividend of 0.145 yuan per share, with a cash dividend rate of 49.5%. CNOOC distributed an interim dividend of 0.59 Hong Kong dollars per share.

(Authors of this section: Li Suxiao, Wang Chao, Reviewers: Yu Ling, Yin Dongqing)

Development Trends of Foreign Oil Companies

In 2023, the oil and gas industry is still in a period of turbulent change, and the development environment faces many uncertainties. Oil companies take advantage of the situation and actively adjust their development strategies. International major oil companies refocus on upstream resource strategies, further upgrade the optimization level of core assets, evaluate low-carbon projects based on investment returns, strive to improve the profitability of low-carbon businesses, continuously optimize organizational structures, and improve management efficiency. National oil companies accelerate the construction of the entire industry chain of upstream and downstream integration, while focusing on the development of low-carbon businesses with high synergy with traditional oil and gas businesses. Independent oil companies continue to accelerate mergers and acquisitions, form scale effects, and focus on the development of low-carbon technologies.

International Major Oil Companies Refocus on Oil and Gas Resource Strategies, Strive to Improve the Profitability of Low-Carbon Businesses, and Continuously Optimize Organizational Structures to Improve Management Efficiency

Increase Oil and Gas Investment and Production Targets, Implement Large-Scale Mergers and Acquisitions, and Refocus on Oil and Gas Resource Strategies

In 2023, international major oil companies generally increased oil and gas investment, with a total year-on-year increase of 18.18%, reflecting the gradually consistent recognition of the strategic status of oil and gas by international major oil companies, that is, the trend of carbon neutrality is irreversible, but oil and gas still play an important role in the global energy system. International major oil companies have returned to upstream resource strategies, focusing more on value growth while increasing investment and production, and are committed to building a low-cost, low-emission, and high-profitability asset portfolio.

European oil companies increase oil and gas investment and adjust aggressive production reduction targets. BP and Shell had previously set aggressive production reduction targets, reduced upstream business investment, increased the sale of oil and gas assets, but the global energy shortage brought about by the Ukraine crisis made the two companies re-examine their strategies and make timely adjustments. In February 2023, BP released a new production plan, adjusting the oil and gas production targets for 2025 and 2030 to 2.3 million barrels/day and 2 million barrels/day, respectively, an increase of 300,000 barrels/day and 500,000 barrels/day compared to before, slowing down the reduction rate, mainly increasing unconventional production in North America and deepwater production in Mexico. It also plans to increase investment in the oil and gas field by $1 billion per year before 2030; In June 2023, Shell released a development strategy adjustment plan, maintaining steady or slight growth in oil and gas production before 2030, canceling the production reduction target of 1% to 2% per year, increasing investment in integrated natural gas and upstream businesses, and the proportion of total investment from 50% in 2025 to about 55% in 2030.

American oil companies implement large-scale mergers and acquisitions, and the optimization level of core assets is further upgraded. In 2023, the merger and acquisition transaction amount of two American oil companies reached 125.2 billion US dollars, setting new highs for the two companies' merger and acquisition amounts in nearly 20 years. Among them, ExxonMobil's acquisition of Pioneer Natural Resources for 59.5 billion US dollars and Chevron's acquisition of Hess Corporation for 53 billion US dollars were the most eye-catching. After the acquisition, ExxonMobil's production increased from 3.7 million barrels/day to 4.4 million barrels/day, and Chevron's production increased from 3 million barrels/day to 3.4 million barrels/day [1]. At the same time, ExxonMobil's reserve replacement rate quickly increased from 58 to 170%, and Chevron's reserve replacement rate increased from 88 to 115%, effectively making up for the hidden danger of weak production growth caused by the continuous decline in upstream investment in previous years. The mega-scale mergers and acquisitions of the two companies can not only quickly increase oil and gas production and expand asset scale, but more importantly, they can further optimize the oil and gas asset portfolio and achieve value growth after the merger.

First, enhance core assets and reduce production costs. After acquiring Pioneer Natural Resources, ExxonMobil further strengthened its leading advantage in the Permian Basin, and Pioneer Natural Resources' oil and gas assets in the Permian Basin have a breakeven point below 35 US dollars/barrel, which will help ExxonMobil maintain low-cost production in the next decade. Chevron's upstream strategic focus is on North and South America, diversifying its asset portfolio. After acquiring Hess, it not only enhances its core assets such as North American shale oil and gas and Gulf of Mexico offshore oil and gas, but also further expands to Guyana's deepwater resources, which are the largest and fastest-growing in the world, which will help the company increase oil and gas production faster in the next decade. **Second, introduce advanced management models and reduce management costs.** Both American independent oil companies acquired by Chevron and ExxonMobil have more streamlined and efficient management models. It is expected that after the merger, they will promote the management efficiency of the two major American oil giants, achieving scale increase and no increase in management costs. **Third, strategically layout low-carbon assets and achieve low-carbon production.** The greenhouse gas emission intensity and methane emission intensity of Pioneer Natural Resources and Hess Corporation are both lower than those of ExxonMobil and Chevron, which will help reduce the latter's emission levels.

The gap in oil and gas development paths and goals between European and American oil companies will widen. Although international oil giants are refocusing on upstream strategies, due to the inconsistency of resource bases and development strategies, the gap in future oil and gas development paths and goals will widen between European oil companies and American companies. Although European oil companies have adjusted their aggressive production targets and increased investment, oil and gas production will not increase significantly, especially for BP. Without major mergers and acquisitions, oil and gas production will still be in a reduction trend in the next decade, but the trend will slow down. American oil companies will have a larger increase in oil and gas production in the future through increased oil and gas investment and large mergers and acquisitions (Fig. 8).

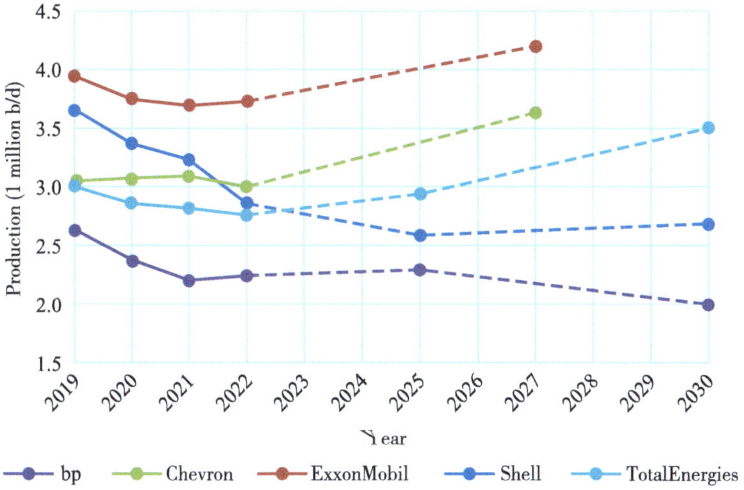

Fig. 8 Forecast of production of international oil giants. *Data source* Energy Intelligence

Improve the Profitability of Low-Carbon Business, Actively Explore New Models of Integration and Synergy Between Low-Carbon Business and Oil and Gas Business, and the Development of Low-Carbon Business Enters a New Stage of Focusing on Benefits

Since 2018, international oil giants have diversified their layout in low-carbon business, and investment has continued to increase significantly. However, since 2023, the growth rate of investment in low-carbon business has slowed down, and the strategic focus of low-carbon business of international oil giants has shifted from speed to benefit, focusing on low-carbon business that can create value for investors.

Pay more attention to the profitability of low-carbon business and adjust the scope of low-carbon business. International oil giants have previously set ambitious capacity targets in multiple low-carbon fields. Starting in 2023, profitability has become the main standard, and renewable energy business has been strategically reviewed again, abandoning some areas with unclear development prospects or lack of competitive advantages. For example, Shell believes that the return on home low-carbon electricity business is lower than expected, and it divested this business in September 2023; CCUS business lacks unique competitive advantages and the technology path is not clear, so it gave up the capacity target set for 2035. ExxonMobil decided to exit the economically poor algae biofuel project that has been operating for 14 years. Chevron stated that it will prioritize investing in CCUS projects in the United States to obtain tax benefits and increase project profitability. Overall, the low-carbon business of international oil giants has begun to enter the "nuclearization" development stage, focusing on advantageous fields, and the development goals are more pragmatic (Table 5).

Table 5 Updated low-carbon business development goals of international oil giants in 2023

Company	Low-carbon business goals
ExxonMobil	CCUS: Global leader in carbon capture
	Biofuels: Production capacity to reach 200,000 barrels/day by 2030
Shell	Biofuels: Annual production of approximately 2 million tons of sustainable aviation fuel by 2025
	Electric vehicle charging stations: 200,000 electric vehicle charging stations globally by 2030
bp	Hydrogen: Annual production to reach 600,000 tons by 2030
	Renewable energy installed capacity: Installed capacity to reach 50 gigawatts by 2030
	Biofuels: Production to increase to 100,000 barrels/day by 2030
	Electric vehicle charging stations: 100,000 electric vehicle charging stations globally by 2030
Chevron	CCUS: Carbon capture capacity to reach 25 million tons/year by 2030
	Biofuels: Production capacity to reach 100,000 barrels/day by 2030
Total	Renewable energy installed capacity: Installed capacity to reach 100 gigawatts by 2030
	Biofuels: Production capacity to reach 5 million tons/year by 2030, with aviation fuel production capacity reaching 1.5 million tons/year
	Hydrogen: Annual production to reach 1 million tons by 2030
	Electric vehicle charging stations: 150,000 electric vehicle charging stations globally by 2025

Source Annual reports of each company

Actively exploring new models of "integrated and synergistic" development of low-carbon business and oil and gas business. Entering 2023, major international oil companies generally recognize that energy transition requires smooth transition and orderly development. Rapidly shifting to renewable energy cannot balance energy security and economy, and they begin to explore new models of "balanced" development of low-carbon business and oil and gas business. **First, they are more inclined to invest in low-carbon businesses that can develop synergistically with traditional oil and gas businesses.** From 2018 to 2021, the main focus of mergers and acquisitions by major international oil companies was on renewable energy power generation, accounting for more than 80% of the total number of mergers and acquisitions. In 2022, the number of mergers and acquisitions of CCUS, biomass energy, and hydrogen energy projects began to surpass renewable energy power generation projects, with each accounting for half. By 2023, the number of mergers and acquisitions of CCUS, biomass energy, and hydrogen energy projects has completely surpassed renewable energy power generation projects, accounting for 60% of the number of low-carbon business acquisitions (Fig. 9). Among them, CCUS projects have become the focus of investment by major oil companies. In 2023, all five major international oil companies invested in CCUS projects, mainly in the

European Union, the United States, Canada, Australia and other regions that already have relatively mature and systematic CCUS development support policies. The most typical example is ExxonMobil's acquisition of Denbury in North America in August 2023. The latter's carbon dioxide pipeline system can connect ExxonMobil's main production areas in North America, refining centers, and carbon storage points to form a pipeline network, enabling ExxonMobil to achieve low-carbon production at a lower carbon reduction cost. Second is **Building industrial clusters for integrated and synergistic development of oil and gas business and low-carbon business.** Major international oil companies integrate oil and gas, renewable energy power generation, hydrogen energy, energy storage, and CCUS for development, establish comprehensive integrated industrial clusters, and achieve point-to-chain, chain-to-cluster development. For example, bp's carbon–neutral industrial cluster in Teesside, UK, includes the UK's largest H2 Teesside blue hydrogen project, the UK's first HyGreen Teesside green hydrogen project, NET power generation project, and NEP CCUS project (plus electricity use side). Total's $27 billion comprehensive energy utilization project in Iraq, which includes the development of the Ratawi oil field, the construction of a central gas processing plant, a seawater desalination project, and a solar power generation project. Overall, the carbon–neutral industrial clusters developed by major international oil companies have three highlights: first, they form a circular use of various types of low-carbon energy; second, they are close to demand, with a broad customer prospect; third, they obtain government subsidies to ensure basic profitability.

Opening up high-growth potential low-carbon new businesses. In 2023, the lithium business became a new area of low-carbon business investment for international oil giants. ExxonMobil believes that with the popularization of electric vehicles, lithium will become a high-growth potential low-carbon investment field

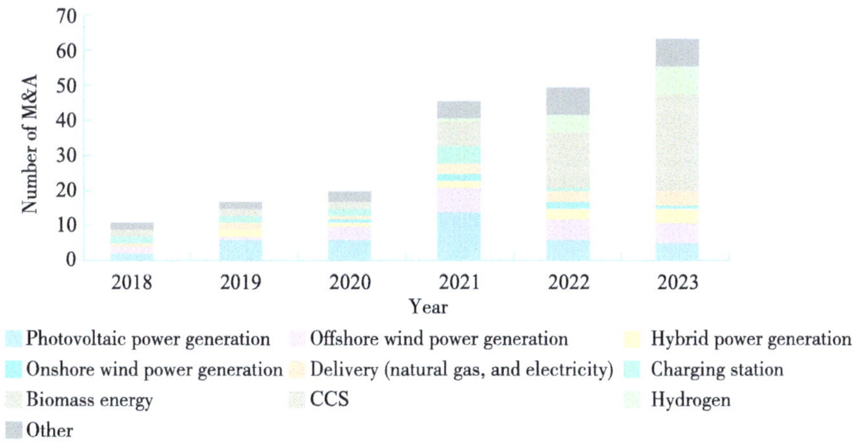

Fig. 9 Mergers and acquisitions of low-carbon businesses by major international oil companies from 2018 to 2023. *Data source* S&P Global

in the next few decades. In May, ExxonMobil invested $100 million to purchase 120,000 acres of lithium mining rights from Galvanic Energy, which has 4 million tons of lithium carbonate equivalent, enough to meet the battery needs of about 50 million electric vehicles. Subsequently, ExxonMobil held talks with Albemarle, the world's largest lithium producer, Tesla, and other car manufacturers, as well as globally renowned battery manufacturers such as Samsung SDI, on lithium metal cooperation. Chevron CEO Mike Wirth stated that lithium extraction aligns with Chevron's "core capabilities," and the company is exploring opportunities to develop its lithium business.

Continuously Innovate and Optimize the Organizational Management System to Improve Organizational Flexibility and Creative Efficiency

In recent years, international oil giants have actively transformed traditional business management structures and business operation models, establishing new management models for integrated energy companies. In 2023, these companies further optimized their organizational structures, showing characteristics of flatter management structures, more optimized decision-making processes, and more agile adaptability. In March, Shell split its original natural gas and renewable energy solutions sector, merging integrated natural gas and upstream businesses, and merging renewable energy and energy solutions businesses with downstream businesses, reflecting Shell's customer-centric energy transition path. After the adjustment, Shell simplified from four major business sectors to three. Shell stated that fewer business sectors will bring stronger cooperation and faster responses, and a flatter structure is expected to reduce Shell's structural operating expenses by $2 billion to $3 billion before 2025. ExxonMobil adjusted its corporate structure in 2022 and moved its headquarters to Houston in the first half of 2023, enhancing management efficiency through more direct communication between the headquarters management department and business departments. ExxonMobil stated that it had achieved its goal of saving $9 billion in structural expenses ahead of schedule in the third quarter of 2023, which will provide important support for the improvement of the company's profitability level.

In addition, to ensure that employees are aligned with the company's transformation strategy and to encourage employees to jointly promote the achievement of transformation goals, international oil giants have increased climate-related performance incentives. For example, Shell implemented the "Shell Energy Transition Road" evaluation system for all employees, adding three new indicators: selling low-carbon products, reducing operational emissions, and cooperating in decarbonization, which together account for 15% of the annual bonus scorecard; the company's long-term incentive plan will increase the proportion of energy transition performance indicators linked to senior employee salaries from 20 to 25% between 2023 and 2025. TotalEnergies formulated a "green allowance" plan, providing an "energy efficiency and transition subsidy" of 2000 euros per person for 35,000 employees in France for five consecutive years, encouraging employees to adopt low-carbon lifestyles such as buying electric vehicles, installing home solar panels, and wind turbines.

National Oil Companies Accelerate the Construction of Integrated Upstream and Downstream Industry Chains, While Focusing on the Development of Low-Carbon Businesses That Have High Synergy with Traditional Oil and Gas Businesses

Firmly Develop Oil and Gas Businesses and Set Ambitious Production Growth Targets

National oil companies generally bear the responsibility of ensuring national energy security, and the development of the oil and gas main business is still the focus of these companies. The stabilization and rebound of oil prices in recent years have further strengthened their confidence in developing oil and gas businesses, and they have put forward ambitious production growth plans (Table 6). **On the one hand, some national oil companies with abundant Chinese resources actively promote international cooperation in Chinese resources**, increasing Chinese energy supply while realizing the monetization of Chinese resources, such as the large-scale tendering of 13 oil and gas fields including the Mansuriya gas field by the Iraqi National Oil Company in June 2023; **On the other hand, some national oil companies with limited Chinese resources actively optimize their international exploration asset portfolios**, such as Petronas of Malaysia, which, while exiting oil and gas field blocks in Myanmar, Vietnam, and Brunei, acquired high-quality oil and gas assets in Mexico, Suriname, and Brazil, shifting its development focus to the Americas.

Implementing Downstream Expansion Strategy

National oil companies, while maintaining stable upstream production, are accelerating their extension to the downstream industry chain and actively promoting the integrated development of refining. In particular, national oil companies from resource countries are further strengthening joint ventures in the high-demand growth market in Asia, with Saudi Aramco being the most outstanding, investing successively in China's integrated refining projects and large-scale refining enterprises (Table 7), to secure long-term crude oil supply orders and a stable downstream chemical market. In addition, national oil companies are accelerating the formation of large-scale refining enterprises and scaling up their refining business. For example, the Abu Dhabi National Oil Company and Austria's OMV plan to merge their two refining companies to form a chemical company valued at over 30 billion US dollars. Indian Oil Corporation plans to invest 6100 billion rupees (approximately 7.4 billion

Table 6 National oil company production growth targets

Company name	Production growth plan
Saudi Aramco	Increase crude oil production capacity from 12 million barrels/day in 2022 to 13 million barrels/day in 2027
Abu Dhabi National Oil Company	Plans to invest $150 billion to increase oil and gas production from 2023 to 2027, raising crude oil production to 5 million barrels/day by 2027
Thailand National Petroleum Company	Plans to achieve a compound growth rate of oil and gas production of 5% from 2023 to 2027
Brazil National Petroleum Company	Plans to increase oil and gas investment to 100 billion US dollars in the next five-year investment plan (2024–2028), a 10% increase from the previous strategic plan (2023–2027), to support the goal of continuous growth of oil and gas production before 2030

Source Various company websites

US dollars) to build a large-scale refining company with a world-class cracking unit, which is the largest investment in the refining business in the history of Indian Oil Corporation.

Table 7 Saudi Aramco's downstream investment cooperation in China in 2023

Time	Specific investment situation
March	Jointly building a refining complex project in Panjin, Liaoning, with a total investment of about 83.8 billion RMB, Saudi Aramco holds 30% of the shares. After completion, it will operate a refinery with a daily processing capacity of 300,000 barrels and a chemical plant with an annual output of 1.65 million tons of ethylene and 2 million tons of p-xylene
July	Acquired 10% of Rongsheng Petrochemical for 24.6 billion RMB. Rongsheng Petrochemical operates a 40 million tons/year integrated refining project, has the world's largest PTA, PX and other chemical product capacity, and is an important producer of polyester, new energy materials, engineering plastics and high value-added polyolefins
September	Signed a cooperation framework agreement with private refining company Oriental Energy, planning to acquire 10% of Oriental Energy's shares. Oriental Energy owns and operates a 16 million tons/year integrated refining unit, a methanol-to-olefins and derivatives combined unit, and a purified terephthalic acid production unit
October	Saudi Aramco signed a memorandum of understanding with Shandong Yulong Petrochemical, planning to acquire 10% of its shares. Shandong Yulong's integrated refining project has a designed crude oil processing capacity of 20 million tons/year and is currently in the final stages of construction. Once completed, it will become a high-end petrochemical, new material, and high-end manufacturing concentration area with international influence

Increase Investment in Low-Carbon Business, Establish a Low-Carbon Operation Management Company, and Focus on Developing Low-Carbon Businesses That Have High Synergy with Traditional Oil and Gas Businesses

In recent years, national oil companies have also begun to announce emission reduction targets, mainly involving Scope 1 and Scope 2. To achieve emission reduction targets and low-carbon development, national oil companies mainly take the following measures. **First, increase investment in low-carbon businesses,** such as the Brazilian National Petroleum Corporation, which has increased the proportion of low-carbon investment from 4% to 6%~15% in its latest expenditure plan (2024–2028), leaving a large flexible space for future low-carbon input growth; Pemex plans to invest 12 billion US dollars in low-carbon investment from 2023 to 2027, a 10% increase from before. **Second, establish an independent low-carbon business company for professional management**, such as the Abu Dhabi National Oil Company and the Abu Dhabi National Energy Company investing in the establishment of Masdar Clean Energy Company, which mainly operates low-carbon businesses such as green hydrogen; Petronas has invested in the establishment of Gentari, a new energy project company focusing on renewable energy power generation, hydrogen, and green travel; Qatar Energy has specifically established a Renewable Solutions Department, focusing on the investment and operation of renewable energy projects. **Third, focus on developing low-carbon businesses such as CCUS and hydrogen energy that have high synergy with traditional oil and gas businesses**. The focus of the National Oil Company's development is still on oil and gas, and the main path to emission reduction is concentrated on decarbonizing oil and gas operations, so it is more inclined to develop CCUS and hydrogen energy.

Independent Oil Companies Continue to Accelerate Mergers and Acquisitions, Forming Scale Effects, While Focusing on the Development of Low-Carbon Technology

Continuously Increase the Intensity of Mergers and Acquisitions to Enhance the Competitiveness of Core Assets

In 2023, independent oil companies generally increased capital expenditure, on the one hand, increased exploration expenditure to ensure synchronous growth of reserves and production; on the other hand, they actively expanded their scale through mergers and acquisitions, with some companies increasing their M&A expenditure to seek scale effects and synergies. For example, in April, Canada's Suncor Energy acquired all assets of Total's Canadian subsidiary for $4 billion, mainly including a 31.23% stake in the Fort Hills oil sands project and a 50% stake in the Surmont oil sands project. After the acquisition, Suncor Energy's core oil sands assets will be further strengthened. In June, U.S. shale oil producer Civitas Resources acquired some shale oil assets in the Midland Basin and Delaware Basin held by NPF Energy Capital for $4.7 billion, with an internal rate of return exceeding 30% for 40% of the assets acquired, significantly enhancing the profitability of its core assets. In December, Chesapeake acquired Southwestern Energy for $17 billion, becoming the largest natural gas producer in the United States.

Focus on the Development of Low-Carbon Technologies with Unique Advantages

In terms of low-carbon business, independent oil companies pay more attention to low-carbon technology, with Western Oil being the most typical. Western Oil is committed to promoting the commercial and large-scale development of Direct Air Capture (DAC) technology. It invested $1 billion at the end of 2022 to build a direct air capture plant in the Permian Basin and plans to establish 70 carbon capture plants worldwide by 2035. In July 2023, Western Oil acquired DAC technology company Carbon Engineering Ltd for $1.1 billion to help it develop carbon capture facilities. In August, Western Oil signed an agreement with the Abu Dhabi National Oil Company to plan a joint venture to establish a direct air capture facility in the UAE. Western's air capture technology is also becoming a new profit point. In September, Western Oil signed an agreement with Amazon to sell it 250,000 tons of carbon credits over the next 10 years.

(Authors of this section: Xia Chuyang, Fan Xugang, Reviewer: Yu Ling)

Development Trends of Chinese Oil Companies

In 2023, the reform of state-owned enterprises entered a new stage focusing on core competitiveness and core functions, with industrial leadership, technological innovation, and safety support as tasks. Chinese oil companies vigorously promoted oil and gas exploration and development and increased reserves and production, continuously deepened the transformation and upgrading of refining, continuously strengthened market marketing, steadily promoted green and low-carbon transformation, increased the intensity of technological innovation, actively carried out the layout of new energy, new materials, and new businesses, and achieved stable and efficient operation of the two major oil and gas industry chains.

Stabilize the Basic Plate of Key Oil and Gas Regions, Increase the Intensity of Tackling "Two Deep and One Unconventional", and the Results of Increasing Reserves and Production are Significant

In 2023, Chinese oil companies continued to increase exploration investment, highlighted efficient exploration and economic recoverable reserves in key oil and gas regions, promoted the implementation of large-scale reserves areas and the preparation of reserve potential areas, fully promoted deep, deep-sea and unconventional oil and gas reservoir exploration, and the results of increasing reserves, production, sales, and efficiency of oil and gas were good.

Strategic Breakthroughs Have Been Made in Key Regional Exploration, and New Fields of "Two Deep and One Unconventional" Are Actively Expanded

Strengthen strategic regional risk exploration, oil and gas rich zone belt exploration and shale oil and gas exploration, accelerate unconventional production capacity construction, accelerate the development towards deep land and deep sea, achieve a number of major breakthroughs in deep and ultra-deep, unconventional oil and gas fields, the construction of national shale oil and gas demonstration zones continues to advance, new areas and new fields continue to make important discoveries, major breakthroughs have been made in deep coalbed methane exploration and development, unconventional oil and gas production continues to grow, becoming an important support for the country's oil and gas reserve increase and production. **PetroChina** discovered a geological reserve of over 100 million tons in Huan County, Gansu Province, marking a major breakthrough in oil exploration in the western fault fracture zone of the Ordos Basin. The shale gas well Zi201 in Neijiang, Sichuan, achieved stable production, becoming the first high-yield industrial gas flow with commercial development value drilled in the ancient Cambrian shale formation worldwide. **Sinopec** The "Asia's deepest well" "Deep Land No.1" Yuejin 3-3XC well was successfully drilled and obtained high-yield oil and gas flow, refreshing the two records of Asia's deepest well inclination and ultra-deep drilling horizontal displacement. The first phase of Bazhong gas field has a proven geological reserve of 305.5 billion cubic meters, showing the good exploration potential of tight sandstone in the northeastern Sichuan region. **CNOOC** Continuously increasing the intensity of oil and gas exploration and development, through the innovation of basin formation mechanisms and understanding of oil and gas reservoir formation models, the world's largest Archean metamorphic rock Bozhong 26–6 oil field was discovered in the southern part of the Bohai Sea, with a cumulative proven and controlled geological reserve exceeding 200 million tons of oil equivalent. The shallow Bohai Sea Qinhuangdao 27–3 oil field has a proven geological reserve of over 100 million tons of oil.

Highlight Effective Development and Strive to Increase Beneficial Output

Taking beneficial development as an important means to improve the quality and efficiency of the company, continuously promoting the demonstration project of beneficial production, focusing on increasing reserves, increasing production, controlling investment, reducing costs, etc., in deepening old areas, expanding shallow layers, tackling deep layers, and insisting on low-cost development strategy and high-quality development mainline in the simultaneous development of oil and gas. Daqing, Shengli and other old oil regions have deepened fine exploration and development, strengthened the application of technology to significantly improve the recovery rate, and the cumulative production of crude oil has once again set a record, reaching an important node. **PetroChina** In the field of unconventional oil and gas, it has built Longdong, Jimusaer, and Gulong national shale oil demonstration zones and bases, and put into operation a 100,000-ton terrestrial shale oil beneficial development demonstration platform in Cangdong Sag. Daqing Oilfield has maintained a stable production of 30 million tons for 9 consecutive years, with a cumulative production of crude oil exceeding 2.5 billion tons, accounting for 36% of the total onshore crude oil production in the country. Through innovative chemical flooding technology to improve the recovery rate, the cumulative production of tertiary oil recovery has exceeded 300 million tons, and the world's largest tertiary oil recovery research and development production base has been built. **Sinopec** Focusing on the "ten trillion cubic meters" reserve target, with the main line of "hard and stable production in old areas, fast production in new areas, breakthrough in difficult to move, and expansion of shale gas", insisting on beneficial development, promoting marine gas, tight gas, shale gas stable production and increase, continuously tackling the beneficial production model of low-grade unused reserves, vigorously promoting the fine water injection adjustment technology during the ultra-high water cut period, innovatively forming low-permeability oil reservoir pressure drive water injection development technology, tackling breakthroughs in offshore, high temperature and high salt, heavy oil reservoir chemical flooding to significantly improve the recovery rate technology. **CNOOC** Once again achieved billion-ton oil and gas exploration new discoveries, opened up new exploration areas such as deep water, deep layers, hidden oil and gas reservoirs, and basin edge depressions. The first phase of the Bozhong 19–6 condensate gas field development project was successfully put into operation, with a cumulative proven geological reserve of natural gas exceeding 2 trillion cubic meters and a condensate oil geological reserve exceeding 200 million cubic meters, marking the entry of China's offshore deep complex buried hill oil and gas reservoir development into a new stage.

Accelerate the Adjustment of the Refining Industry Structure, Highlight the Development of Quality and Efficiency, and Comprehensively Promote Transformation and Upgrading

By 2023, Chinese oil companies will fully enhance the scale level of "refining integration", vigorously promote "oil reduction and chemical increase" and "oil reduction and special increase", transform from fuel-type refineries to refining-integrated refineries, increase the output of chemical raw materials and fine chemicals, and increase the supply of efficient characteristic products and high value-added products such as paraffin, asphalt, lubricating oil, base oil, low-sulfur marine fuel oil, high-end special materials, high-end polyolefins, high-end new materials, etc.

Strengthen the Synergy of Refining Integration and Production and Sales Integration, and Accelerate the Optimization of Supply-Side Structure

Adhere to market orientation, focus on the adjustment of refining product structure with "oil transformation, oil to special" as the main, strengthen the synergy of refining integration and production and sales integration. **PetroChina** Continuously optimize the structural layout, steadily and orderly promote the construction of major transformation and upgrading projects, and promote the existing refining and chemical integration enterprises towards differentiation and high-end development. The ethylene unit of the ten-million-ton refining and chemical integration project—Guangdong Petrochemical refining and chemical project successfully produced qualified products and entered the full trial production stage. **Sinopec** The refining and chemical business flexibly adjusts its operating strategy, strengthens structural optimization, insists on the optimization of production and sales integration, effectively coordinates the rhythm of "oil transformation" and "oil to special", and accelerates the construction of world-class bases. **CNOOC** The adjustment and transformation of the refining and chemical industry structure continues, further optimizing the layout of the refining and chemical industry base, the Huizhou Petrochemical quality upgrade project was put into operation, and the production capacity of aromatics and other chemical products was greatly increased.

Steadily Increase the Proportion of High Value-Added Chemical Products, Promote High-Quality and High-Efficiency Development

Adhere to the "basic + high-end" and "chemical + material" development direction, aim at basic low-cost, high-end high value-added, strengthen advantageous businesses, develop high-end businesses, exit disadvantageous businesses, through changing methods, adjusting structures, driving innovation and connotative development, accelerate the improvement of the comprehensive competitiveness of the chemical sector. **PetroChina** Combining its own technical and market advantages, it accelerates the transformation from "fuel type" to "chemical products and organic materials type". The first production line of Fushun Petrochemical's 5000 tons/year general hydrogenation catalyst project successfully trial-produced the first batch of PHF-151 hydrogenation refining catalysts, marking a breakthrough in the construction of China Petroleum's largest general hydrogenation catalyst project. **Sinopec** strengthens cost control, closely integrates production, sales, research and application, steadily increases the proportion of high value-added products, accelerates the extension of the high-end graphite material industry chain and the landing of key projects in the high-end layout of the washing chemical characteristic industry chain. **CNOOC** focuses on the production of high-end asphalt, special lubricating oil base oil, solvent oil and other characteristic oil products, strengthens the high-end lubricating oil base oil, special oil industry chain, uses its own excellent low-temperature cycloalkane base oil and good viscosity-temperature wax base oil and other resource advantages, develops high-end lubricating oil, and achieves breakthroughs in high-end internal combustion engine oil, industrial transformer oil and various types of process lubricating oil.

Complementing and Mutually Beneficial with Resource Countries, Refining and Chemical International Cooperation Continues to Deepen

Using advantages in production technology and engineering construction, combined with the local capabilities and resource advantages of resource countries, to promote mutual complementarity and seek win–win development. In 2023, **Sinopec** and Kazakhstan National Petroleum and Natural Gas Company signed a key terms agreement on the polyethylene project in the Atirau region of Kazakhstan in Xi'an. **PetroChina** successfully completed the modernization project of the Chimkent refinery in Kazakhstan. The quality of the factory's oil products, production efficiency and environmental protection level have greatly improved, becoming the most advanced modern refinery in Central Asia with an annual refining capacity of 6 million tons, contributing to the stability of the local finished oil market.

Highlighting "Innovation is the First Driving Force for Development", Increasing the Intensity of Tackling Key Problems, and Making New Breakthroughs in Scientific and Technological Innovation

In 2023, Chinese oil companies grasp the trend of technological revolution and industrial transformation, fully build national strategic scientific and technological forces and energy and chemical innovation highlands, accelerate the creation of original technology source areas, optimize the layout of key scientific and technological tackling projects, solve "neck-sticking" technical problems, and achieve a number of important stage results in scientific and technological tackling, results promotion and application, and Chineseation of technical equipment, and substantial progress has been made in tackling key core technologies.

Strengthening the Joint Innovation Mechanism and Tackling Core Technologies, the Independent Ability of Oil and Gas Exploration and Development Has Significantly Improved

Through independent research and development, collaborative research and development and other ways, to create an innovation platform, optimize the research and development mechanism. The Deep Oil and Gas Innovation Alliance, jointly formed by Chinese related universities, research institutes, other enterprises, etc., aims to create an innovation joint body with deep integration of industry, university, research and application, promote the scientific and technological innovation development of deep oil and gas exploration and development, and provide theoretical, technical and talent support for the low-cost, sustainable development and utilization of deep oil and gas. A number of major scientific and technological tackling projects have been implemented, the intensity of major equipment Chineseation tackling has been continuously increased, and a number of key technologies have been targeted to form a high-level scientific and technological self-reliance and self-improvement to promote high-quality development. **PetroChina** Tuha Oilfield has carried out a large number of experiments to improve the recovery rate for the main contradiction of ultra-deep heavy oil water drive development, and innovatively formed the theoretical system of ultra-deep heavy oil gas injection throughput development and the "profile control + foam drive + gas injection throughput" multi-technology coordinated development technology system. **Sinopec** The "Deep Land No.1" Yuejin 3-3XC well was drilled to a depth of 9432 m, refreshing the two records of the deepest well slant depth and ultra-deep layer drilling horizontal displacement in Asia. **CNOOC** The "Deep Sea No.1" ultra-deepwater large gas field was successfully put into production, China's marine oil and gas development achieved a historic leap to 1500 m ultra-deepwater, China's independently designed and built Asia's first deepwater conductor platform "Sea Base No.1" was successfully installed, the first set of Chinese shallow water, deepwater subsea wellhead and oil tree respectively completed subsea installation, the world's first large-scale offshore heavy oil thermal recovery development oil field—Lvda 5–2 North Phase I project was successfully put into production, the world's first offshore shale oil Weiye-1 well drilling and fracturing test was successful, major breakthroughs were made in marine oil and gas development technology tackling.

Strengthening the Deep Integration of Digital Technology and Business, Significantly Improving Operational Management Efficiency

Chinese oil companies accelerate the promotion of intelligent production and digital management, and widely apply intelligent technology to all major links of the oil and gas industry chain. Through data sharing, business collaboration and intelligent application construction, they gradually move towards the stage of intelligent oil and gas fields. In terms of oil and gas exploration, they create a new work mode of "business management process, collaborative work platform, and result sharing network". **PetroChina** Implement remote supervision and command of shale gas wells, and construct a new fracturing collaboration model of "indoor decision-making + real-time field adjustment". Build a digital twin delivery of models, data, and documents centered on factory objects, and create a ground engineering information perception system + process control system. **Sinopec** The first intelligent oil depot management system pilot construction project—Zhanjiang Donghai Island Commercial Storage Intelligent Oil Depot Management System has completed acceptance review and fully launched. Focus on the digital transformation of the entire industry. **CNOOC** Promote the development of intelligent oilfield groups represented by Enping oilfield group, accelerate the greening and intelligent development of offshore oil and gas fields through intelligent oilfield construction, "shore carbon into the sea" and other means.

Adhere to the Integrated Development of Traditional Oil and Gas and New Energy, and Strive to Achieve Green and Low-Carbon Transformation

By 2023, Chinese oil companies will base on resource endowment and the advantages of the oil and gas industry chain, coordinate oil and gas supply security and green and low-carbon development, accelerate the formation of green and low-carbon production methods, and help the country accelerate the construction of a clean, low-carbon, safe and efficient modern energy system.

Implement Green Production, Explore and Promote CCUS and Carbon Neutral Projects

Adhere to the implementation of the national "dual carbon" requirements, vigorously promote green and low-carbon production, actively implement energy-saving and emission-reduction renovation projects, adjust and optimize corporate green and low-carbon energy use, open a new mode of CCUS business operation, and effectively open up the entire industry chain of CCUS business. **PetroChina** Create three super energy basins with carbon dioxide flooding and carbon sequestration of tens of millions of tons in the Ordos Basin, Junggar Basin, and Songliao Basin. The first CCUS training base in China with "practical operation" as its feature has been built in Jilin Oilfield to accelerate the application of CCUS technology and talent training. **Sinopec** The "Qilu Petrochemical—Shengli Oilfield CCUS Project" captures the carbon dioxide emitted by Qilu Petrochemical and transports it to Shengli Oilfield for oil displacement through pipelines, realizing long-distance high-pressure pipeline transportation technology and carbon dioxide displacement oil and sequestration collaborative technology, and establishing a three-dimensional monitoring network for tracking. **CNOOC** In cooperation with international oil companies, it carries out the Daya Bay CCS/CCUS cluster research project, actively explores the new model of offshore "negative carbon" industry. China's first million-ton offshore carbon dioxide sequestration demonstration project—Enping 15–1 oilfield carbon dioxide sequestration demonstration project was officially put into use in the Pearl River Estuary, marking the initial formation of a complete set of drilling and completion technology and equipment system for offshore carbon dioxide injection, sequestration, and monitoring in China.

Develop New Energy Business According to Local Conditions, and Accelerate the Integrated Development of Oil and Gas and New Energy

Chinese oil companies adhere to ecological priority and diversified integrated development, vigorously promote wind power and photovoltaic power generation in onshore oil and gas mining areas and surrounding areas, coordinate the promotion of offshore wind power and oil and gas exploration and development, accelerate the improvement of new energy development and utilization capabilities in the upstream of oil and gas, and actively promote the construction of green oil and gas field demonstrations. **PetroChina** A batch of new energy projects such as the 1.1 million kilowatt photovoltaic power generation project in the Kashgar region and the restart of the Geermu engine in Qinghai Oilfield and supporting new energy projects have started construction. A number of projects have been completed, such as the million-kilowatt photovoltaic power station in Xinjiang, large-scale centralized wind power in Jilin, deep geothermal heating in Beijing, and large-scale green hydrogen in Yumen. The scale of new energy development and utilization has achieved a historic breakthrough. **Sinopec** The first 10,000-ton photovoltaic power generation direct green hydrogen production project in China—Xinjiang Kuche Green Hydrogen Demonstration Project has been fully completed and put into operation, which can reduce carbon dioxide emissions by about 485,000 tons per year. Shengli Oilfield has built the first energy system with integrated source, network, load, and storage in the oil and gas field with independent intellectual property rights, reducing carbon dioxide emissions by about 730,000 tons per year, and opening a new practice of "controlling energy, reducing costs, increasing green, reducing carbon, and improving efficiency" in the entire industry chain. **CNOOC** In accordance with the strategy of "differentiated development, coordinated development, and integrated development", China's first deep offshore floating wind power platform "CNOOC Guanlan" was successfully incorporated into the Wenchang oilfield group grid, marking a major progress in China's deep offshore wind power key technology.

(This section was written by: Wen Shuhan, Reviewed by: Yin Dongqing)

Analysis of Chinese and Foreign Carbon Markets

In 2023, the global consensus on carbon reduction was further consolidated, and the impact of the carbon market continued to strengthen. The EU carbon price hit a record high, breaking through 100 euros/ton for the first time. The construction of China's carbon trading system has made positive progress, the National Carbon Emission Trading System has generally operated smoothly, and the National Voluntary Greenhouse Gas Emission Reduction Trading Market officially started. In the future, the scale of Chinese and foreign carbon markets will continue to expand, their role will be further enhanced, and the oil and gas industry will face both opportunities and challenges.

International Carbon Market Analysis

In 2023, the influence of the international carbon market continued to strengthen while major countries and regions increase their efforts to control carbon dioxide emissions. The global carbon market scale further expanded, and the EU carbon price hit a historical high. The EU Carbon Border Adjustment Mechanism (CBAM) officially entered the transition phase, and the carbon tariff process in some countries accelerated.

X. Qian et al., *Global Oil and Gas Industry Development Report: 2023*,
https://doi.org/10.1007/978-981-95-2375-7_12

Analysis of International Low Carbon Policies

The International Community's Requirements for Carbon Dioxide Reduction Have Become Stricter

Global climate change has been intensifying, and rapid emission reduction became a consensus in the international community. The 28th Conference of the Parties (COP28) to the United Nations Framework Convention on Climate Change conducted the first Global Stocktake of the Paris Agreement. To achieve the goal of limiting warming to 1.5 °C, all sectors worldwide need to take rapid, substantial, and immediate emission reduction actions, reducing greenhouse gas emissions by 43% by 2030 and 60% by 2035 based on 2019 levels, and achieving net-zero emissions by 2050. The International Energy Agency (IEA) updated the "Net Zero Emissions Roadmap: Global Pathways to Achieve the 1.5 °C Target" report, emphasizing the urgency of low-carbon development according to the 1.5 °C target and calling for a faster response from the global energy industry.

Major countries, regions, and sectors propose higher emission reduction targets. In 2023, the global total carbon dioxide emissions were about 40.9 billion tons, a slight increase of 0.5% year-on-year, of which fossil fuel-related carbon dioxide emissions were 36.8 billion tons, an increase of 1.1% year-on-year. Major countries, regions, and sectors have formulated and implemented action plans for greenhouse gas emission reduction targets. The EU proposed the "2040 Climate Target", planning to reduce greenhouse gas emissions by 90–95% (compared to 1990) by 2040, and it was expected to enter the legislative process in the first half of 2024. France issued the "2030 Greenhouse Gas Emission Reduction Action Plan", planning to reduce emissions by 50% based on 1990 levels. The Ukrainian government issued the "Long-term Low Greenhouse Gas Emission Development Strategy", planning to achieve carbon neutrality by 2050. Australia reformed the "Safeguard Mechanism", requiring companies with carbon dioxide emissions exceeding 100,000 tons/year to reduce emissions by 30% (compared to 2005) by 2030, covering sectors including mining, oil and gas extraction, manufacturing, transportation, waste disposal, and some power companies. The International Maritime Organization announced that the international shipping industry would achieve net-zero greenhouse gas emissions around 2050. The U.S. Environmental Protection Agency proposed to fully reduce the carbon emissions of new light-duty vehicles by 2032, reducing the average greenhouse gas emissions by 56% (compared to 2026), and planned to achieve net-zero emissions in the transportation industry by 2050. As part of the EU's "Fit for 55" policy, from 2030 onwards, ships over 5000 tons in EU, Norway, and Iceland (EEA) ports would achieve net-zero emissions at berth. New Zealand announced that by 2030, farm emissions would be reduced by at least 30% compared to 2005 (Table 1).

Countries have been actively taking action to reduce emissions. The European Union proposed the Green Deal Industrial Plan and issued directives to improve energy efficiency, aiming to accelerate carbon neutrality through measures such as

Table 1 Emission reduction plans of major countries/regions/sectors in 2023

Country/region/industry	Planned time	Emission reduction target	Status
EU	2040	90–95% reduction compared to 1990	In progress
France	2030	50% reduction compared to 1990	Enacted
Ukraine	2050	Achieve net-zero emissions	Enacted
High-emission companies in Australia	2030	30% reduction compared to 2005	Enacted
International shipping industry	2050	Achieve net-zero emissions	Enacted
Light-duty vehicle industry in the U.S.	2032	56% reduction compared to 2026	Enacted
Some ships in EEA ports	2030	Achieve net-zero emissions	Enacted
New Zealand farming industry	2030	Reduce emissions by 30% compared to 2005	Enacted

enhancing the competitiveness of European green industries and reducing primary energy consumption. France passed the Green Industry Act to promote climate transition and announced to phase out coal-fired power generation by 2027. Canada announced that it would gradually phase out inefficient fossil fuel subsidies from 2024, provide at least 15% tax relief for investments in clean energy and clean technologies, and invest 40–50 billion Canadian dollars in clean electricity, green infrastructure, and low-carbon technologies over the next decade.

The EU's Carbon Border Adjustment Mechanism Officially Entered the Transition Phase, and Some Countries Accelerated the Process of Carbon Tariffs

On October 1, 2023, the EU's Carbon Border Adjustment Mechanism (CBAM) officially entered a three-year transition phase, covering six major sectors including steel, aluminum, cement, fertilizers, electricity, and hydrogen. After the EU proposed the CBAM, some countries also accelerated the pace of carbon tariff policy. The UK announced that it would implement the CBAM mechanism from 2027, covering imported products including steel, iron, aluminum, fertilizers, hydrogen, ceramics, glass, and cement. At the end of 2023, the Clean Competition Act (CCA), proposed by Democratic senators in 2022, completed the second reading. The CCA is a form of CBAM, but it targets not only importers of carbon-intensive products but also Chinese producers in the United States, taxing only the part that exceeds the carbon emission industry baseline, aiming to enhance the competitiveness of the US clean manufacturing industry and reduce carbon emissions. The Australian government initiated an assessment of whether to introduce CBAM, and it was expected to form an assessment report in the third quarter of 2024. Canada conducted a public consultation on CBAM in 2021 and was exploring the formulation of related policies.

The United States and Japan required some Chinese sectors to collect and disclose carbon emission data. US bipartisan senators proposed the PROVE IT Act, which would require the US Department of Energy to collect average emission intensity data from the production process of some goods within two years of the act taking effect and establish a public database, involving multiple sectors such as aluminum, steel, plastics, and cement. The Japanese Ministry of Economy, Trade, and Industry planned to require electric vehicle manufacturers to calculate and disclose the carbon dioxide emissions generated by battery manufacturing from 2024.

Developing Countries Intended to Accelerate Greenhouse Gas Emission Reductions, but Lacked Technical and Financial Support

According to IEA estimates, between 2024 and 2030, emerging markets and developing economies other than China need to invest 80–100 billion US dollars annually to achieve global net-zero emissions by 2050. African countries passed the Nairobi Declaration and Call to Action on Climate Change at the first African Climate Summit, calling on developing and developed countries to work together to reduce greenhouse gas emissions and urging developed countries to fulfill their climate commitments. The UAE announced that it would provide 4.5 billion US dollars to support the development of clean energy in African countries. Indonesia and Vietnam were in talks with the Just Energy Transition Partnership (JETP) to seek funding from the G7 to help the two countries reduce carbon emissions in the power industry and achieve peak carbon emissions in power ahead of schedule.

Analysis of the International Carbon Trading Market

The Global Carbon Market Continued to Expand

In 2023, two new carbon emission trading markets (ETS) were added globally, in Washington State, USA, and Indonesia, bringing the total to 29 carbon markets at different levels of jurisdictions. The Washington State carbon market in the United States includes transportation, construction, industry, and power generation sectors, covering 63% of total greenhouse gas emissions. The Indonesian carbon market has been in the trial operation stage, covering only coal-fired power plants from 2023 to 2024. As of the end of 2023, there were 8 carbon markets under construction globally, including Pennsylvania, New York, Colombia, Ukraine, EU ETSII, Turkey, Vietnam, and Sakhalin in Russia.

With the Implementation of Multiple Policies, EU Carbon Prices Hit a Record High

On April 25, 2023, the EU Council officially passed the "Fit for The relevant amendments to the 55" package plan, the shipping industry would gradually be included in the carbon market, with 40% of emissions included in the quota management by 2024, reaching 70% by 2025, and fully included by 2026. The free allowances in avaiation sector would gradually phase out from 2024 to 2026, and all would be allocated by auction from 2027. Affected by factors such as carbon market reform and energy price fluctuations, the EU carbon price broke through 100 euros/ton for the first time in history in February, and the overall trend for the year was highly volatile, with an average price of 92.21 dollars/ton, an increase of 7.6% year-on-year. And the price range was 72.28–106.82 dollars/ton, the price difference narrowed by 27.1% compared to the previous year (Fig. 1). The trading activity of the carbon market has declined, with an average cover ratio of 1.93, a decrease of 7.2% year-on-year. The annual futures trading volume was 7.47 billion tons, basically unchanged as the previous year.

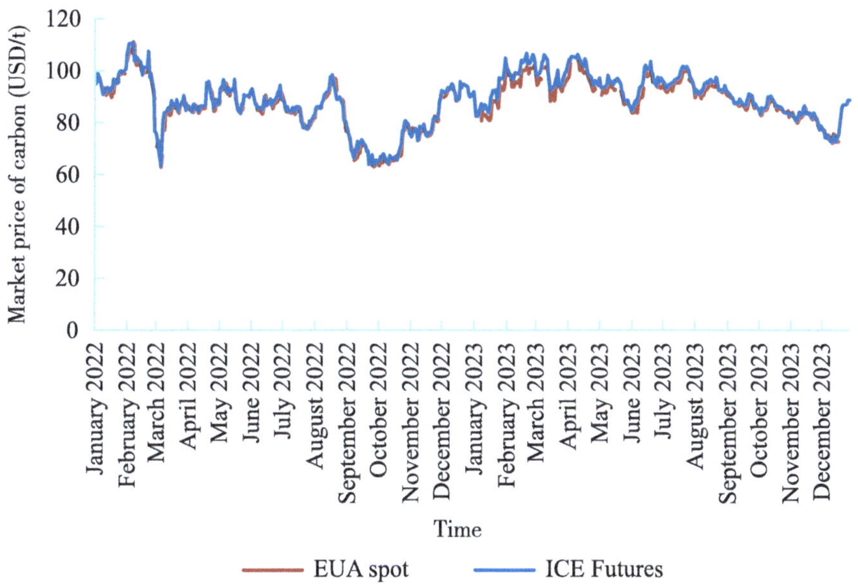

Fig. 1 EU carbon market price trend 2022–2023. *Data source* Reuters, European Energy Exchange (EEX)

Fig. 2 UK carbon market price trend 2022–2023. *Data source* Reuters

UK Clarified Issuance of Additional Allowances, Carbon Price Dropped Significantly

In 2023, the UK carbon price fell all the way after a brief rise at the beginning of the year, with an average annual price of 66.15 dollars/ton, a year-on-year decrease of 29.2%. The price range was 41.77–94.23 dollars/ton (Fig. 2). The sharp drop in UK carbon prices was due, on one hand, to a decrease in natural gas consumption in the power industry, which reduced emission reduction demand. On the other hand, the UK announced carbon emission trading system reform details in July, confirming the additional issuance of 53.5 million tons of allowances from 2024 to 2027.

North American Carbon Market Prices Varied

The United States and Canada have not yet established a national carbon market. Instead, some states and provinces established two regional carbon emission trading systems, the Western Climate Initiative (WCI) and the Regional Greenhouse Gas Initiative (RGGI). Due to differences in covered sectors, climate policies, and emission reduction targets, each market showed different trends (Fig. 3).

WCI California-Quebec carbon price hit a record high. WCI includes four markets in the US states of California, Washington, and the Canadian provinces of Quebec and Nova Scotia. The California and Quebec markets were linked in 2013, covering multiple sectors such as power generation and power imports, large industry, fuel supply, and natural gas transportation. The annual average price of the

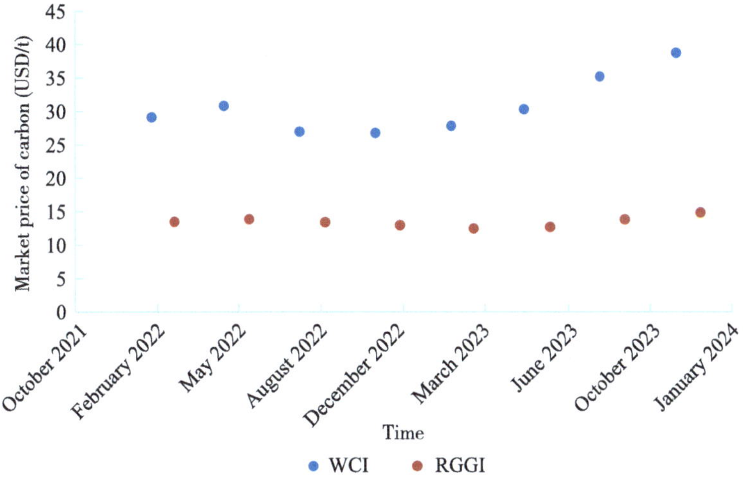

Fig. 3 Price trends of major carbon markets in North America 2022–2023. *Data source* WCI, RGGI

WCI California-Quebec carbon market was 33.03 dollars/ton, a year-on-year increase of 16.1%, and the highest price of the year was 38.73 dollars/ton, a record high. The rise in carbon prices was mainly due to stricter climate policy goals. California passed a roadmap for carbon neutrality by 2045 at the end of 2022, and carbon emissions would be reduced by at least 40% from 1990 levels by 2030. Around this goal, California has made new adjustments to the carbon market and allowance supply, planning to reduce the allowance budget from 2025 to 2030, which supported the carbon price. At the same time, the Canadian federal government proposed a higher carbon price target, setting it at 48 dollars/ton in 2023, increasing by 11 dollars/ton each year, and reaching 126 dollars/ton in 2030. If the carbon pricing system of each province did not reach the federal government's minimum benchmark price, an additional federal carbon tax would need to be paid. Newfoundland and Labrador, Nova Scotia, and Prince Edward Island have implemented a fuel additional carbon tax from July 2023. 2023 is the last year of the "Cap and Trade" system in the WCI Nova Scotia carbon market, with only two auctions held, and high-emission industrial users would implement the "Output-based Pricing System" from 2024. The WCI Washington carbon market was launched in January 2023, and the state regulatory authorities intended to connect with the California-Quebec carbon market under the WCI framework.

RGGI carbon price slightly increased. In 2023, RGGI held 4 auctions, with an average price of 13.49 dollars/ton, a slight increase of 0.2% year-on-year. RGGI only covers the power sector. As the proportion of renewable energy generation continues to increase, the carbon dioxide emissions of the power industry continue to decrease, reducing the demand for carbon emission allowances. RGGI participants included 12 states in the United States, among which the Pennsylvania state court ruled that RGGI tax revenue was illegal, and the state did not conduct quota trading. Virginia decided to withdraw from RGGI before the end of 2023 by vote. In addition, North Carolina's plan to join RGGI fell through, as the law prohibited the state from implementing a cap-and-trade mechanism for the power sector.

Carbon Prices in South Korea and New Zealand Generally Fell

The annual average price of the South Korean carbon market was 8.76 dollars/ton, a year-on-year decrease of 52.7%. In 2023, affected by the oversupply of allowances, the carbon price in South Korea fell all the way from the beginning of the year, dropping below 6 US dollars/ton at its lowest (Fig. 4). In late September, the government decided to issue a series of new market regulations, relaxing market participation and transfer rules, and reducing government involvement. As a result, the carbon price briefly rebounded to 9 US dollars/ton. However, due to weak demand, the market turned downward again, and the decline continued until the end of the year. In order to further boost the market, the government planned to increase the number of market makers. In addition, South Korea was also considering revising its carbon market policy to better respond to CBAM.

The annual average price of the New Zealand carbon market is 38.38 US dollars/ton, a year-on-year decrease of 24.0%, showing a trend of falling first and then

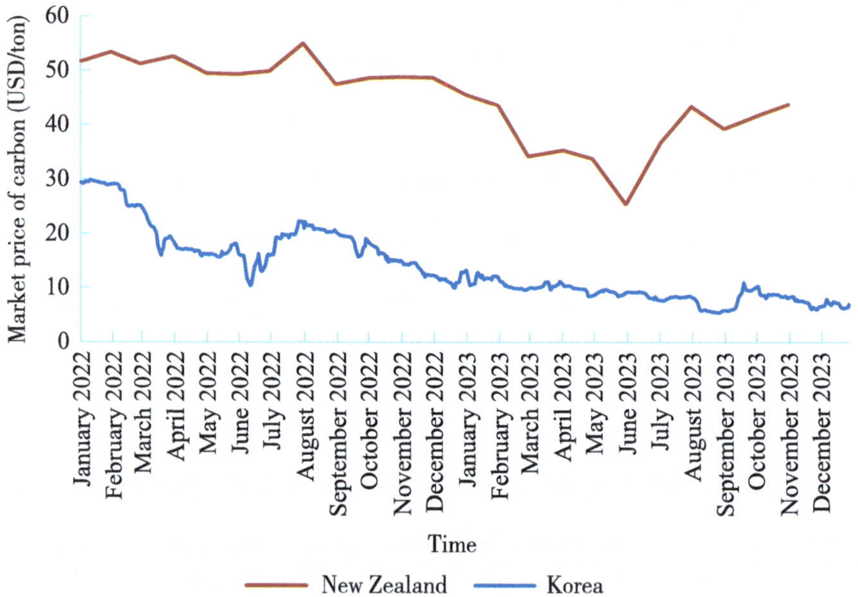

Fig. 4 Price trend of the carbon market in New Zealand and South Korea from 2022 to 2023. Data source: Reuters, Korea Exchange

rising. At the beginning of 2023, the carbon price in New Zealand continued to fall from the previous year. In March, the auction failed once because the clearing price was lower than the floor price set by the government. In June, due to the uncertainty of the general election and carbon market reform, the carbon price reached the lowest of the year at 25.3 US dollars/ton. In July, the government proposed to reconsider the carbon price control mechanism, and the carbon price quickly rebounded, returning to above 40 US dollars/ton. Since then, the market has generally been in a wait-and-see state, and the carbon price fluctuated around 40 US dollars/ton.

Outlook for the International Carbon Market

The scale of the international carbon market will continue to expand. First, the number of countries and regions covered by carbon markets will further increase. The Mexico's carbon market, after three years of trial operation and transition period, was originally scheduled to officially operate in 2023, but will be postponed to 2024 due to legislative process issues. In October 2023, the Tokyo Stock Exchange in Japan began operating the carbon credit trading market, planning to implement ETS in 2026. Brazil, Malaysia, Thailand, Pakistan, India, Chile and other countries have proposed ETS construction plans. Second, the sectors covered by carbon markets will continue to grow. The EU ETS will further cover the shipping sector and gradually phase out free allowances under the ETS framework. The EU ETS II, to be implemented from 2028, will be expanded to the construction and transportation sectors.

The oil and gas industry will face greater emission reduction challenges in the future. The EU and North American WCI carbon markets have respectively included refining and oil and gas supply sectors in the coverage. Canada plans to build a federal oil and gas carbon market, and the specific rules will be open for consultation in 2024 and implemented in 2025. According to the current framework draft, this carbon market will cover direct emissions from oil and gas upstreams and LNG facilities, including crude oil production, open-pit oil sands mining, natural gas production and processing, and LNG production. It will cover carbon dioxide, methane, nitrogen oxides and other greenhouse gases, with a preliminary target of reducing emissions by 35–38% by 2030 compared to 2019. In addition, some countries and regions have introduced policies to control carbon emissions from local oil and gas production and consumption. Alberta in Canada has set emission benchmarks for large emitters including oil and gas companies, and will gradually tighten the benchmarks from 2023 to 2030. Norway, Argentina, Colombia and other countries also levy additional carbon taxes on different oil and gas products. The EU plans to control methane emissions from oil and gas imports from 2030.

(This section was written by Chen Yuanyuan and reviewed by Fan Hui)

Analysis of China's Carbon Market

In 2023, the National Carbon Market operated smoothly overall, the price discovery function initially appeared, oil and gas companies actively carried out forestry carbon sink projects, and strengthened carbon asset management. The State Council reviewed and approved the "Interim Regulation on the Administration of Carbon Emission Trading (Draft)", and an important step was taken in the construction of the carbon trading system. The national trading market for voluntary greenhouse gas emission reduction was officially launched. It is expected that the coverage of the carbon market will further expand in the future, and the "green electricity-carbon emission" effective connection mechanism will gradually form, helping China's green and low-carbon transformation.

Operation of China's Carbon Market

The National Carbon Market Operates Smoothly, and Prices Rise Significantly

In 2023, the national carbon market operated for 242 trading days, with a total annual trading volume of 212 million tons, an average daily trading volume of 875,800 tons, which is 4.2 times that of the previous year (Fig. 5). Among them, bulk agreement trading is the main trading method in the national carbon market, accounting for 83.5% of the total trading volume, and the listed agreement trading volume accounts for 16.5% of the total trading volume.

In 2023, the average trading price of the national carbon market was 68.15 yuan/ton, a year-on-year increase of 23.2%. After July, the price overall rose significantly, and the single-day closing price broke through 80 yuan/ton for the first time since the operation of the national carbon market, reaching a maximum of 81.67 yuan/ton. The price increase was driven by companies' concentrated performance on the one hand, and the Ministry of Ecology and Environment required 95% of the key emission units in each administrative region to complete their performance before November 15, and all to be completed before December 31; on the other hand, the carbon emission quota transfer plan was implemented, the national baseline was lowered, and the quota supply was generally tightened.

Regional Pilot Carbon Market Prices and Transaction Volumes Vary

In terms of price, except for Guangdong, Hubei, and Chongqing, the average transaction price of other regional pilot carbon markets has increased year-on-year. In 2023, the average transaction price of the Beijing Environment Exchange was 113.26 yuan/ton, ranking first among all regional pilot carbon markets, with a year-on-year

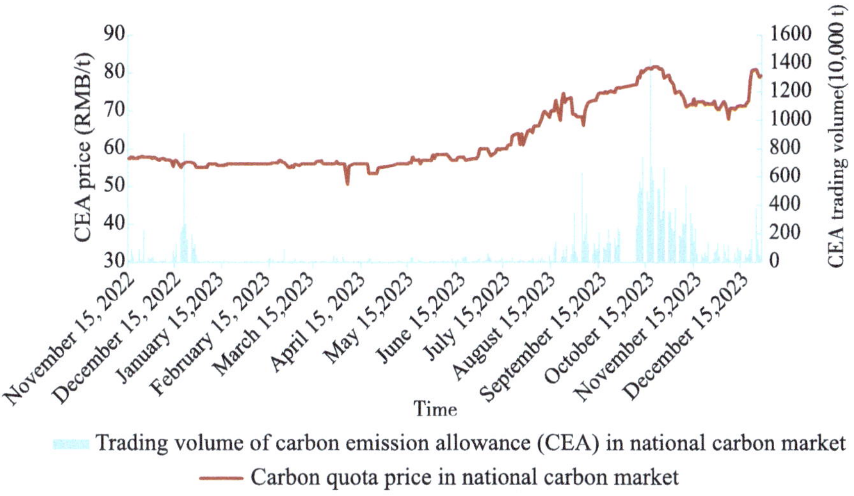

Fig. 5 National carbon market carbon quota price and trading volume. *Source of information* Shanghai Environment and Energy Exchange

increase of 17.5%, mainly due to the reduction of 500,000 tons of carbon emission quotas supplied by paid competitive bidding in 2022 compared to 2021. The price range of other regional pilot carbon markets is between 30 and 76 yuan/ton. In terms of transaction volume, the total transaction volume of the 8 regional pilot carbon markets increased by 26.1% year-on-year, with Shanghai, Tianjin, Chongqing, and Fujian seeing an increase in transaction volume, while other regions saw a decrease. The Fujian Strait Resources and Environment Exchange had the largest increase in transaction volume, reaching 242%, mainly due to reasons such as lowering transaction fees (Fig. 6).

Regional pilot carbon markets are actively exploring the expansion of new trading varieties and industries. The Tianjin Emissions Exchange is exploring the first national data product transaction to deal with the EU's "carbon tariff". The Guangdong Provincial Department of Ecology and Environment issued the "Guangdong Province 2023 Carbon Emission Quota Allocation Plan", further expanding the coverage of the carbon market, including the ceramics, transportation, and data center industries, further promoting the local key industries to shift towards clean and low carbon.

Further Improvement of National Carbon Market Policy

On March 15, 2023, the Ministry of Ecology and Environment issued the "2021 and 2022 National Carbon Emission Trading Quota Total Setting and Allocation Implementation Plan (Power Generation Industry)", compared with the first compliance

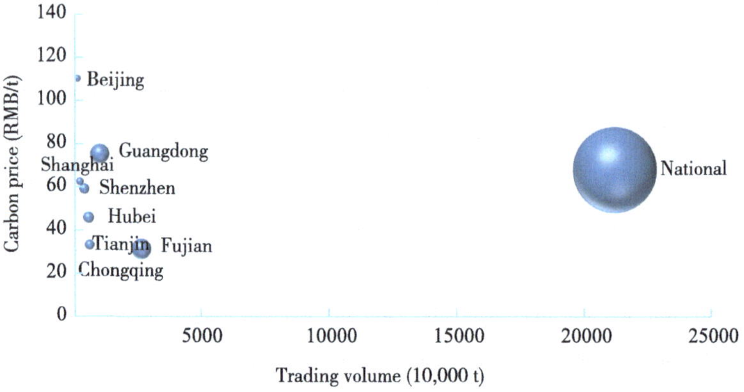

Fig. 6 Comparison of trading conditions between regional pilot carbon markets and the national carbon market (bubble size represents transaction amount) *Source of information* Compiled based on data from exchanges such as the Shanghai Environment and Energy Exchange, Shenzhen Emission Rights Exchange, Beijing Environment Exchange, etc.

period (2019 and 2020), adjustments were made in terms of annual quota management, balance value, benchmark value, correction coefficient, etc. First, annual quota management is implemented. Unlike the previous compliance period where two years were combined for compliance and the same quota allocation benchmark value was used, 2021 and 2022 adopted differentiated quota allocation benchmark values, based on the actual emissions of the previous year to determine the benchmark value for the second year, making the benchmark value more in line with the actual situation of the industry. Quotas were issued and compliance was carried out separately for 2021 and 2022. Second, the balance value was introduced for the first time as the basis for setting the benchmark value. The proportion of units with measured carbon content of coal in 2020 has increased significantly, and the calculated result of the overall carbon emission intensity of the industry is about 10% lower than the corporate emission data used when the benchmark value was calculated, resulting in a significant difference between the benchmark values for 2021 and 2022 compared to 2019 and 2020. This quota allocation plan introduces a balance value. The balance value is the carbon emission intensity value corresponding to the balance of the carbon emission quota of various types of units for power and heat supply and their verified emission volume (quota to be paid), and it has become a reference for setting the benchmark values for power and heat supply. Third, the benchmark values for coal-fired power and heat supply units were reduced. Based on the balance values of various types of units in 2021, the benchmark values for coal-fired power units were reduced by 6.29–18.82%, and those for heat supply units were reduced by 11.83–12.3%. Fourth, the applicable range of the load factor correction coefficient for units was adjusted. Consistent with conventional coal-fired power units, the quota allocation for coal-fired cogeneration units also uses the load factor correction coefficient, and quota compensation is given for low-load operation of cogeneration units. Fifth, a flexible mechanism for advancing the 2023 quota and a personalized relief mechanism were added. For key emission units with a quota gap rate of more than 10%, if they are unable to complete compliance on time due to operational difficulties by purchasing quotas, they can advance some quotas from the pre-allocated quotas for 2023 to complete compliance, and the advance amount should not exceed 50% of the quota gap. For enterprises that undertake major livelihood security tasks, if they still find it difficult to complete compliance after implementing the compliance exemption mechanism and flexible mechanism, a personalized relief plan will be coordinated and studied.

The National Trading Market for Voluntary Greenhouse Gas Emission Reduction Officially Launched

Since 2023, the work to restart the national voluntary greenhouse gas emission reduction market has been accelerated. On October 19, the state issued the "Greenhouse Gas Voluntary Emission Reduction Trading Management Measures (Trial)", which comprehensively regulated the approval and registration of CCER projects, verification and registration of emission reductions, trading and related activities, and subsequently issued the afforestation carbon sink and other 4 "Measures for the Administration of voluntary Greenhouse Gas Emission Reduction Trading" "Greenhouse Gas Voluntary Emission Reduction Project Approval and Emission Reduction Verification Implementation Rules" and other documents, laying the institutional foundation for the start of the CCER market.

On January 22, 2024, the National trading market for voluntary greenhouse gas emission reduction was officially launched. The newly issued CCERs are centrally traded at the Beijing Green Exchange, which helps to increase the concentration and liquidity of transactions. The existing CCER emission reductions are still traded at local pilot carbon exchanges and will become invalid from January 1, 2025. The methodology for new CCERs adopts a "top-down" management model, which is uniformly selected and published by the Ministry of Ecology and Environment, and no longer requires developers to apply for record. The first batch of methodologies includes afforestation carbon sinks, grid-connected solar thermal power generation, grid-connected offshore wind power generation, and mangrove forest construction projects, reflecting the state's support for the development of high-cost renewable energy and the transformation of forestry ecological value. Overall, the restart of CCER will help promote industries and companies that make significant contributions to carbon reduction and increase, such as renewable energy and forestry, to participate in greenhouse gas emission reduction actions, and will help promote the formation of a national carbon market system that complements and connects compulsory and voluntary carbon markets. Fourth, the green electricity market and regional carbon markets strengthen their connection. In 2023, the four pilot carbon markets in Tianjin, Beijing, Shanghai, and Hubei achieved a connection with the green electricity market. On March 28, Tianjin issued a notice on the verification and fulfillment of the 2022 carbon emission report, clarifying that key emission units can apply for a deduction of the green electricity volume purchased into the grid when calculating the net purchase and use of electricity. On April 19, Beijing issued a notice on the management of carbon emission units and the pilot work of carbon emission rights trading in 2023, clarifying that the green electricity carbon emissions purchased and used by key carbon emission units through market means are calculated as zero. On June 8, Shanghai issued a notice on adjusting the calculation method of green power carbon emissions in the purchased electricity of carbon trading companies in the city, clarifying that the carbon emissions of purchased green electricity are calculated as zero. On November 6, Hubei Province issued the 2022 Carbon Emission Rights Allocation Plan, and emission control companies can use the emission reduction corresponding to the green electricity trading certificate jointly certified by the Hubei Power Trading Center and the Hubei Carbon Emission Rights Trading Center to offset actual carbon emissions. The green electricity market and the national carbon market need to be further connected. According to the Notice on Doing a Good Job in the Reporting and Verification of Greenhouse Gas Emissions of Some Key Industry Enterprises from 2023 to 2025, enterprises that purchase and use green electricity through market transactions need to separately report this part of electricity consumption and provide relevant supporting materials, and the corresponding emissions are temporarily calculated based on the average carbon emission factor of the national power grid. Fifth, oil and gas companies strengthen carbon asset management and accelerate the layout of forestry carbon sink projects Oil and gas companies strengthen the development and management of carbon assets. In 2023, China National Petroleum Corporation(CNPC) Jilin Oilfield's waste heat utilization project was packaged into international carbon assets, and Sinopec Carbon Science Company's upstream carbon reduction project was officially approved by the German Federal Environment Agency, obtaining the "entry ticket" for German carbon trading. Oil and gas companies actively carry out forestry carbon sink projects. In April, CNPC purchased 220 tons of carbon sink from the Xiaohai Mangrove Ecological Restoration Project from the Wanning City Government of Hainan Province for carbon neutrality at the international large-scale event held in Hainan Province, which has practical significance for realizing the ecological value of mangroves, exploring blue carbon trading and mangrove protection. In August, CNOOC signed a forestry carbon sink purchase agreement with Zhuozi County Xingnong Agricultural Investment Co., Ltd. in Inner Mongolia. CNOOC will purchase about 500,000 tons of CCER developed by Zhuozi County according to the carbon sink afforestation project methodology to offset the carbon emissions produced by its gas-fired power plants. In October, Sinopec offset the annual carbon emissions of the Dongying Crude Oil Depot through the carbon-sink generated by the Shengda Ecological Forest Farm Carbon-Sink Afforestation Project in Shengli Oilfield. The Dongying Crude Oil Depot was also certified as the first "carbon neutral" crude oil depot in China. Sixth, Outlook for China's Carbon Market On January 5, 2024, the State Council reviewed and approved the "Interim Regulations on Carbon Emission Rights Trading (Draft)", which clarified the basic system of China's carbon emission rights trading management, and put forward clear requirements for improving data quality management and strengthening legal responsibilities, which will promote the continuous standardized development of the national carbon market. The future development of China's carbon market will show the following trends: First, the industry coverage will gradually expand. The Ministry of Ecology and Environment has organized special discussions on expanding the industry coverage of the national carbon market, further refined the reporting and verification requirements for emissions from industries such as cement and electrolytic aluminum, and the signal for expansion is becoming clearer. Second, the "Green Electricity-Carbon Market" connection mechanism will be further strengthened. Opening up the connection channel between the electricity and carbon markets will help form a system synergy and jointly promote carbon reduction. Third, the carbon market is expected to gradually introduce a paid allocation mechanism for carbon quotas. Globally, the transition from free allocation to paid allocation of carbon quotas is a trend. Some Chinese pilot carbon markets have explored the implementation of auction mechanisms. Against the backdrop of the accelerated construction of the national carbon market, oil and gas companies should continue to promote energy conservation and emission reduction, implement clean energy substitution, accelerate industrial transformation and development, and further strengthen the tackling of low-carbon, zero-carbon, and negative-carbon related technologies. On the other hand, they should pay attention to the market opportunities for the development of potential CCER projects such as CCUS, offshore wind power, and solar thermal power generation for offset mechanisms, and at the same time, leverage the mutual recognition and interconnection mechanism of the carbon electricity market to tap the energy-saving and carbon-reducing potential of corporate green electricity projects.

(Section written by: Xiang Chao Reviewer: Fan Hui)

Energy Transition and New Industries

Development of the Hydrogen Energy Industry

New Situations and Requirements for the Development of Hydrogen Energy in China

The Development of Clean Hydrogen Has Become a Global Consensus, and International Competition and Cooperation Face New Situations

With the global consensus on building a safe and stable green energy system, investment and innovation in the clean hydrogen industry continue to increase. According to the "Hydrogen Energy Future Development Trend Research Report" released by the International Hydrogen Energy Committee, the global demand for hydrogen energy will increase tenfold by 2050. Based on the huge application potential of hydrogen energy and the enormous value of its industrial chain, its development and utilization have not only become an important path for many countries to achieve energy transition, but also risen to an important field of international competition. Developed countries and regions such as Europe, Japan, and the United States are focusing on the layout of the clean hydrogen industry. Developing regions such as Africa and Latin America are actively promoting the landing of clean hydrogen projects based on their resource endowment advantages. Hydrogen energy is gradually developing into a new highlight of global energy transformation and cooperation.

The Risk of Technical Equipment Shortcomings Still Exists, and Industrial Safety Faces New Challenges

China's hydrogen energy industry started late and has a relatively weak foundation. The core technology design, component manufacturing, control systems, etc. of

X. Qian et al., *Global Oil and Gas Industry Development Report: 2023*,
https://doi.org/10.1007/978-981-95-2375-7_13

hydrogen energy utilization have not yet been fully realized. There are still certain shortcomings in the manufacturing of key components such as hydrogen liquefaction, hydrogen storage containers, and fuel cells. The level of autonomy in raw materials such as bipolar plates, proton exchange membranes, and membrane electrodes is low, and product performance needs to be improved. In the face of international competition, technology blockades, and trade controls, China needs to concentrate its efforts to strengthen the autonomy and controllability of key technologies and equipment, improve industrial supporting capabilities, and ensure the safety of the supply chain.

The Standard and Evaluation System Lags Behind, and Market Promotion Faces New Tasks

China's hydrogen energy standard system has not yet fully covered the "production-storage-transportation-use" links. There are still some gaps in the construction of hydrogen refueling stations, hydrogen storage and transportation, and hydrogen energy applications, and it cannot meet the actual needs of all links in the upstream, middle and downstream of the industrial chain. It is particularly important to strengthen the formulation and implementation of standards for special equipment and applications in the hydrogen energy industry and improve the overall technical level and quality of standards.

At the industry level, technology research and development lags behind market demand, product performance indicators are uneven, and accurate and fair third-party testing and evaluation are lacking. The interactive pattern of "technology guides standards, and standards regulate technology" has not yet been formed. Accelerating the establishment and improvement of the hydrogen energy standard and testing and evaluation system, promoting the update and iteration of hydrogen energy products and technologies, and improving the industrial development environment are new tasks facing the development of hydrogen energy in China.

Overview of China's Hydrogen Energy Industry Development in 2023

Hydrogen energy industry innovation resources will further gather, with the goal of industrial chain scale construction and major project demonstration and promotion, the hydrogen energy industry will gradually be built into an important tool to promote the construction of modernization with Chinese characteristics and help achieve the "dual carbon" goal.

Overview of China's Hydrogen Energy Supply and Demand

According to statistics from the China Hydrogen Energy Alliance Research Institute, as of June 2023, China's hydrogen production capacity is about 49 million tons/year, mainly concentrated in the northwest, east, and north regions, an increase of 1.4% from the end of 2022. Fossil energy hydrogen production is still the mainstream hydrogen production method in China, accounting for 77.9% of the total hydrogen production capacity; Hydrogen production capacity from water electrolysis accounts for about 1.3% (Fig. 1). As of December 2023, China has completed and operated 58 renewable hydrogen projects, an increase of 22 from the end of 2022, with a total project installed capacity of 654.4 megawatts. The technology route is mainly alkaline electrolytic water hydrogen production, accounting for 95.4%; the power source is mainly photovoltaic, accounting for 83.5%; the application direction is mainly refining, accounting for 40.1%. According to the utilization rate of China's hydrogen production project capacity in the first half of 2023, it is estimated that China's hydrogen production will reach 35.7 million tons or more in 2023.

Hydrogen energy has a wide range of uses. In the industrial field, hydrogen as an important chemical raw material can be used in the production process of synthetic ammonia, methanol, refining, coal-to-oil gas, steel, etc.; in the construction field, hydrogen fuel cells generate electricity for buildings, and waste heat can be recovered for heating and hot water; in the transportation field, hydrogen is used as a fuel cell for passenger cars, and can also be used as a fuel internal combustion engine power in aircraft, ships and other non-land transport carriers; in the power field, hydrogen as a new type of energy replaces coal and other non-renewable energy for power generation, while completely subverting the traditional power generation system, it also eliminates carbon emissions from the power source, and can be used for peak regulation of renewable energy power systems.

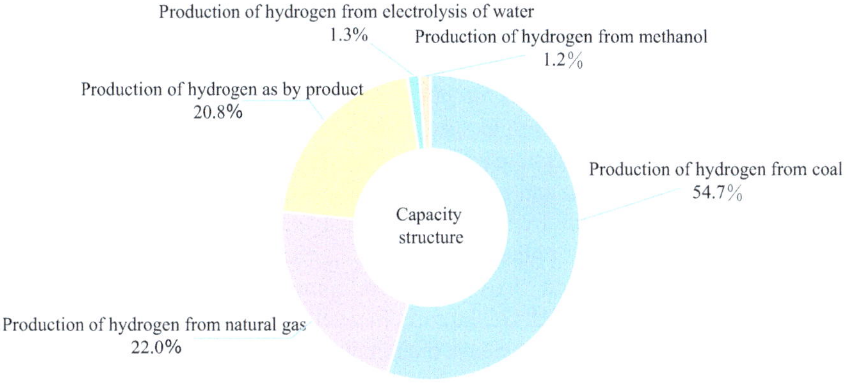

Fig. 1 Structure of China's hydrogen production capacity (as of the first half of 2023). *Data source* China Hydrogen Energy Alliance, CNPC Economic Research Institute

With the continuous expansion of hydrogen energy application scenarios, the demand for hydrogen energy in our country will greatly increase. It is estimated that by 2030, the proportion of hydrogen energy in the domestic terminal energy system will reach 5%, and by 2050, it will reach at least 10%, with a hydrogen demand of nearly 60 million tons.

Significant Breakthroughs in Technological Innovation

In recent years, our country has made significant breakthroughs in scientific research on hydrogen energy industry, and the key technologies such as hydrogen preparation, storage and transportation, refueling, and fuel cell system integration have been continuously improved. Many key technologies and equipment have been independently developed, and a relatively complete hydrogen energy innovation chain has been basically constructed.

According to the "China Hydrogen Energy and Fuel Cell Industry Data Manual 2022", among the 55 subdivided technical directions in the six major technical fields of hydrogen production, hydrogen storage and transportation, hydrogen refueling, fuel cells, frontier cross, and hydrogen safety and quality control, our country has 2 technologies in the leading stage, 19 technologies in the parallel stage, and 34 in the following stage. Among the 53 subdivided technologies in the parallel and following stages, the gap of 33 subdivided technologies has narrowed, and the gap of 5 technologies represented by vehicle-mounted type III bottles, fuel cell systems, air compressors, and bipolar plates has rapidly narrowed. The scale of hydrogen refueling infrastructure construction leads the world, the construction of pipeline infrastructure is orderly advanced, and the innovation of hydrogen energy technology is booming.

Industrial Layout and Green Hydrogen Project Overview

In August 2023, six departments of our country jointly issued the "Hydrogen Energy Industry Standard System Construction Guide (2023 Edition)", systematically constructing a standard system for the whole industry chain of hydrogen production, storage, transportation, and use, covering five subsystems of foundation and safety, hydrogen preparation, hydrogen storage and transportation, hydrogen refueling, and hydrogen energy application. At present, our country's hydrogen energy industry has gradually formed "five major development regions in the east, west, south, north, and middle", which are the early fuel cell vehicle research and development and demonstration area in the east, the renewable energy hydrogen production and fuel cell stack research and development area in the west, the large-scale demonstration and hydrogen refueling network planning mature area in the south, the fuel cell stack and key parts research and development area in the north, and the fuel cell parts research and development and bus large-scale demonstration area in the middle, and gradually radiates to the surrounding areas, the industrial agglomeration effect is initially apparent.

With the rapid growth of our country's installed capacity and power generation of renewable energy, green electricity hydrogen production has gradually become an important choice for regions rich in new energy resources, and renewable energy hydrogen production projects have accelerated their landing. According to the statistics of the Hydrogen World Big Data Platform, as of December 2023, our country has built and operated 58 renewable hydrogen projects; the total project scale has reached 654.4 terawatts, an increase of 26.3% year-on-year.

The key green hydrogen projects in our country are listed in Table 1.

Hydrogen Storage, Transportation and Refueling

Compared with other gases and liquids, hydrogen as an energy carrier faces challenges of low energy density, hydrogen embrittlement and safety issues. Its safe and efficient storage and transportation are key to the industrial scale development. In the development of the hydrogen energy industry, the storage and transportation of hydrogen is the key bridge connecting the hydrogen production end and the demand end, so efficient and low-cost hydrogen storage and transportation technology is a necessary guarantee for large-scale hydrogen use. According to the storage state of hydrogen, the storage and transportation methods can be divided into gaseous storage and transportation, low-temperature liquid storage and transportation, organic liquid storage and transportation and solid storage and transportation. At present, high-pressure gaseous hydrogen storage and low-temperature liquid hydrogen storage have entered the commercial application stage, while organic liquid hydrogen storage and solid material hydrogen storage are still in the technology research and development stage (Table 2). From the overall development trend, according to the prediction of the China Hydrogen Energy Alliance report, hydrogen energy storage and transportation will follow the technology development direction of "low pressure to high pressure" and "gaseous to multiphase", gradually improving the storage and transportation capacity of hydrogen.

Hydrogen refueling stations are specialized places for refueling hydrogen fuel for fuel cell vehicles. As a key link in serving the commercialization of hydrogen energy transportation, it is an important infrastructure for the development of the hydrogen energy industry. In terms of scale, the number of hydrogen refueling stations in China is increasing year by year (Fig. 2). As of December 2023, 429 hydrogen refueling stations have been built nationwide, 274 are in operation, covering 29 provinces, with the top five provinces being Guangdong, Shandong, Hebei, Henan, and Zhejiang, accounting for 45.8% of the total. The cumulative supply capacity of operating hydrogen refueling stations has reached 208,000 kg/day. According to the announced planning targets, by 2025, China will have built at least 762 hydrogen refueling stations, with a total of 88,000 fuel cell vehicles, and the scale of the hydrogen energy industry will be close to 700 billion yuan.

Table 1 Key green hydrogen projects in our country

Company name	Project name
State Power Investment Corporation Limited	Jilin Dadian Integrated Demonstration Project of Hydrogen Production and Ammonia Synthesis from Wind and Photovoltaic Power Generation
China Petrochemical Corporation	Kuche Green Hydrogen Demonstration Project in Xinjiang Ordos Wind and Photovoltaic Power-to-Hydrogen Demonstration Project
China National Petroleum Corporation	Integrated Project of Wind and Photovoltaic Power-to-Hydrogen and Ammonia Synthesis in Xinghe, Ulanqab; Renewable Energy Hydrogen Production Project in Yumen Oilfield
National Energy Group	Integrated Wind and Photovoltaic Power-to-Hydrogen and Ammonia New Type Energy Storage Demonstration Project
China Energy Construction Corporation Limited	Yumen Integrated Wind, Photovoltaic, Hydrogen Storage and Hydrogen Energy Comprehensive Utilization Project Urat Middle Banner Integrated Wind and Photovoltaic Hydrogen Production and Ammonia Demonstration Project Balin Left Banner Green Hydrogen-based Chemical Industry Base Demonstration Project Songyuan, Jilin Annual Production of 45,000 Tons of Electrolyzed Water Hydrogen, 200,000 Tons of Flexible Synthetic Ammonia and 20,000 Tons of Green Methanol Integrated Project
China National Coal Energy Group Co., Ltd	Ordos 100,000 Tons/Year Liquid Sunlight Methanol Technology Demonstration Project 500,000 tons/year off-grid wind and Photovoltaic hydrogen production and green ammonia synthesis technology demonstration project
Guofu Hydrogen Energy, Longyuan New Energy and China Machinery Equipment	Fengzhen Wind and Photovoltaic Hydrogen Production Integration Project in Inner Mongolia
China Energy Construction Group	Manzhouli 60,000 tons/year Wind and Photovoltaic Hydrogen Production Integration Demonstration Project in Inner Mongolia Chifeng Wind and Photovoltaic Hydrogen Production Integration Demonstration Project
China Huaneng Group Co., Ltd.	Pengzhou 13 MW Hydroelectric Hydrogen Production Technology Innovation Demonstration Project; Chifeng Wind and Photovoltaic Storage High Proportion Coupling Green Electricity Hydrogen Production Demonstration Project in Inner Mongolia
China Yangtze Power Co., Ltd.	China Three Gorges Green Electricity Green Hydrogen Demonstration Station, the first domestic river port-based hydrogen production and refueling integrated project

(continued)

Table 1 (continued)

Company name	Project name
Shuimu Mingtuo (Damao) Hydrogen Energy Technology Co., Ltd.	Phase I construction of 150,000 cubic meters/hour regenerated water electrolysis hydrogen production and 1800 tons/day dynamic green synthesis ammonia system, supporting 1.5 million kilowatts wind farm and demonstration area overall public auxiliary engineering, completed and put into operation in 2024
China General Nuclear Power Group Co., Ltd., Jidaoneng Energy	Wind and Photovoltaic Hydrogen Production Million Tons Green Methanol Project
Xintian Green Energy, Dongfang Electric, Datang Group	Phase I is the construction of hydrogen production project in the park, hydrogen fuel cell demonstration production line and green electricity hydrogen production supporting wind power generation project; Phase II is the construction of hydrogenation station and supporting integrated energy station project, rural revitalization distributed wind power project, rural revitalization photovoltaic power generation project
CRRC Wind Power	Phase I plans 300 sets of large square electrolytic water hydrogen production equipment, green hydrogen green methanol 50,000 tons/year

Sales of Hydrogen Fuel Cell Vehicles

At present, the number of fuel cell vehicles in China has exceeded 10,000, and the overall application is showing a rapid expansion trend. According to data from the China Association of Automobile Manufacturers, in 2022, the sales volume of hydrogen fuel cell vehicles in China reached more than 5000, about 4 times that of 2017, showing a rapid growth trend (Fig. 3). The "Energy Saving and New Energy Vehicle Technology Roadmap 2.0" compiled by the China Society of Automotive Engineers predicts that by 2025, the number of hydrogen fuel cell vehicles in China will increase to 100,000, and by 2035, it will reach about 1 million.

Hydrogen Energy Industry Policy

International Hydrogen Energy Policy

With the increasing urgency of global energy low-carbon transformation needs, countries around the world have all focused their energy transformation on the field of hydrogen energy. According to incomplete statistics, at least 14 countries around the world have issued policies related to hydrogen energy. In addition to Japan and South Korea, the United States, the United Kingdom, Russia and other countries have also issued a series of hydrogen energy-related policies. The total number of hydrogen

Table 2 Comparison of hydrogen storage technologies

Storage method	Core technology	Advantages	Disadvantages	Technology maturity
Gaseous storage	High pressure compression	Lower cost, normal temperature operation, low hydrogen storage energy consumption, fast hydrogen charging and discharging speed	Hydrogen storage density is small, storage container volume is large, there is a risk of hydrogen leakage and container explosion	The technology is mature and currently the most widely used
Low-temperature liquid storage	Low-temperature insulation	High energy density, high volume density, short refueling time	High cost, high refrigeration energy consumption, high insulation requirements	The technology is mature and mainly used in fields such as aviation
Organic liquid storage	Organic hydrogen storage medium	High hydrogen storage density, high stability, good safety, convenient transportation, hydrogen storage medium can be recycled multiple times	High cost, high dehydrogenation temperature, high energy consumption, low hydrogen purity	No major technical obstacles
Solid state storage	Physical or chemical adsorption of hydrogen	Good safety, high hydrogen storage density, high hydrogen purity, convenient transportation, fast hydrogen charging and discharging	High cost, constraints on hydrogen storage	Still in the stage of technological improvement, has been demonstrated in distributed power generation, wind power hydrogen production

Source "World Energy Blue Book", Tsinghua University Institute of Nuclear Energy and New Energy Technology, KPMG Analysis

energy policies issued by 14 countries around the world has reached more than 40 (Table 3).

Japan, as a representative of hydrogen energy development in Asia, determined early on to develop hydrogen energy and updates it almost every year. For Western countries such as the United States and the United Kingdom, their hydrogen energy policies have distinct regional characteristics. Some regions with better energy development conditions will introduce some separate hydrogen-related policies to boost

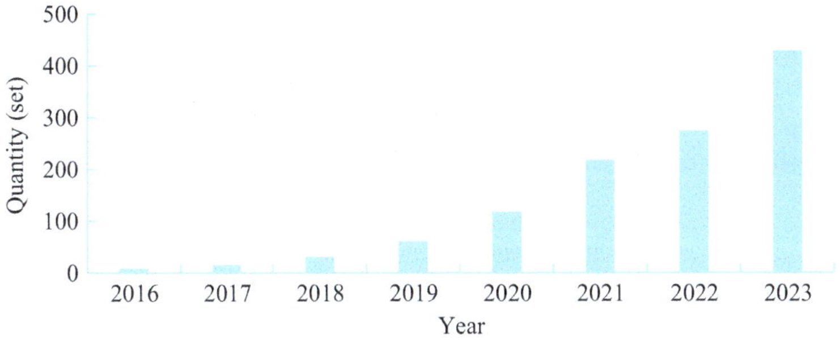

Fig. 2 Number of hydrogen refueling stations built in China. *Data source* China Hydrogen Energy Alliance, PetroChina Economic Research Institute

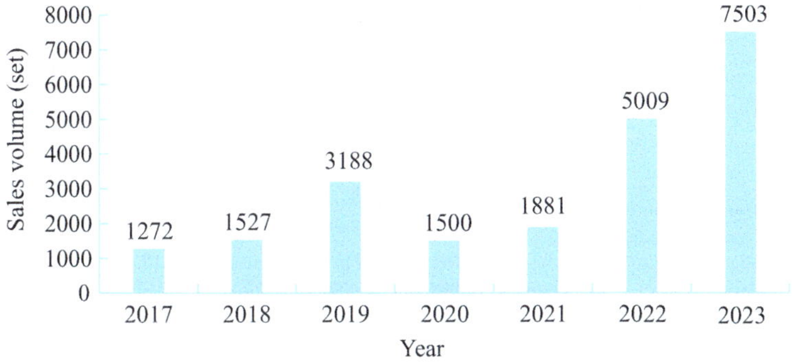

Fig. 3 Annual sales of fuel cell vehicles in China. *Source* China Association of Automobile Manufacturers, PetroChina Economic Research Institute

local hydrogen industry development. Looking at the overall hydrogen energy target planning of Western countries, some goals seem somewhat overly optimistic. For example, the "U.S. Hydrogen Energy Development Plan" mentions "reducing the cost of electrolytic cells to $300/kW, extending the operating life to 80,000 h, achieving a system conversion efficiency of 65%, and reducing the price of hydrogen for industrial and power sectors to $1/kg". Russia's hydrogen energy policy formulation has a strong specificity. Unlike other countries that vigorously develop green hydrogen, Russia mainly focuses on developing blue hydrogen.

Domestic Hydrogen Energy Policy

The National Development and Reform Commission released the "Medium and Long-term Plan for Hydrogen Energy Industry Development (2021–2035)" on March 23, 2022, reflecting the policy's positioning for the future development of the

Table 3 Hydrogen energy development policies by country

Country	Policy	Content	Time	Issuer
Japan	The Strategic Road Map for Hydrogen and Fuel Cells (2014)	Stage 1: Increase the installed fuel cell capacity in Japan to 1.4 million units by 2020 and 5.3 million units by 2030; Stage 2: (2020-2030 overseas purchase price down to 30 yen hydrogen/m^3 Stage 3: zero carbon hydrogen production, transportation and storage	2014	JH2A/Fuel Cell Strategic Association
	The Strategic Road Map for Hydrogen and Fuel Cells (2016)	The use of hydrogen fuel cells has been fully accelerated, and the technical standards have been revised to meet international standards	2016	METI
	Basic Hydrogen Strategy	By 2030, the commercial power generation of hydrogen fuel cells will reach 1 GW, and the cost will be controlled within 17 yen/kW. By 2050, FCVs will be fully popularized, fuel vehicles will be stopped for sale, power generation capacity will be increased to 15-30 GW, and the cost will be reduced to 12 yen/(kW h).	2017	Japanese government
	Development strategy for hydrogen and fuel cell technology	The technology development strategy focuses on industries related to fuel cells, hydrogen preparation and storage and transportation, and hydrogen fuel power generation	2019	Japanese government
	The Strategic Road Map for Hydrogen and Fuel Cells (2019)	The R&D of 10 key projects in three technical fields: fuel cell, hydrogen energy supply chain, and hydrogen production from electrolytic water	2019	METI
	The 6th Strategic Energy Plan	By 2030 and 2050, clean hydrogen/ammonia power generation will account for 1% and 10% of the total electricity supply, and hydrogen will be used as the main power source to combat climate change and carbon neutrality by 2050, and the demand for hydrogen and ammonia will reach 3 million tons/year by 2030	2021	Japanese government
	Revised the 2017 Basic Hydrogen Strategy	A six-fold increase in hydrogen use to 12 million tons by 2040; The public-private sectors will jointly invest 15 trillion yen over the next 15 years to promote hydrogen applications, and nine technologies, including fuel cells, electrolytic water hydrogen production equipment, are listed as "strategic areas."	2023	Japanese government

(continued)

Table 3 (continued)

Country	Policy	Content	Time	Issuer
U.S	New York State fuel cell subsidy policy	To provide subsidies of $2,000 for leasing or buying and selling new energy vehicles, including hydrogen fuel cell vehicles	2016	Government of New York
	The U.S. Hydrogen Economy Roadmap: Executive Summary	By the end of 2022, the total segmented hydrogen market will reach 12 million tons, and the number of hydrogen fuel cell vehicles will reach 50,000 and hydrogen energy material handling vehicles will reach 50,000. By 2025, the total demand for hydrogen energy will be 13 million tons, 125,000 hydrogen fuel vehicles will be put into operation in the field of material handling, and 200,000 hydrogen fuel cell vehicles of all types will be on the road. By 2030, the hydrogen market will exceed 17 million tons, with 5.3 million hydrogen-fueled vehicles and 5,600 hydrogen fueling stations in the United States. By 2050, the revenue of the hydrogen energy industry will reach $750 billion	2019	FCHEA
	The Hydrogen Energy Plan development plan	By 2030, the technical and economic indicators of hydrogen energy development: (1) the cost of electrolyzer is reduced to $300/kW, the operating life is 80,000 hours, the system conversion efficiency is 65%, the price of hydrogen in the industrial and power sectors is reduced to $1/kg, and the price of hydrogen in the transportation sector is reduced to $2/kg; (2) The cost of hydrogen transmission and distribution in the transportation sector in the early market is reduced to $5/kg, and the cost of hydrogen transmission and distribution in the eventually expanded high-value product market is reduced to $2/kg; (3) The cost of the vehicle hydrogen storage system reaches $8/(kW h) under the energy density of 2.2 kW h/kg and 1.7 kW h/L, and the cost of high-strength carbon fiber for hydrogen storage tanks reaches $13/kg	2020	DOE

(continued)

Table 3 (continued)

Country	Policy	Content	Time	Issuer
	Fuel cell Electric Trucks: A Vision for freight movement in California and Beyond	By 2035, 200 hydrogen refueling stations will be built and 70,000 hydrogen heavy trucks will be on the road	2021	California Battery Partnership

hydrogen energy industry. It also clarifies the application scenarios and fields encouraged by the policy, outlines the medium and long-term blueprint for hydrogen energy, helps to strengthen investors' confidence in the development of the hydrogen energy industry, and boosts the long-term expectations of industry participants. While the state attaches great importance to the hydrogen energy industry, local governments actively respond. The local "Fourteenth Five-Year" plans that have been released have planned for the development of hydrogen energy, and some regions have released special plans for the hydrogen energy industry. From the overall content, the key focus is on renewable energy hydrogen production, the development of the entire hydrogen energy industry chain, and hydrogen fuel cell vehicles (Table 4).

Prospects for the Key Directions of China's Hydrogen Energy Industry Construction

Extensively Carry Out Domestic and Foreign Exchanges and Communication, Explore New Models and New Forms of International Cooperation in Clean Hydrogen, and Ensure Industrial Competitive Advantages

The European Union, the United States, Japan and other countries and regions have fully promoted the future development strategy of clean energy and accelerated the deployment of the clean hydrogen industry system. To achieve the healthy development of China's hydrogen energy industry, under the guidance of the new energy security strategy of "four revolutions and one cooperation", we should strengthen exchanges and cooperation with domestic and foreign leading enterprises in the fields of hydrogen-based energy trade, international standard setting, key technology research and development, and transnational demonstration project construction, explore cooperation opportunities and formulate a systematic cooperation framework. Relying on extensive deployment in scenario development, demonstration scale and commercial exploration, we explore diversified cooperation models. In key areas such as efficient hydrogen production from renewable energy, hydrogen liquefaction, and hydrogen fuel cells, we should increase the layout and continuously

Table 4 China national and regional hydrogen policies

Time	Policy	Content
May 2012	12th Five-Year Plan for the Development of Strategic Emerging Industries	FCVs are included in the key development direction of the national development of strategic emerging industries
May 2015	Made in China 2025	Continue to support the development of the FCV industry and focus on improving core technology innovation
Jun. 2016	China Energy Technology Innovation Action Plan 2016–2030	Clearly outline the renewable energy hydrogen production, hydrogen energy and fuel cell technology innovation as the key development content
Nov. 2016	13th Five-Year Plan for the Development of Strategic Emerging Industries	Promote R&D and industrialization of FCVs, and realize mass production and large-scale demonstration application by 2020
Apr. 2017	Mid- to Long-term Development Plan for the Automotive Industry	Gradually expand the pilot demonstration map for FCVs
Jun. 2020	Energy Work Guidelines 2020	Formulate and implement plans for the development of the hydrogen energy industry, organize research on key technologies and equipment, and actively promote application demonstrations
Sept. 2020	Notice on the Development of Fuel Cell Vehicle Demonstration Applications	The demonstration city will be "awarded instead of compensated", and the demonstration will be carried out focusing on the technical research and industrial application of key components
Mar. 2021	Outline of the 14th Five-Year Plan for Economic and Social Development and Long-range Objectives through 2035 of the People's Republic of China	In the field of cutting-edge technology and industrial change such as hydrogen energy and energy storage, we will organize the implementation of future industry incubation and acceleration plans, and plan the layout of a number of future industries

(continued)

Table 4 (continued)

Time	Policy	Content
Mar. 2022	The Medium and Long-Term Plan for the Development of Hydrogen Industry (2021–2035)	Clarify the energy attributes of hydrogen, put forward the basic principles of the development of hydrogen energy industry, the goals of each stage of the development of hydrogen energy industry, and deploy important measures to promote the high-quality development of hydrogen industry
Jan. 2023	Guiding Opinions on Accelerating the Development of Electronics Industry	Accelerate the research on efficient hydrogen production technology, and promote the R&D of hydrogen storage materials, hydrogen storage containers and vehicle hydrogen storage systems; Support the development and application of system integration technology for hydrogen production, hydrogen storage and hydrogen burning
Apr. 2023	Energy Work Guidelines 2023	Accelerate research on new key energy storage technologies and green hydrogen production, storage and transportation technologies, and promote the large-scale application of energy storage and hydrogen energy. Strengthen research on new power systems, energy storage, hydrogen energy, pumped storage, CCUS and other standard systems
Jun. 2023	On the National Key R&D Program "Hydrogen Energy Technology" and other 7 key special project application guidelines for 2023	Focusing on the three technical directions of hydrogen energy green production and large-scale transfer system, safe storage and rapid transmission and distribution system, and convenient transformation and efficient power system, 19 tasks are planned to be started and the state will allocate RMB340 millionl
Jul. 2023	Guiding Catalogue for Industrial Restructuring 2023 (draft for commet)	Hydrogen energy technology and application; Efficient and economical hydrogen production, hydrogen transport and high-density hydrogen storage technology development and application and equipment manufacturing, hydrogen refueling stations and vehicle clean alternative fuel filling stations, mobile new energy technology development and application, a new generation of hydrogen fuel cell technology research and development and application, renewable energy hydrogen production, liquid, solid and gaseous hydrogen storage, pipeline trailer hydrogen transport, pipeline hydrogen, hydrogen refueling stations, hydrogen coupling and other hydrogen energy technology application

(continued)

Table 4 (continued)

Time	Policy	Content
Aug. 2023	Guidelines for Establishing the Standards System on Hydrogen Energy Industry (2023)	The standard system of the whole industrial chain of hydrogen energy production, storage, transport and use has been established, covering five sub-systems of basic and safety, hydrogen preparation, hydrogen storage and transport, hydrogen filling, and hydrogen energy application, which are further decomposed according to technology, equipment, system, safety, and testing, forming 20 second-level sub-systems and 69 third-level sub-systems
Oct. 2023	Green Aviation Manufacturing Industry Development Outline (2023–2035)	Promote research and development demonstration exploration of hydrogen aircraft
Oct. 2023	Guiding Opinions of the National Development and Reform Commission and other Departments on Promoting Green Innovation and High-quality Development of the Oil Refining Industry	Green hydrogen refining, carbon dioxide capture, utilization and storage (CCUS) and other technologies have been industrialized, large-scale demonstration and verification, and a number of green and low-carbon benchmark enterprises can be built for reference and replication, and support the national carbon emissions to peak before 2030
Dec. 2023	High-Speed Action Initiative on Jointly Building Hydrogen Energy in China	Support high-speed operating vehicles and high-speed hydrogen refueling stations, such as reducing high-speed tolls and optimizing the construction and operation costs of hydrogen refueling stations
Dec. 2023	Guidance Catalogue for Industrial Restructuring (2024)	Hydrogen technology and application: Renewable energy hydrogen production, hydrogen transport and high-density hydrogen storage technology development and application and equipment manufacturing, hydrogen refueling stations and vehicle clean alternative fuel filling stations, mobile new energy technology development and application, a new generation of hydrogen fuel cell technology research and development and application, renewable energy hydrogen production, liquid, solid and gaseous hydrogen storage, pipeline trailer hydrogen transport, pipeline hydrogen, hydrogen refueling stations, Popularizing and applying hydrogen technologies such as hydrogen-electric coupling

(continued)

Table 4 (continued)

Time	Policy	Content
Jan. 2020	Action Plan for the Development of Hydrogen Energy Industry in Tianjin (2020–2022)	By 2022, the total output value of hydrogen energy will exceed RMB15 billion, at least 10 hydrogen refueling stations will be built, and more than 1,000 hydrogen FCVs such as logistics vehicles, forklifts and buses will be promoted
Jun. 2020	Shandong Medium to Long Term Development Plan for Hydrogen Energy Industry (2020–2030)	By 2025, 10,000 FCVs and 100 hydrogen refueling stations will be built. From 2026 to 2030, a total of 50,000 FCVs and 200 hydrogen fueling stations will be built
Sept. 2020	Sichuan Hydrogen Industry Development Plan (2021–2025)	By 2025, the application scale of FCVs (including heavy trucks, medium and light logistics, buses) will reach 6,000 units and 60 hydrogen refueling stations
Jul. 2021	Hebei "14th Five-Year" Hydrogen Energy Industry Development	By 2025, 100 hydrogen refueling stations will be built, 10,000 FCVs will be built, the annual output value of the hydrogen industry chain will reach RMB50 billion, and the green hydrogen production capacity will reach 100,000 tons/year
Aug. 2021	Beijing Hydrogen Energy Industry Development Implementation Plan (2021–2025)	By 2025, the cumulative scale of the hydrogen industry chain will reach more than RMB100 billion, and strive to build 37 new hydrogen refueling stations, and the cumulative promotion of FCVs will exceed 10,000
Feb. 2022	Inner Mongolia "14th Five-Year" for Hydrogen Energy Development	By 2025, 60 hydrogen refueling stations will be built, 5,000 FCVs will be promoted, hydrogen production capacity will be 1.6 million tons/year, green hydrogen will account for more than 30%, and the total output value will reach RMB100 billion
Jun. 2022	Shanghai Medium and Long-Term Plan for the Development of Hydrogen Energy Industry (2022–2035)	By 2025, about 70 hydrogen refueling stations will be built, the number of FCVs will exceed 10,000, and the industrial scale of the hydrogen industry chain will exceed RMB100 billion
Jul. 2022	Guizhou "14th Five-Year" Hydrogen Energy Industry Development Plan	At the end of the "14th Five-Year Plan", the demonstration operation of FCVs exceeded 1,000. The total investment scale of the hydrogen energy industry to exceed RMB10 billion, and 15 hydrogen refueling stations will be built (including comprehensive energy stations containing oil and gas hydrogen).
Jul. 2022	Shanxi Medium and Long-Term Plan for the Development of Hydrogen Energy Industry (2022–2035)	By 2025, the number of FCVs will reach more than 10,000, and a number of hydrogen refueling stations will be deployed and built, leading the country in application scale

(continued)

Table 4 (continued)

Time	Policy	Content
Aug. 2022	Liaoning Hydrogen Energy Industry Development Plan (2021–2025)	By 2025, the output value will reach RMB60 billion, and the number of FCVs in the province will reach more than 2,000 and more than 30 hydrogen refueling stations
Aug. 2022	Shaanxi "14th Five-Year" Hydrogen Energy Industry Development Plan	By 2025, about 100 hydrogen refueling stations will be built and put into operation, and about 10,000 FCVs will be promoted, and the scale of the whole industry chain will reach more than RMB100 billion
Sept. 2022	Henan Medium and Long Term Plan for the Development of Hydrogen Energy Industry (2022–2035)	By 2025, the total output value of the hydrogen industry will reach RMB100 billion, FCVs will to exceed 5,000, the terminal price of hydrogen will be reduced to less than RMB30/kg, and 3 to 5 green hydrogen demonstration projects will be built
Oct. 2022	"Hydrogen Power Jilin" Medium and Long-term Development Plan (2021–2035)	By 2025, with the "hydrogen power Jilin" project, the hydrogen production capacity of renewable energy will reach 60,000 to 80,000 tons/year, and the output value will reach 10 billion scale
Nov. 2022	Ningxia Hydrogen Energy Industry Development Plan	By 2025, the amount of hydrogen produced from renewable energy will reach more than 80,000 tons, more than 10 hydrogen refueling stations will be constructed, more than 500 hydrogen fuel cell heavy trucks will be built, and the amount of hydrogen produced from renewable energy will reach more than 300,000 tons by 2030
Nov. 2022	Anhui Medium and Long-Term Plan for the Development of Hydrogen Energy Industry	By 2025, fuel cell system production capacity will reach 10,000 units/year, FCVs capacity of 5,000 units/year, the number of hydrogen refueling stations 30, the total output value will reach RMB50 billion
Nov. 2022	Hunan Hydrogen Energy Industry Development Plan	By 2025, 10 hydrogen refueling stations will be built, 500 hydrogen FCVs will be promoted, and hydrogen energy infrastructure will be gradually improved
Dec. 2022	Fujian Hydrogen Energy Industry Development Action Plan (2022–2025)	By 2025, the application scale of FCVs in the province will reach 4,000, and to build more than 40 types of hydrogen refueling stations, achieving an output value of more than RMB50 billion

(continued)

Table 4 (continued)

Time	Policy	Content
Jan. 2023	Guiding Opinions of the General Office of the Gansu Provincial People's Government on the Development of Hydrogen Energy Industry	By 2025, Gansu will build a hydrogen production and storage base with hydrogen production capacity of about 200,000 tons/year; Reduce carbon emissions by about 2 million tons/year; More than 20 hydrogen energy enterprises have been introduced and cultivated, with an annual output value of RMB10 billion
Jan. 2023	Qinghai Medium and Long-Term Plan for the Development of Hydrogen Energy Industry (2022–2035)	By 2025, the production capacity of green hydrogen to reach 40,000 tons, the construction of green hydrogen demonstration projects not less than 5, the number of FCVs in operation not less than 150, the number of hydrogen energy heavy trucks in mining areas not less than 100, the construction of 3 to 4 hydrogenation demonstration stations, and the output value of the whole green hydrogen industry chain is RMB 3.5billion
Jan. 2023	Jiangxi Medium and Long Term Plan for the Development of Hydrogen Energy Industry (2023–2035)	By 2025, renewable energy hydrogen production will reach 1,000 tons/year, about 500 FCVs, put into operation a number of hydrogen-powered ships, and build 10 hydrogen refueling stations
Feb. 2023	Notice of Guangdong Province on organizing the implementation of the Measures for Promoting the Development of Hydrogen Energy Industry in 2023 (the first batch)	Hydrogen refueling station subsidy: For hydrogen refueling stations with a hydrogen sales price of no more than RMB30/kg in 2022, an operation subsidy of RMB15/kg will be provided
Mar. 2023	Notice on the Technology List of the Major Demonstration Project of New Energy Equipment Scientific and Technological Innovation in Inner Mongolia in 2023	Two hydrogen energy projects that adapt to a wide range of power fluctuations, namely, the development and demonstration application of key equipment for thousand-standard square-level electrolytic water hydrogen production, the industrialization technology development and application of solid hydrogen storage materials, will receive subsidies of RMB10 million and RMB6 million respectively
Apr. 2023	Notice on organizing the implementation of the measures for promoting the development of hydrogen energy industry in Guangdong in 2023 (the third batch)	Maximum subsidy for fixed refueling stations: RMB5 million, and maximum subsidy for skid-mounted refueling stations: RMB2.5 million

(continued)

Table 4 (continued)

Time	Policy	Content
Jun. 2023	Implementation Opinions on Comprehensively Promoting the Construction of "Electric Fujian" (2023–2025)	Hydrogen fuel cell ship single ship subsidy of up to RMB15 million: hydrogen FCV enterprises with sales volume of 50 or more in a year will be rewarded with RMB2 million; Hydrogen fuel cell system RMB20,000/set (set) reward, a single maximum of RMB5 million
Jun. 2023	Luoyang City 17 measures for hydrogen industry development	The maximum one-time reward of hydrogen energy enterprises is RMB10 million: the subsidy of no more than 30% for technical transformation projects is not more than RMB10 million
Jul. 2023	Karamay Hydrogen Energy Industry Development Action Plan (2023–2025)	Form a hydrogen energy supply system based on industrial by-production hydrogen and nearby use of renewable energy hydrogen production, "green hydrogen" production capacity reaches more than 100,000 tons/year, and the actual hydrogen production cost is less than RMB18/kg; Cultivate and introduce 20 high-quality enterprises in key areas, and strive to exceed RMB30 billion in the scale of related industries in the city
Jul. 2023	Measures for the Safety Administration of Hydrogen Energy Industry of Hebei Province (Trial)	Renewable hydrogen production does not require a safe production permit for hazardous chemicals
Aug. 2023	Huzhou Hydrogen Energy Industry Development Plan (2023–2035)	By 2027, it will cultivate and introduce more than 50 enterprises in related fields of hydrogen energy industry, cultivate more than 2 listed enterprises, and the annual output value of hydrogen energy and related industries will exceed RMB10 billion; More than 5 independent hydrogen refueling stations and integrated energy supply service stations will be built, and more than 5 iconic hydrogen energy application scenarios with demonstration synergies will be created. The "Hydrogen Corridor around Taihu Lake" with the influence of the Yangtze River Delta initially established.
Sept. 2023	Guangxi Medium and Long-Term Plan for the Development of Hydrogen Energy Industry (2023–2035)	The supply capacity of high-purity hydrogen to reach 2,000 tons, 10 hydrogen refueling stations, and 500 hydrogen FCVs
Oct. 2023	Shanghai's Plan for Further Promoting New Infrastructure Construction (2023–2026)	Support offshore wind power for hydrogen production

(continued)

Table 4 (continued)

Time	Policy	Content
Oct. 2023	Measures for Promoting High-quality Development of Hydrogen Energy Industry in Gansu Province "Interim" (Draft for comment)	For green hydrogen manufacturing enterprises that have signed medium and long-term gas supply contracts with hydrogenation stations in the city, the factory price does not exceed RMB30/kg, and the hydrogen supply amount of hydrogen supply stations will be given RMB10/thousand restraint hydrogen subsidies, and the maximum for a single enterprise is not more than RMB700,000
Nov. 2023	Jilin Hydrogen Energy Industry Safety Management Measures (Trial)	Green hydrogen production projects such as electrolytic water hydrogen production (renewable energy sources such as solar energy and wind energy) and their hydrogen production and hydrogenation integrated stations do not need to be built in the chemical park. Green hydrogen production does not require a safe production permit for hazardous chemicals
Nov. 2023	Opinions on Accelerating the Innovation and Development of Hydrogen Energy Industry in Guangdong Province	By 2025, more than 10,000 FCVs will be promoted, the annual hydrogen supply capacity will exceed 100,000 tons, and more than 200 hydrogen refueling stations will be built
Dec. 2023	Tibet's Hydrogen and Oxygen Industry Development Plan (2024–2030) (Draft for Comment)	By 2025, renewable energy hydrogen production capacity will reach 10,000 tons/year, oxygen production will reach 80,000 tons/year, green hydrogen production cost will be reduced to RMB18/kg, and 10 demonstration projects such as renewable energy hydrogen production, oxygen production, and municipal natural gas hydrogen blending will be built; By 2027, renewable energy will produce 30,000 tons of hydrogen per year, 240,000 tons of oxygen per year, and the production cost of green hydrogen will be reduced to RMB15/kg; By 2030, the hydrogen and oxygen industry will reach the domestic advanced level
Dec. 2023	Tibet's Hydrogen and Oxygen Industry Development Plan (2024–2030) (Draft for Comment) Jilin's Notice on Issuing the Implementation Plan of the New Circuit for the Pre-emptive Layout of the Hydrogen Energy Industry and the New Energy Storage Industry	By 2025, renewable energy hydrogen production capacity will reach 200,000 tons/year, the installed capacity of new energy driven by hydrogen energy industry projects will exceed 5 million kilowatts, and the capacity of hydrogen-based green energy such as green synthetic ammonia, green methanol and green aviation coal will reach 1 million tons

lead domestic technology and industry leaders to expand international markets and enhance the international competitiveness of China's hydrogen energy industry.

Build a Quality Assurance System for Hydrogen Energy Products, Improve Equipment and Quality Inspection, and Support the Market-Oriented Development of the Industry

With the formulation and certification of hydrogen energy standards as the focus, establish and improve the measurement—inspection—certification standards and system for key technical equipment in the industrial chain, and actively guide the industry to carry out high-quality technical innovation and standard formulation. In terms of standard setting, benchmarking the revision work of international hydrogen energy related standards, combing the gaps in the standard system of China's hydrogen energy industry, jointly carrying out the revision of related hydrogen energy standards with multiple industry entities, and building a standard system for multiple application scenarios such as hydrogen-electric coupling, green hydrogen chemical industry, and hydrogen metallurgy. In terms of inspection and certification, develop rapid and efficient hydrogen fuel inspection and measurement methods, establish a hydrogen fuel quality evaluation process, establish a national hydrogen fuel quality inspection center in a timely manner, form a complete hydrogen energy technology equipment inspection and certification system, a full-cycle hydrogen quality control system, and achieve hydrogen energy standard governance and innovation iteration coordination. In terms of market development, relying on the hydrogen energy big data platform, fully grasp the development of the hydrogen energy industry in various places and the operation of enterprises, establish a safety management system and safety maintenance operation record database covering the entire life cycle of hydrogen, and ensure safe development.

Safely and Orderly Promote the Demonstration of Hydrogen Energy Projects, Plan the Construction of Supply Chain Infrastructure, and Reduce the Cost of Industrial Development

Fully considering the current situation that the technology of the entire hydrogen energy industry chain in our country needs to be fully improved, the early stage should be mainly pilot demonstrations, and gradually expand the application scale. Drawing on international demonstration experience, support the construction of a batch of large-scale renewable energy hydrogen production comprehensive application demonstration bases, reduce the cost of hydrogen production through scale; deploy a batch of "renewable energy hydrogen production + storage and transportation + comprehensive application" full-chain major engineering demonstrations around renewable hydrogen coupled coal chemical industry, hydrogen-electric fusion and other fields, to solve the problems of insufficient upstream and downstream linkage and unclear consumption scenarios; rely on project layout to promote the construction of hydrogen refueling stations in key areas, and clarify the infrastructure approval management system from the perspective of energy supervision, continuously optimize the infrastructure network layout, improve the convenience of facility applications, and reduce the cost of the hydrogen energy supply chain.

(This section was written by: Lan Mengtong Wang Yuelin, Reviewer: Gao Hui)

Development of Energy Storage Industry

The rapid scaling of renewable energy has driven substantial demand for energy storage solutions. As a critical enabler of China's "interaction of load and storage between source and network" integrated power system, energy storage enhances grid flexibility and stability while addressing challenges in the energy trilemma (security, affordability, sustainability). On the generation side, energy storage systems optimize efficiency; on the consumer side, they act as "power reservoirs" enabling peak cut. Since 2023, various places in China have accelerated the implementation of energy storage projects. Energy storage technology has achieved diversified development, with breakthrough technologies emerging continuously and it making substantial progress. The energy storage industry is still undergoing transformation, and new technologies, models, forms, and trends are reshaping the industry development model. FDI, self-owned brands and new cross-industry players have begun to confront each other, and industry competition is becoming increasingly fierce. The scale of energy storage operation has grown exponentially, the scale of bidding has rolled out new heights, and the development of the energy storage industry has become mature and diversified.

Emerging Trends and Requirements

Energy Storage as a Strategic Growth Engine

By December 2023, China had built and put into operation new power storage projects with a total installed capacity of 31.39 GW/66.87 GWh(+45% YoY),marking the first year where pumped hydro fell below 60% of new installations. As early as 2023 H1, the newly added operating volume of the energy storage industry had exceeded the total for 2023. From January to June 2023, the newly added operating installed capacity of new energy storage in our country was about 8.63 GW/17.72 GWh, equivalent to the total installed capacity of previous years. The rapid growth of the national energy storage installed capacity verifies the large demand for flexible resources under the power system with a high proportion of wind and light power generation.

The "14th Five-Year Plan" has proposed that China should make the energy storage industry a new economic growth point, fully utilize the advantages of energy storage, especially electrochemical energy storage, and achieve technological and industrial overtaking. Facing the urgent demand for commercialization and scale development of energy storage during the "15th Five-Year Plan" period, energy storage industry as a whole still faces many challenges. In the future, as an important part of the power system, the energy storage industry will complete the transition from initial commercialization to scale development as soon as possible.

Technology R&D: Key to Global Competitiveness

As a core technology to promote reform of the energy system, energy storage is not only a high-tech, high-growth potential emerging industry, but also a key support for building a clean, low-carbon, safe and efficient a system for new energy sources and building the new electricity system. The construction of energy storage is related to the timely realization of the carbon peaking and carbon neutrality goals. Facing the competition of world energy green technology, it is urgent to accelerate the construction of a new energy storage technology innovation system. New energy storage is one of the breakthroughs to create new forms of energy industry and create a new economic engine. Energy storage can improve the level of new energy development and utilization and promote the sustained and healthy development of the new energy industry. In a "dual circulation" development pattern, the layout of the new energy storage industry faces major opportunities.

With the deepening of carbon peaking and carbon neutrality goals strategy, China's energy storage industry urgently needs to face multiple challenges. Global trend of low-carbon is accelerating the transformation of energy structure, and the related international competition is becoming more and more intense. Argentina, Chile, Bolivia and Brazil are analyzing the creation of "Lithium OPEC" which may rewrite the global lithium resource supply pattern, and the new EU battery regulation (Regulation 2023/1542) are gradually building green barriers for China new energy industry. With global energy transition and development of the renewable energy international competition is becoming more and more intense. Strengthening the R&D of energy storage technology will help seize the commanding heights of new energy technology and enhance China's competitiveness and influence in the global new energy field.

The energy storage industry is in a critical period of transformation, and technology R&D needs strong industrial chain support. Strengthening the development of the energy storage industry can drive the development of upstream and downstream industries, including battery materials, energy storage devices, grid equipment, etc. As an important path to enhance the global competitiveness of renewable energy industry chain, Energy storage technology R&D will directly affect China's level and influence in the global renewable energy field.

Overview of China's Energy Storage Industry Development in 2023

Proportion of Energy Storage Installed in China in 2023 by Technology

At present, various energy storage technologies are flourishing. Electrochemical energy storage is developing uniquely, and the scale of new energy storage in 2023 has further increased. From the distribution of new energy storage technologies that have been put into operation, the proportion of lithium-ion battery energy storage has further increased, jumping from 94.5% in 2022 to 97.4% in 2023, lead carbon battery energy storage accounts for 0.5%, CAES accounts for 0.5%, the flow battery energy storage accounts for 0.4%, and other new energy storage technologies account for 1.2% (Fig. 4).

The lithium iron phosphate batteries are still the mainstream. CAES, flow battery energy storage, flywheel energy storage system (FESS) and other technologies are developing rapidly, and non-lithium energy storage technology applications are gradually increasing: the first flywheel fire storage frequency regulation project, the first flywheel + lithium battery mixed storage frequency regulation project, and the user side largest single lead carbon battery project have been put into operation in succession; 300 MW power level compressed air is accelerating layout, various flow battery

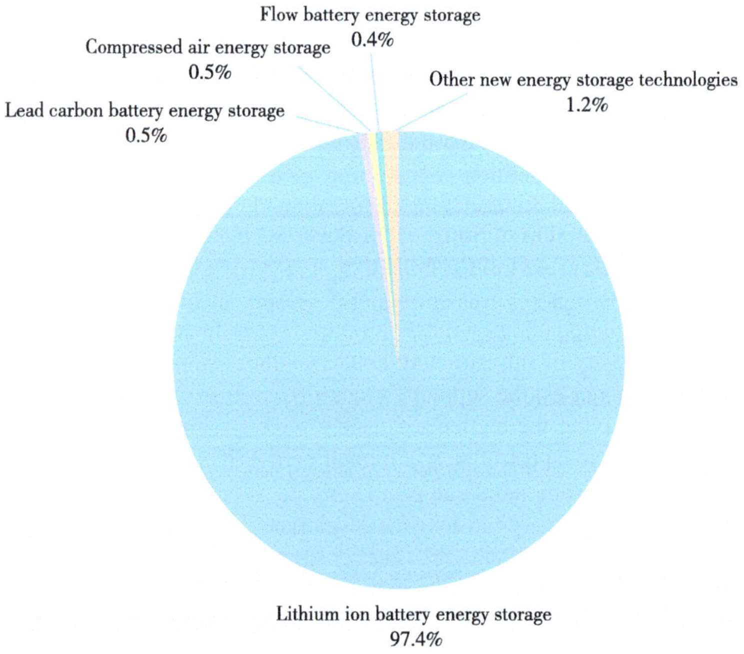

Fig. 4 Cumulative installed capacity of new power storage projects in operation in China (by technology)

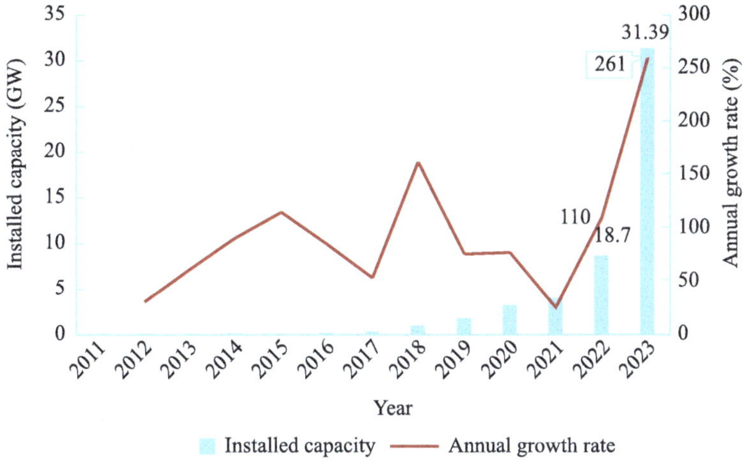

Fig. 5 Total installed capacity of new energy storage projects in operation nationwide (2011–2023)

technology routes and 100 MW level sodium battery projects are included in the provincial demonstration project list.

Scale of Electrochemical Energy Storage Installed in China in 2023

By the end of 2023, the total installed capacity of new energy storage projects that have been built and put into operation nationwide has reached 31.39 GW/66.87 GWh, with an average storage time of 2.1 h. The newly added installed capacity in 2023 is about 22.6 GW/48.7 GWh, an increase of more than 260% compared to the end of 2022 (Fig. 5), nearly 10 times the installed capacity at the end of the "13th Five-Year Plan" period.

In 2023, China's new energy storage market continued to develop rapidly, with the number of projects increasing by 46% compared to 2022, and the scale of new additions growing by 65% over the previous year. The growth rate of projects at the hundred-megawatt level was significant. By the end of December 2023, more than 100 hundred-megawatt projects were put into operation, an increase of approximately 370% YoY (Fig. 6).

Latest Breakthroughs in Electrochemical Energy Storage Technology

2023 was a year of vigorous development for China's electrochemical energy storage technology. In terms of material system technology, non-lithium electrochemical energy storage technology accelerated its development, and industrial breakthroughs were achieved in segmented technology fields such as sodium-ion batteries, condensed batteries, and solid-state batteries. The disruptive innovation of

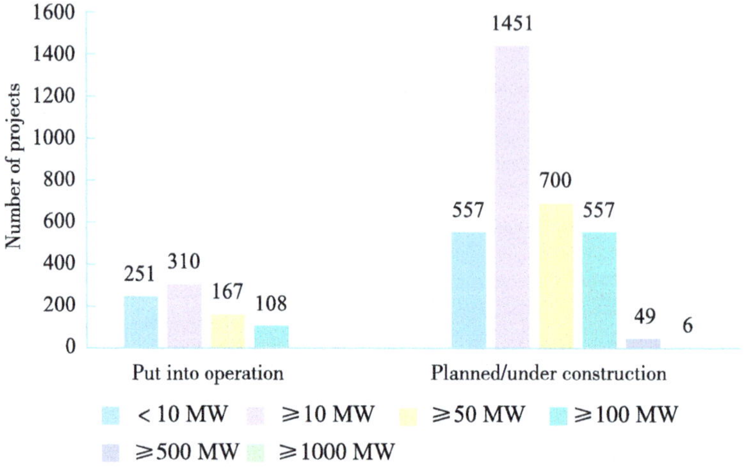

Fig. 6 Distribution of project numbers by power level (January-December 2023)

the electrochemical system of power batteries represented by solid-state batteries is expected to completely solve the problems of endurance, energy replenishment, and safety of power batteries. Currently, the first generation of semi-solid batteries jointly developed by Qingtao Energy Development Co., Ltd. and SAIC Group will mass-produced in their first-generation semi-solid-state batteries in Q3 2024. CATL is advancing the commercialization of sodium-ion batteries, lithium manganese iron phosphate (LMFP) batteries, and condensed-state batteries.

In terms of structural technology innovation, China's battery industry has achieved leapfrog progress through differentiated innovations For instance, the self-developed "Gold Brick" battery by Extreme Krypton allows the energy density of lithium iron phosphate battery packs to increase by over 10% compared to traditional ternary lithium battery packs, with battery volume utilization rate reaches 83.7%, achieving a global advantage in technology; Leapmotor has adopted battery-chassis integration technology, achieving a volume utilization rate of up to 79 and 15% improvement in system lightweighting. From the layout of batteries to the composition of batteries, China's electrochemical energy storage has firmly occupied a leading position in global electrochemical energy storage technology through differentiated innovation.

In the exploration of energy storage technology, a new type of energy storage technology has emerged that relies on existing technical capabilities and system integration capabilities-mobile energy storage. This technology relies on Vehicle-to-Grid (V2G) technology. Electric vehicles had a stock of more than 13 million in 2022, with a sales penetration rate of 25.6%. It is expected that under the backdrop of the rapid development of the new energy vehicle market, the stock of new energy vehicles will reach 100 million by 2030, and even climb to 500 million by 2060. Against this backdrop, mobile energy storage technology related to new energy vehicle batteries has begun to gain traction.

Table 5 lists the electrochemical energy storage technology situation in 2023.

Table 5 Overview of electrochemical energy storage technology in 2023

Category	Name	Technical highlights	Application cases
Electrochemical Energy Storage	Condensed Battery	High Specific Energy + High Safety	CATL collaborates with COMAC to develop civilian electric manned aircraft
	Chocolate Swap Battery Module	Small Modular Swap System	CATL begins mass production of Chocolate Swap Battery Modules
	Kirin Battery	Refreshes the global record for battery system integration	2023 Extreme Krypton 001 Car Model
	Sodium-ion Battery	Over 90% discharge retention rate in − 20 °C low temperature environment	CATL and Chery jointly launch battery brand ENER-Q
	Lithium Manganese Iron Phosphate Battery (M3P)	Low Cost, High Safety, High Energy Density	2023 Tesla Model 3 66 kWh Version
Compressed Air Energy Storage (CAES)	300 Megawatt Non-Reheat Compressed Air Energy Storage System	Core technology independently developed, 100% domestically produced equipment	China Energy Construction Gansu Jiuquan 300 Megawatt Compressed Air Energy Storage Power Station Demonstration Project
Molten Salt Energy Storage	High and Low Temperature Dual Tank Molten Salt Energy Storage System	Anhui Provincial Energy Bureau's reply agrees that the molten salt energy storage project of the State Energy Group Anhui Company is a technological innovation project	State Energy Group Suzhou Power Plant molten salt heat storage system renovation project
Flywheel energy storage (FESS)	FESS joint frequency regulation	Participating in the grid primary frequency regulation together with thermal power	Huaneng Laiwu Power Plant

The Development of New Long-Duration Energy Storage Technology Has Begun

Long-duration energy storage is an energy storage system that can achieve daily, monthly, and even seasonal charge and discharge cycles based on ordinary energy storage systems. It has the ability to provide stable power support for the power system and improve the system's reliability and flexibility. Long-duration energy storage has a significant advantage in smoothing the grid dominated by renewable energy, can adjust the fluctuation of renewable energy generation over a longer time dimension, avoid grid congestion, increase the absorption capacity of renewable energy, and can better achieve power shifting, transferring the power of renewable energy generation systems to peak power demand periods, which is conducive to the peak-valley arbitrage mechanism of industrial and commercial energy storage, and meets the needs of future power system reform.

By 2023, new non-lithium long-duration energy storage technology applications have gradually broken through, with the first flywheel fire storage frequency regulation project, the first supercapacitor + lithium battery mixed storage frequency regulation project, and the largest single lead carbon battery project on the user side being put into operation one after another; 300 MW of CAES completed reverse power delivery, and various long-duration energy storage technology routes were included in the provincial demonstration project list. Latest long-duration energy storage technology development routes continue to emerge, and government has promoted large-capacity, medium-long time scale energy storage technology demonstrations, and actively deployed the research and development of various long-duration energy storage technologies such as all-vanadium flow batteries, iron-chromium flow batteries, CAES, molten salt heat storage, hydrogen energy storage, etc.

By the end of 2023, CAES, molten salt heat storage, flywheel energy storage and other long-duration energy storage technology pilot projects have made breakthrough progress, becoming the highly watched long-duration energy storage technologies, among which, the development progress of compressed air energy storage is eye-catching, and the overall industry development has gradually shifted from technology to actual landing projects, with traditional energy companies with geographical advantages betting on compressed air energy storage. In the second half of 2023, Zhongning County, Ningxia, launched the first 100 MW artificial underground compressed air energy storage project in China. The project excavated a ring-shaped gas storage tank at a depth of 150 m underground, using surplus electricity to drive the compressor, compressing the air to a high-pressure state, and storing it in the gas storage system. The project's total designed gas storage volume is 100,000 m^3, equivalent to the volume of 40 standard swimming pools.

The Development of Energy Storage Business Models is Being Explored

New energy storage business models can be divided into three types according to different electricity target customers: energy storage on the generation side (source), user-side energy storage (load), and grid-side energy storage (grid). On the source side, centralized renewable energy grid connection generates smooth power output, reducing the need for wind and light abandonment; on the grid side, the instability of electricity generates power peak regulation, system frequency regulation, and other auxiliary operation needs; on the load side, the peak-valley arbitrage, self-use backup, mobile portability, and other needs of end users give birth to various energy storage applications.

Since 2023, the new shared energy storage model has developed rapidly, and the shared energy storage new model may become the mainstream domestically. Looking at the structure of bidding projects, the generation side, energy storage batteries, and shared energy storage account for a larger proportion, of which shared energy storage's bidding volume increase from January to April 2023 tends to stabilize, accounting for about 23% cumulatively, and energy storage battery bidding is concentrated in March and April, with shared energy storage plus energy storage batteries accounting for 52% in total.

The current mainstream business models focus on achieving integrated development of "interaction of load and storage between source and network, the coordination and complementarity of multiple energy sources, and the smart regulation of energy demand." to fully tap the potential value of new energy storage, so that energy storage can better serve power consumption, auxiliary services, spot market, capacity compensation, capacity leasing, and peak-valley electricity price adjustment, etc., which are urgent pain points that the electricity market needs to solve.

The energy storage business model is hot, and the trend of energy storage industry financing and listing is obvious. Since 2023, there have been several financing events of energy storage companies with amounts over 100 million yuan. By the end of 2023, the total financing amount reached 53.4 billion yuan (only counting financing events that disclosed specific amounts, excluding upstream material financing events), involving fields such as sodium-ion batteries, flow batteries, PCS, portable/household storage systems, energy storage safety, battery recycling, intelligent manufacturing, virtual power plants, AI digitization, etc. (Table 6). However, with the introduction of new energy storage policies, the pace of company listings has slowed down.

Energy Storage Industry Policy

By the end of 2023, a total of 647 energy storage-related policies have been issued by the central and local governments (Fig. 7), including 60 central policies. Guangdong, Zhejiang, Shandong, and Jiangsu have issued more policies.

From a local perspective, 24 provinces (regions, cities) across the country have clearly defined the scale of new energy storage development, with a total demand for new energy storage exceeding 60 million kilowatts. The "14th Five-Year Plan" and medium-to-long-term new energy storage development goals and key tasks in various places all require energy storage: by 2025, the planned installed capacity of 27 provinces will reach 84 gigawatts, among which Inner Mongolia, Henan, Guangdong, Hubei, Guangxi and other places have further increased their energy storage installation targets (Fig. 8).

From the perspective of the national overall layout, since 2021, the National Energy Administration and relevant departments have continuously issued documents such as "Guiding Opinions on Accelerating the Development of New Energy Storage", initially constructing an industry development policy system guided by top-level design, supported by project management, dispatching and use, market

Table 6 Listing situation of some energy storage companies in 2023

Company	Business field	Status	Login section	Fundraising
Haibosichuang	Energy Storage System	IPO review has been inquired	Shanghai Stock Exchange STAR Market	783 million yuan
Ailuo Energy	Energy Storage System	Listed in January 2024 (code: 688717.SH)	Shanghai Stock Exchange STAR Market	Issue price 55.66 yuan/share, issue quantity 40 million shares
Ruipu Lanjun	Battery	Listed in December 2023 (code: 00,666.HK)	Hong Kong Stock Exchange	1.83 billion Hong Kong dollars
Mengguli	Battery	Listed in August 2023 (code: 301487.SZ)	Shenzhen Stock Exchange GEM	Issue price 5.32 yuan/share, issue quantity 58 million shares
Wotai Energy	Energy Storage System	IPO review has been inquired	Shanghai Stock Exchange STAR Market	1 billion yuan
Ailuo Energy	Energy Storage System	Listed in January 2024 (code: 688717.SH)	Shanghai Stock Exchange STAR Market	Issue price 55.66 yuan/share, issue quantity 40 million shares
Honeycomb Energy	Battery	Withdrawn A-share IPO listing application	Shanghai Stock Exchange STAR Market	15 billion yuan
Ates	Photovoltaic Energy Storage	Listed in June 2023 (code: 688472.SH)	Shanghai Stock Exchange STAR Market	Issue price 11.1 yuan/share, issue quantity 541 million shares
Ligao New Energy	BMS	IPO accepted	Shenzhen Stock Exchange GEM	734.79 million yuan
Company	Business area	Status	Login section	Fundraising
Huasu Technology	BMS	Listed in March 2023 (code: 301157.SZ)	Shenzhen Stock Exchange GEM	Issue price 56.5 yuan/share, issue quantity 15 million shares
Shuangdeng	Battery	IPO accepted	Shenzhen Stock Exchange GEM	1.575 billion yuan
First Flight New Energy	Photovoltaic Inverter	Review passed	Shenzhen Stock Exchange GEM	3.512 billion yuan
Gurui Watt	Photovoltaic Inverter	Passed hearing	Hong Kong Stock Exchange	1 billion yuan

(continued)

Table 6 (continued)

Company	Business area	Status	Login section	Fundraising
Power King Shares	Battery	Listed in September 2023 (code: 831627.BJ)	Beijing Stock Exchange	138 million yuan
Guohong Hydrogen Energy	Fuel Cell	Listed in December 2023 (code: 09663.HK)	Hong Kong Stock Exchange	1.563 billion Hong Kong dollars
Sanjing Electric	Photovoltaic Inverter	Approved	Shanghai Main Board	1.204 billion yuan
Xuchang Intelligent	Energy Storage System	Registration Submitted	Beijing Stock Exchange	300 million yuan

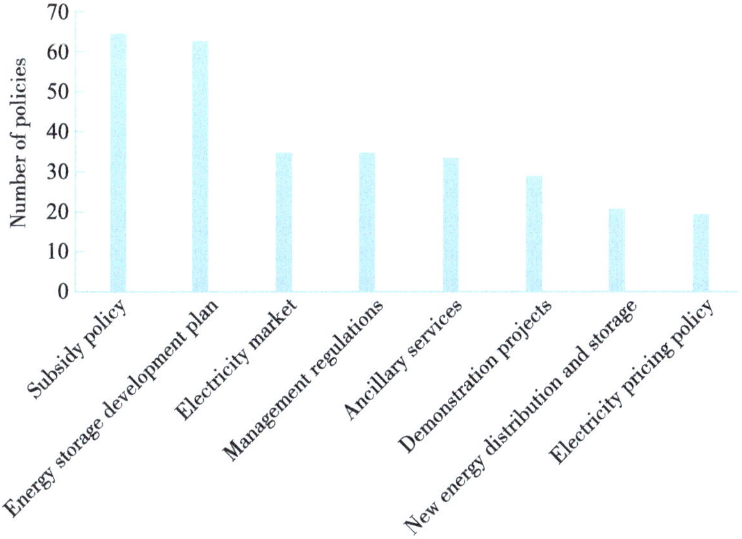

Fig. 7 Number of important energy storage policies in 2023

mechanisms, pricing and safety management and other special policies, and guaranteed by pilot demonstrations, talent training, industry-education integration, and standard system common policies. In February 2023, the National Energy Administration issued the "Technical Guidelines for Planning New Energy Storage for Cross-Provincial Power Allocation from renewable Energy Bases" (Draft for Comment), which is the first planning technical document issued at the national level to guide the allocation of energy storage scale at new energy bases, clarifying the important principles of energy storage ratio, key parameters to be considered in capacity calculation, etc., forming a more scientific and economical energy storage ratio scheme,

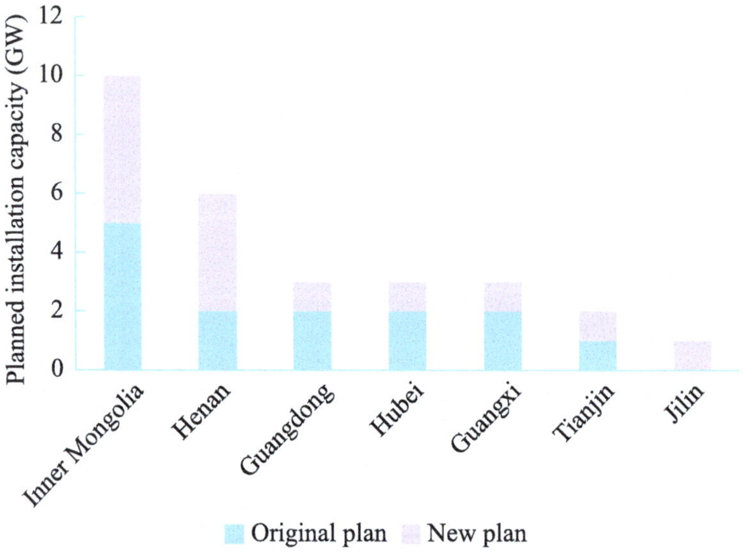

Fig. 8 New installed capacity planning situation in 2023

which is more conducive to the orderly development of the energy storage industry. At the same time, the "Guiding Opinions on Strengthening the Stability Work of the renewable Power System" also pointed out that the construction of energy storage should be scientifically arranged to form power regulation capabilities in multiple time scales and application scenarios.

At present, China's energy storage policies roughly include top-level design, discipline construction, industry norms, grid operation, project management, implementation plans, and scientific and technological breakthroughs. Generally speaking, the future policy release tends to combine China's national conditions, strengthen policy support, cultivate business models for different application scenarios, promote shared energy storage projects as market entities to participate equally in transactions, access standards, settlement rules, and revise peak-valley price differences in areas with commercial promotion conditions. Policies for the integrated development of source, network, load, and storage. It is worth noting that the second batch of energy storage demonstration applications has received widespread attention and has become a weathervane for the national development of energy storage. At present, there are 56 energy storage demonstration applications with an installed capacity of 8.2 GW/29.9 GWh. The second batch of energy storage demonstration applications shows a clear trend towards diversification of technology routes, with long-duration energy storage projects such as compressed air as the main body.

Table 7 lists the important policies of China's energy storage industry in 2023.

Table 7 Overview of important policies in China's energy storage industry in 2023

Date	Policy	Core content
January 6	The General Office of the National Energy Administration issued a notice on publicly soliciting opinions on the "Blue Book on the Development of renewable Power Systems (Draft for Comment)"	"Strengthening the construction of the energy storage scale application system" is listed as a key task for the overall development of the renewable power system
January 17	The Ministry of Industry and Information Technology and other six departments issued the "Guiding Opinions on Promoting the Development of the Energy Electronics Industry"	Develop new energy storage batteries that are safe and economical. Research breakthroughs in key technologies such as ultra-long-life high-safety battery systems, large-scale high-capacity efficient energy storage, and mobile energy storage for transportation tools, and accelerate the development of new types of batteries such as solid-state batteries, sodium-ion batteries, and hydrogen energy storage/ fuel cells. Establish a distributed photovoltaic cluster supporting energy storage system; accelerate the research and development of photovoltaic products and energy storage systems suitable for intelligent microgrids
January 18	National Energy Administration issued "Key Points of Energy Supervision Work in 2023"	Accelerate the construction of auxiliary service markets, establish a special work mechanism for power auxiliary service markets, study and formulate power auxiliary service pricing methods, establish and improve the auxiliary service sharing mechanism involving users, promote marketization of varieties such as frequency modulation and standby, and continuously guide new entities such as virtual power plants and new energy storage to participate in system regulation. In terms of stabilizing system safety and stable operation, explore the promotion of "source-network-load-storage" collaborative governance
February 3	Ministry of Industry and Information Technology, Ministry of Transport and other eight departments jointly issued "Notice on Organizing the Pilot Work of Comprehensive Electrification of Vehicles in Public Areas"	Support the innovation of business models such as battery swapping, financing leasing, and vehicle-electricity separation; accelerate the pilot application of "light-storage-charging-discharging" integration; explore the implementation path of renewable energy vehicles participating in the electricity spot market, and improve the trading mechanism of green electricity storage and discharge

(continued)

Table 7 (continued)

Date	Policy	Core content
March 31	National Energy Administration issued "Several Opinions on Accelerating the Development of Energy Digitalization and Intelligence"	Accelerate the development of renewable energy microgrids and highly reliable digital distribution systems, and improve the intelligent and efficient configuration and operation optimization control level of user-side distributed power sources and new energy storage resources
Time	Policy	Core content
April 24	National Energy Administration issued a notice on publicly soliciting opinions on "Guiding Opinions on Strengthening the Stability Work of the New Power System (Draft for Soliciting Opinions)"	Scientifically arrange energy storage construction. Build energy storage as needed, orderly construct pumped storage, and actively promote the construction of new energy storage; diversify the scientific configuration of energy storage, fully play the advantages of various new types of energy storage such as electrochemical energy storage, compressed air energy storage, flywheel energy storage, hydrogen energy storage, heat (cold) energy storage, explore new scenarios for energy storage integration development, and improve the safety guarantee level and comprehensive efficiency of the power system
May 12	The Comprehensive Department of the National Energy Administration issued the "Notice on Further Doing a Good Job in the Planning and Construction of Pumped Storage"	According to the needs of renewable energy development and power system operation, scientifically plan, reasonably layout, and orderly construct pumped storage. For provinces that have been thoroughly demonstrated and have no demand gap, they will not be added to the regulations for the time being
May 15	National Development and Reform Commission issued "Notice on the Capacity Price of Pumped Storage Power Stations and Related Matters" (Development and Reform Price [2023] No. 533)	The capacity price of 48 pumped storage power stations that are in operation and planned to be put into operation before the end of 2025 is determined. It is required that the local development and reform commissions should strengthen the supervision of the implementation of the capacity price of pumped storage power stations

(continued)

Table 7 (continued)

Time	Policy	Core content
May 17	National Development and Reform Commission and National Energy Administration issued "Implementation Opinions on Accelerating the Construction of Charging Infrastructure to Better Support the Rural Revitalization of New Energy Vehicles"	Encourage the research of key technologies such as two-way interaction between electric vehicles and the grid (V2G), coordinated control of light storage and charging, explore the construction of charging infrastructure that provides photovoltaic power generation, energy storage, and charging integration in rural areas with low charging pile utilization rate. Implement peak-valley time-of-use electricity price policy, encourage users to charge during low-valley periods
May 24	The Equipment Department of the Ministry of Industry and Information Technology announced that the "Management Measures for the Recycling and Utilization of Power Batteries for New Energy Vehicles" will be released soon	The intensity of regulatory constraints will be strengthened. In line with the needs of industrial development, we will accelerate the development of a batch of standards such as recycling service points, and improve the level of standard support. We will fully leverage the role of the national industrial finance platform to guide financial institutions to comprehensively use various financial tools such as green credit and green funds to support the development of the power battery recycling industry. We will continue to promote the application of advanced technologies and equipment for the comprehensive utilization of retired batteries, increase supply–demand docking, and promote the large-scale application of advanced technologies
September 15	The National Development and Reform Commission and the National Energy Administration issued a notice on the issuance of the "Basic Rules for the Electricity Spot Market (Trial)"	The document clarifies the operating regions of the spot market, and the costs of auxiliary services are shared by both the supply and demand sides based on the principle of fairness and reasonableness. At the same time, each province (region, city)/region should explore the establishment of a market-oriented capacity compensation mechanism in accordance with the overall deployment of the state and the actual needs, to incentivize investment and construction of various types of power sources, ensure sufficient system power generation capacity, adjustment ability and operational safety. Regions that carry out the spot market should coordinate market price limits, market settlement, power generation cost surveys, etc. with the capacity compensation mechanism. When conditions permit, the establishment of a capacity market can be explored

Outlook for the Key Directions of China's Energy Storage Industry Construction

Focus on the Development of New Types of Electrochemical Energy Storage, and Strengthen the Global Leading Position

China is the country with the fastest growth and the most rapid technological development in the global electrochemical energy storage scale. It is expected that the annual new installed capacity in 2024 will exceed 35 gigawatts, continuing the high-speed development trend of annual new installed capacity exceeding the cumulative installed capacity. A group of companies led by BYD and CATL are performing brilliantly in the electrochemical energy storage industry. The basic tone of China's electrochemical energy storage entering the "spring" has never changed. The progress of energy storage technology and the demand of the market are further expanding, and China's energy storage industry is showing a good trend of diversified development. Among them, new types of electrochemical energy storage led by sodium-ion batteries and all-vanadium redox flow batteries are eye-catching.

Sodium-ion batteries are important materials to break the bottleneck of China's key lithium resources constraints. The industry consensus is to start mass production of sodium-ion batteries in 2024. This battery can be used in small passenger cars, and it has great potential for application in the energy storage industry in the future. The flow battery is currently the best choice for the long-cycle energy storage industry. Its life can currently reach up to 15 years. With the development of technology, it is expected that the life can be increased to 20 years by 2030. Flow batteries have the characteristic of long electrolyte life and have significant competitiveness in medium and long time scale adjustment scenarios greater than 4 h.

The Business Model of Energy Storage is Relatively Single, and the Market Operation Mechanism Urgently Needs to Be Improved

Currently, the business model of energy storage is relatively simple and still heavily dependent on the policy of matching of wind power and solar. The market mechanisms and profit mechanisms for energy storage on the new energy side and the grid side are still not sound, cost guidance is difficult, the utilization rate of energy storage systems is not high, and the return level of energy storage investors is generally low. Energy storage is a provider of flexible power supply, but the market value that truly determines the flexibility of the power system and the market mechanism for its realization are not perfect. The policy mechanism that truly promotes the healthy development of the energy storage industry needs to be gradually established and improved in the process of comprehensively deepening the reform of the power system.

Currently, the production capacity of the energy storage industry is rapidly expanding, and the competition in the industry is intensifying. The market has insufficient understanding of the risks of operating energy storage power stations, and there are problems with profitability, and industry reshuffling is imminent. Looking at the whole year of 2023, the tendering volume of battery systems increased by 168% YOY the tendering volume of energy storage systems increased by 306% year-on-year, and the EPC tendering volume increased by 81% YOY, all of which were far higher than the same period in 2022. The scale of bidding has surged, and corporate competition has become fierce. At the same time, affected by the supply chain linkage, the price of battery-grade lithium carbonate continued to decline in 2023, with the largest drop exceeding 80%. The annual price range of battery-grade lithium carbonate was 96,000 to 510,000 yuan/ton, with an average price of 226,500 yuan/ton, a year-on-year decrease of 53%. The year-end average price of battery-grade lithium carbonate has fallen below 100,000 yuan/ton, a drop of more than 80% compared with the highest price of 600,000 yuan/ton. Affected by this, the average bid price of the domestic market energy storage system continued to decline, falling to 0.79 yuan/(watt·hour) by December 2023, nearly halved compared with the beginning of 2023, and even a quote below 0.6 yuan/(watt·hour) appeared.

Combining Comprehensive Recycling Technology, Gradually Forming a Complete Battery Recycling Business Mechanism

In 2023, the wave of retirement of renewable energy vehicle batteries has arrived. It is predicted that the amount of retired power batteries in 2023 can reach a huge 590,000 tons. If retired batteries are allowed to be recycled and reused in traditional ways, it will cause serious waste of resources and environmental pollution. At the same time, retired batteries do not mean waste batteries, they still contain a large amount of important imported strategic resources such as lithium, nickel, and cobalt.

Perfecting the technology for the tiered use of power batteries and accelerating the construction of a new energy storage recycling system can effectively alleviate the supply constraints of scarce heavy metal resources in our country. Moreover, the environmental hazards of waste lithium batteries are significant, and without recycling, they will cause serious pollution. The centralized harmless treatment of waste lithium-ion batteries and the recycling of their metal materials is an important measure to ensure human health and sustainable environmental development. More importantly, renewable and low-carbon has become a new high ground in international industrial competition, and the recycling of old batteries has become an indispensable link in the lithium battery industry chain. Battery recycling can significantly reduce the carbon emissions of the lithium battery life cycle and is also beneficial to cost reduction and efficiency enhancement of the entire industry chain, including renewable energy vehicles.

(This section was written by: Lu Changbo, Reviewed by: Gao Hui)

Development of the New Energy Vehicle Market

In 2023, China's economy continued to recover and improve, laying a solid foundation for the vigorous development of new energy vehicles. Despite the decline in subsidies for new energy vehicles, the country and various regions have introduced a series of policies to promote car consumption, driving the rapid growth of new energy vehicle sales on a high base, and the market penetration rate has accelerated to over 30%. In 2024, with the continuous efforts of technological progress and policy support, the infrastructure for new energy vehicles continues to improve, private consumption willingness further strengthens, the sales scale of new energy vehicles continues to expand, but the sales growth rate has slowed down, and the market penetration rate has exceeded 35%.

Review of the Domestic New Energy Vehicle Market

The New Energy Vehicle Industry is Developing Rapidly, Entering a New Stage of Market-Driven Development

In 2023, China's new energy vehicle market maintained a good development trend. Although the country lowered the subsidy standards for new energy vehicles, with the introduction of a series of policies to promote new energy vehicle consumption, various regions actively improved the development plans for new energy vehicles, launched consumer subsidies and car-to-countryside activities, and strengthened support measures for the construction of charging and battery swapping infrastructure, ensuring the continuous growth of domestic new energy vehicle sales. The main features are as follows: First, the sales of new energy vehicles grew rapidly on a high base. As the production capacity of new energy manufacturers expanded, market competition intensified, driving the overall decline in the prices of new energy vehicles, further enhancing market competitiveness, and pushing sales to 9.5 million units, a year-on-year increase of 37.9%, accounting for more than 60% of global new energy vehicle sales, ranking first in the world for nine consecutive years (Fig. 9). Second, the internationalization of new energy vehicles accelerated, and exports of new energy vehicles grew explosively, the annual export volume reached 1.2 million units, a year-on-year increase of 77.2%, all record highs, and maintaining the position of the world's largest new energy vehicle exporting country for three consecutive years. Third, the development of new energy vehicles has entered a new stage of market-driven development, and the market penetration rate has continued to rise. The annual market penetration rate exceeded 31.6%, an increase of 6 percentage points from 2022, breaking through the 30% industry outbreak threshold. By the end of 2023, the number of new energy vehicles in use reached 20.41 million, accounting for 6.1% of the total number of vehicles in use, an increase of 2 percentage points from the previous year.

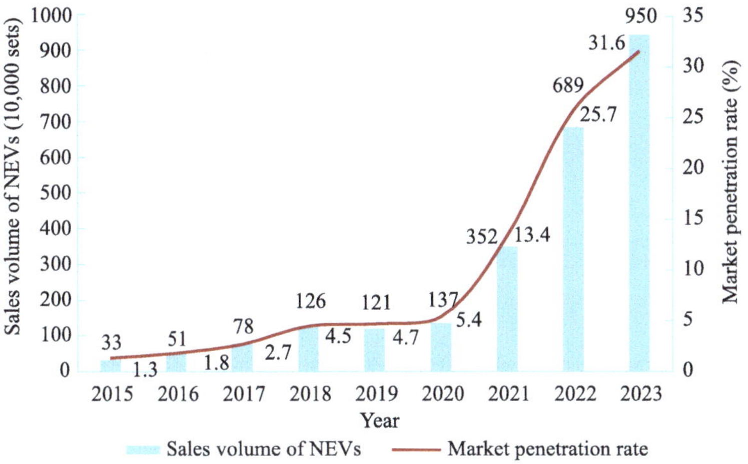

Fig. 9 Sales and market penetration rate of new energy vehicles in China. *Data source* Chinese Passenger Car Association (referred to as the CPCA), China Association of Automobile Manufacturers (referred to as the CAAM)

Passenger Cars Are Gradually Concentrated in the Private Sector, and the Proportion of Light Commercial Vehicles Has Significantly Increased

The continuous upgrading of battery technology and the continuous improvement of infrastructure have reduced consumers' mileage anxiety about buying new energy vehicles, promoting rapid growth in private consumption and becoming the main driver of the growth in sales of new energy passenger vehicles. In 2023, the sales of new energy passenger vehicles exceeded 9 million units, a year-on-year increase of 38.4%, a significant decrease in growth rate compared to 2022, and the market penetration rate reached 34.7%, an increase of 7 percentage points from 2022. Looking at the usage, private non-business use dominates the development. In 2023, the consumption ratio increased to 89.8%, an increase of 1.3 percentage points from 2022 (Fig. 10).

The consumption of new energy commercial vehicles has returned to rationality, and the proportion of light commercial vehicles continues to increase. Because the cruising range of new energy vehicle power batteries cannot meet the needs of large and medium-sized buses, and they do not have an advantage in power compared to traditional fuel vehicles, the consumption of large and medium-sized buses has returned to rationality, and the growth rate of new energy commercial vehicles has dropped significantly. In 2023, the sales of new energy commercial vehicles will be close to 450,000 units, a year-on-year increase of 33%, a decrease of 50 percentage points from 2022, and the market penetration rate is about 11.1%, an increase of 0.8 percentage points from 2022. Looking at the types of vehicles, the proportion of light new energy commercial vehicles continues to expand. In 2023, the proportions of

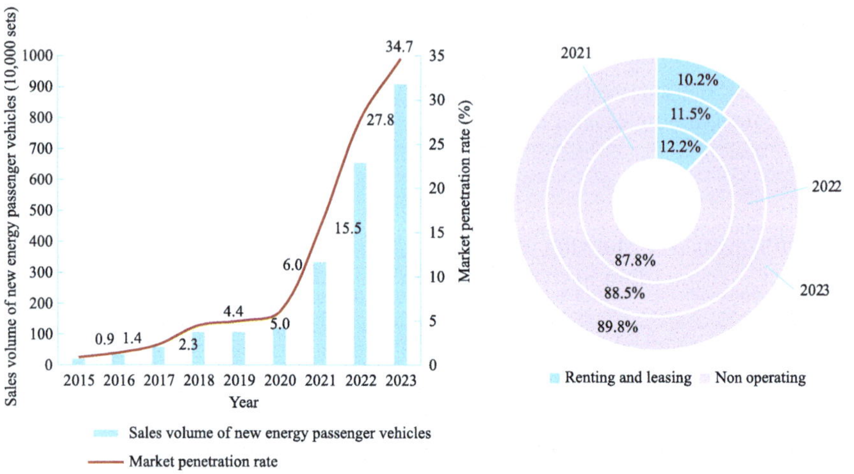

Fig. 10 Sales and Usage Structure of New Energy Passenger Vehicles in 2023. *Data source* CPCA, CAAM, China Automotive Technology & Research Center (referred to as the CATARC)

light passenger cars and light trucks increased to 44.5% and 33.9% respectively, an increase of 5.4 percentage points and 3.5 percentage points from 2022. However, the sales proportions of large buses and medium-sized buses dropped to 4.2% and 4.1% respectively, a decrease of 3.9 percentage points and 6.8 percentage points from 2022 (Fig. 11).

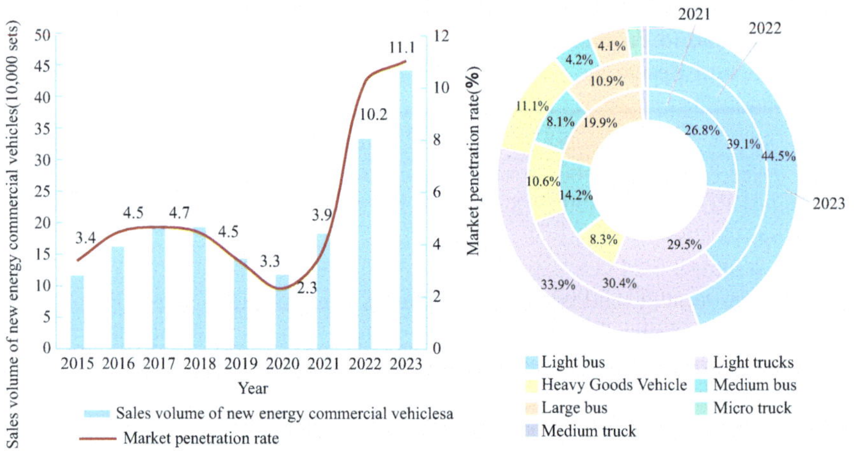

Fig. 11 Sales and structure of new energy commercial vehicles in 2023. *Data source* CPCA, CAAM, CATARC

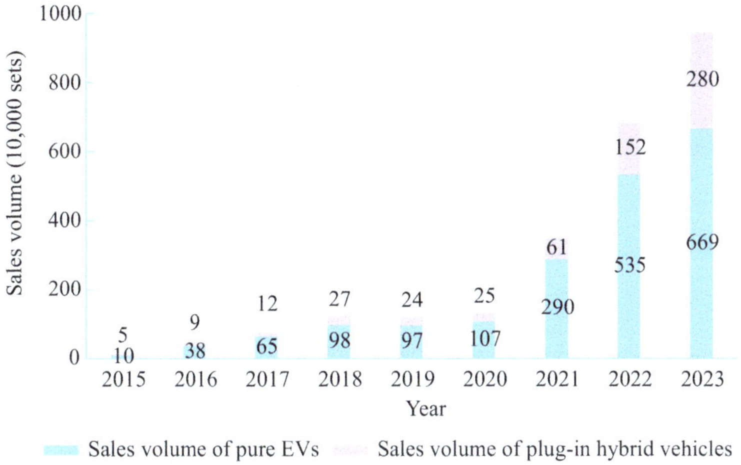

Fig. 12 Sales of new energy vehicles by power type. *Data source* CAAM

Pure Electric Models Still Dominate, Plug-In Models Account for Nearly 30%

In 2023, with the continuous refinement of application scenarios, hybrid vehicles are flexible in use, have a high market acceptance, and have a good market popularization prospect, gradually becoming an important engine for the diversified development of new energy vehicles. Looking at the types of power, pure electric vehicles still dominate the market despite a decline in their share. In 2023, the sales volume of pure electric models is about 6.69 million units, a year-on-year increase of 25%, accounting for 70.5% of the sales volume of new energy vehicles, a decrease of 7.3 percentage points from 2022 (Fig. 12); hybrid vehicles are growing rapidly and quickly seizing the incremental market, with a sales volume of about 2.8 million units in 2023, a year-on-year growth rate of 84.7%, accounting for 30% of the sales volume of new energy vehicles, contributing 49.3% of the increase in sales volume of new energy vehicles; fuel cell vehicles are still in the pilot stage, with products and application scenarios gradually enriching, and the promotion scope is still concentrated in pilot cities, with a sales volume of 7503 units in 2023, a year-on-year increase of 50%.

The Concentration of Passenger Car Brands is High, and a "One Super, Many Strong" Competition Pattern Has Formed

The concentration of the passenger car market remains high, with the top ten new energy vehicle manufacturers selling over 5 million units, accounting for about 56%. Looking at the brands, a "one super, many strong" market pattern has initially formed. BYD's sales reached 3.01 million units, far ahead of other OEM brands, with a market

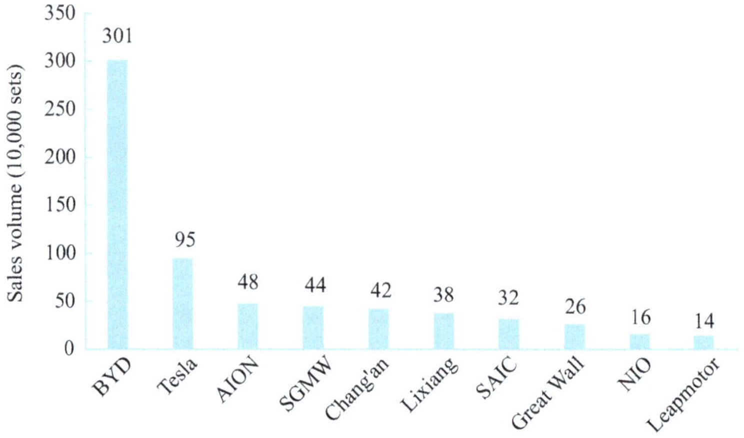

Fig. 13 Sales of top new energy passenger car manufacturers in 2023. *Data source* CAAM, Official Websites of Various Companies

share of 33.4%. In addition, Tesla, Aiways, General Motors Wuling, and Changan ranked high in sales, with sales of 950,000 units, 480,000 units, 440,000 units, and 420,000 units respectively (Fig. 13).

Rapid Growth in the Construction of Charging Facilities, Steady Progress in the Construction of Battery Swap Facilities

China's charging infrastructure industry continues to grow rapidly, strongly supporting the rapid development of China's electric vehicle market on a large scale. The market competition concentration is high, and the vehicle-to-pile ratio remains stable. In 2023, the increase in charging infrastructure was 3.39 million units, a year-on-year increase of 30.6%, and the sales volume of new energy vehicles was 9.5 million units, with a pile-to-vehicle increase ratio of 1:2.8. Among them, the increase in public charging piles was 929,000 units, a year-on-year increase of 42.7%, and the increase in private charging piles built with vehicles was 2.458 million units, a year-on-year increase of 26.6%. As of December 2023, the total number of charging infrastructures nationwide was about 8.6 million units, a year-on-year increase of 65%, gradually forming a virtuous cycle of mutual promotion between new energy vehicles and charging infrastructure. Among them, the stock of public charging infrastructure was about 2.73 million units (Fig. 14), a year-on-year increase of 51.7%, of which DC charging piles were 1.2 million units, AC charging piles were 1.52 million units, and the average monthly increase in public charging piles in 2023 was about 77,000 units; the stock of private charging infrastructure was 5.64 million units, a year-on-year increase of 75.2%. As of the end of 2023, there were a total of 3,550 battery swap facilities nationwide, a year-on-year increase of

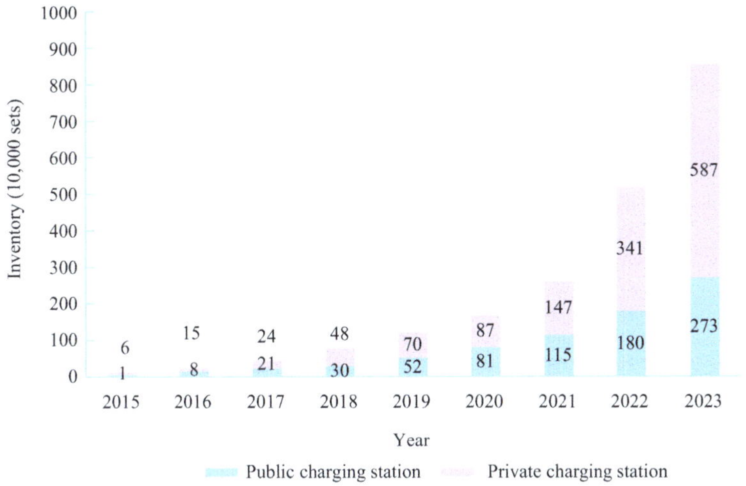

Fig. 14 National electric pile stock situation since 2015. *Data source* China Charging Alliance

79%, of which NIO operated more than 2,300 battery swap stations, accounting for more than 60%.

New Energy Vehicle Development Policy in 2023

In 2023, around the key directions of new energy vehicle development planning, consumer promotion, and infrastructure construction, a total of more than 600 policies related to new energy vehicles were introduced nationwide, mainly local (including city and district) policies. The policy-making body sinks to make policy-making closer to the market and consumers, better promoting the development of new energy vehicles.

New Energy Vehicle Industry Development Policy

At the national level, the policy is mainly to improve, subsidies gradually decrease and exit, the industry development enters the market-driven stage, and during the year, around the theme of steady growth in the automotive industry, policies such as new energy vehicles going to the countryside, tax benefits and exemptions were introduced, which played a stabilizing role in guiding the rapid development of new energy vehicles (Table 8).

Table 8 Overview of important national new energy vehicle policies in 2023

Date	Policy name	Core content
June 12	Notice on Carrying Out New Energy Vehicle Going to the Countryside Activities in 2023 by the General Office of the Ministry of Industry and Information Technology and other five departments	The activities are carried out in a combination of "offline + online". Offline mainly includes the launch ceremony + series of tours in advantageous regions, demonstration activities in characteristic regions, supplemented by various activities actively carried out in various places; "online" activities are carried out by e-commerce and internet platforms according to the on-site activity arrangements, setting up online publicity columns, carrying out "online" promotions, live car sales and other activities, participating in the whole process and continuously carrying out new energy to the countryside, realizing the linkage between offline and "online"
June 19	"Announcement on Continuing and Optimizing the Policy of Exemption from Vehicle Purchase Tax for New Energy Vehicles"	From January 1, 2024 to December 31, 2025, new energy vehicles are exempted from vehicle purchase tax, and the tax exemption for each new energy passenger vehicle does not exceed 30,000 yuan; from January 1, 2026 to December 31, 2027, new energy vehicles are subject to half of the vehicle purchase tax, and the tax reduction for each new energy passenger vehicle does not exceed 15,000 yuan
July 6	Revision of the Parallel Management Measures for Average Fuel Consumption of Passenger Vehicle Enterprises and New Energy Vehicle Points	Adjust the calculation method of new energy vehicle points, which will be implemented from August 1, 2023, and the average score of standard models of new energy passenger vehicles will be reduced by about 40%, and the calculation method and score limit of points have been adjusted accordingly
September 1	Notice of the Ministry of Industry and Information Technology and other seven departments on Issuing the Work Plan for Stabilizing Growth in the Automobile Industry (2023–2024)	Support to expand the consumption of new energy vehicles. Implement the existing preferential policies such as the exemption from vehicle purchase tax for new energy vehicles, do a good job in the liquidation and audit of new energy vehicle subsidy funds, and actively expand the proportion of personal consumption of new energy vehicles. Improve infrastructure construction and operation. Implement the guiding opinions on further building a high-quality charging infrastructure system, optimize the supporting environment. Encourage all regions to scientifically predict the charging demand of new energy vehicles, do a good job in the layout planning of urban and surrounding county and rural public charging networks, and promote the orderly development of charging facility layout construction and supporting power grid expansion and transformation

Data source Various ministries' websites

At the local level, provinces, cities, and autonomous regions have clarified the development goals of new energy vehicles, issued plans to promote the consumption of new energy vehicles, vigorously promoted the activities of new energy vehicles going to the countryside, and strengthened the subsidy policy for oil-to-electricity conversion (Tables 9 and 10).

Infrastructure Policy

Focusing on infrastructure construction, the state has issued support policies in various aspects such as planning, construction, and operation (Table 11), providing support for the construction of new energy vehicle infrastructure.

Local governments around the country, in line with national policy measures and in response to the infrastructure construction needs of their regions, have gradually improved construction planning, construction subsidies, and facility operation incentives in various fields (Table 12), ensuring the continuous advancement of new energy vehicle infrastructure construction and providing support for the rapid development of new energy vehicles.

Outlook for the New Energy Vehicle Market in 2024

Domestic Production Capacity Continues to Expand, the Process of Internationalization Slows Down, and the Competitive Landscape is Deeply Adjusted

With the rapid growth of domestic new energy vehicle sales, the explosive growth of market participants under the boost of capital, driving the rapid expansion of new energy vehicle production capacity, it is expected that the cumulative production of new energy vehicles will exceed 30 million units by July 2024. The market competition is more inward, and the internationalization process of new energy vehicles may slow down under the game of big countries, affecting the industry to enter a period of deep reshaping.

Sales Continue to Grow Rapidly, and the Expansion of the Base Leads to a Decline in Growth Rate

With the continuous support of government policies, the continuously growing new energy vehicle industry has become a new driving force for China's economic growth. Enterprises have made huge investments in the field of R&D, promoting the continuous progress of new energy vehicle technology, and the rapid iteration of automobile intelligence and electrification technology. As consumers' attention to environmental

Table 9 Overview of local new energy vehicle development planning policies in 2023

Date	Province	Policy name
January 2	Tianjin	"Implementation Plan for Green and Low-Carbon Development in the Field of Transportation in Tianjin"
January 3	Gansu	"Guiding Opinions on Hydrogen Energy Industry Development"
January 12	Qinghai	"Three-Year Action Plan for Hydrogen Energy Industry Development in Qinghai Province (2022–2025)"
January 12	Qinghai	"Medium and Long-Term Plan for Hydrogen Energy Industry Development in Qinghai Province (2022–2035)"
January 29	Zhejiang	"Action Plan for Accelerating the Development of New Energy Vehicle Industry in Zhejiang Province"
March 27	Zhejiang	"Implementation Opinions on Promoting the High-Quality Development of New Energy in Zhejiang Province (Draft for Soliciting Opinions)"
April 12	Guizhou	"Plan for the Construction of New Energy Power Battery and Material Research and Development Production Base in Guizhou Province (2022–2030)"
April 21	Shanxi	"Action Plan for Hydrogen Energy Industry Chain in Shanxi Province in 2023"
April 25	Shanghai	"Implementation Rules for Special Funds for Demonstration Application of Fuel Cell Vehicles in Shanghai"
April 26	Shanxi	"Notice on Matters Related to Electric Vehicle Charging and Replacement Service Fees and Electricity Prices"
May 9	Chongqing	"Action Plan for Doubling the Development of Intelligent Networked New Energy Vehicle Parts Industry in the Western Chongqing Region (2023–2027)"
May 25	Henan	"Henan Province's 14th Five-Year Plan for the Development of New Energy and Renewable Energy"
May 28	Anhui	"Plan for the Construction of New Energy Vehicle and Intelligent Networked Vehicle Industry Ecosystem"
June 28	Shanghai	"Top-level Design Scheme for Shanghai Smart Bus"
Time	Province	Policy name
July 12	Hainan	"Hainan Province Vehicle Networking Industry Development Plan"

(continued)

Table 9 (continued)

Time	Province	Policy name
July 20	Shanghai	"Shanghai Transportation Field Hydrogen Energy Promotion Application Scheme (2023–2025)"
August 8	Hainan	"Hainan Province's Medium and Long-term Action Plan for the Promotion of New Energy Vehicles (2023–2030)"
August 29	Xinjiang	"Several Policy Measures of the Autonomous Region to Support the Construction of Hydrogen Energy Industry Demonstration Zone"
September 1	Guangxi	"Guangxi Hydrogen Energy Industry Medium and Long-term Development Plan (2023–2035)"
September 28	Guizhou	"Guiding Opinions on Accelerating the Construction of 'Electric Guizhou'"
September 28	Tianjin	"Tianjin's Implementation Plan for Accelerating the Development of New Energy and Intelligent Networked Vehicle Industry (2023–2027)"
October 13	Anhui	"Anhui Province's High-quality Charging and Replacement Service System Construction Plan (2023–2027)"
October 20	Guizhou	"Implementation Opinions on Accelerating the Construction of Guizhou New Energy Battery and Material Research and Development Production Base (Draft for Soliciting Opinions)"
October 27	Guizhou	"Guiding Opinions on Accelerating the High-quality Development of New Energy Vehicle Industry and Promoting the Construction of 'Electric Guizhou'"
November 6	Guangdong	"Notice of Guangdong Province on Accelerating the Innovation and Development of Hydrogen Energy Industry"
November 13	Hubei	"Hubei Province's Implementation Plan for the Transformation and Development of the Automobile Industry (2023–2025)"

Data source Provincial government websites

protection and sustainability increases, the price and performance of new energy vehicles are increasingly attractive to consumers, and the domestic market is expected to continue to grow. It is estimated that by 2024, the growth rate of China's new energy vehicle sales will further decline, and the annual sales are expected to exceed 11.5 million units, a year-on-year increase of 20%.

Increase the Construction of Supporting Facilities to Promote High-Quality Development of the Industry

It is expected that more policies will be introduced around high-quality development in 2024. The intensity of infrastructure construction has increased. As the market penetration rate increases, the support for charging infrastructure construction will also strengthen, solving the bottleneck of new energy vehicle use. Emission standards and environmental protection requirements are improved, and more consumers

Table 10 Overview of local new energy vehicle consumption promotion policies in 2023

Time	Province	Policy name
January 9	Tianjin	The second phase of "Jin Le Gou" consumer coupon distribution activity
January 28	Shanxi	"Notice on Further Stimulating Automobile Circulation and Expanding Automobile Consumption"
February 3	Guizhou	"Several Measures of Guizhou Province to Stimulate Automobile Circulation and Expand Automobile Consumption"
February 6	Henan	"Notice on Further Stimulating Automobile Circulation and Expanding Automobile Consumption"
February 8	Tianjin	"Notice of the Municipal Transportation Commission on the Allocation of Additional Quotas for Individual Car Lottery in Our City in 2023"
February 20	Guangdong	"Implementation Plan of Guangdong Province to Further Stimulate Automobile Circulation and Expand Automobile Consumption"
March 15	Shanghai	"Notice on the Operation Procedure Related to the Implementation Measures for Encouraging the Purchase and Use of New Energy Vehicles in Shanghai in 2023"
March 21	Beijing	"Beijing's Plan to Encourage Automobile Replacement and Consumption"
March 27	Anhui	"Notice on Several Measures to Support and Expand Automobile Consumption"
April 12	Beijing	"Notice on the Declaration of Fuel Cell Vehicle Demonstration Application Project in Beijing for 2022–2023"
May 6	Shanghai	"Notice on Adjusting Some Contents of the Detailed Rules for the Implementation of Subsidies to Promote Automobile Consumption in Shanghai"
May 30	Shanghai	"Several Policy Measures to Support the Development of Private Investment in Shanghai"
June 1	Chongqing	"Chongqing's 2023 Financial Subsidy Policy for Charging and Changing Infrastructure"
June 2	Hebei	"Opinions on Accelerating the Construction of Charging Infrastructure in Rural Areas and Promoting the Implementation of New Energy Vehicles and Rural Revitalization"
June 7	Fujian	"Implementation Opinions on Fully Promoting the Construction of 'Electric Fujian' (2023–2025)"
July 6	Hunan	"Specific Measures for Administrative and Public Institutions in Hunan Province to Promote New Energy Vehicles"

(continued)

Table 10 (continued)

Time	Province	Policy name
August 2	Shandong	"2023 New Energy Vehicle Going to the Countryside Activity Plan in Shandong Province"
August 13	Jiangxi	Jiangxi Province "Hundred Cities Linkage" Automobile Consumption Season
August 14	Zhejiang	"Zhejiang Province New Energy Vehicle Rural 'Ten Actions' List"
August 31	Hunan	"Several Policy Measures for Hunan Province to Restore and Expand Consumption"
September 4	Tianjin	"Tianjin City's Implementation of the 'Measures to Restore and Expand Consumption' Specific Measures Division Scheme"
September 7	Shanghai	"Several Measures to Revitalize Automobile Circulation and Expand Automobile Renewal Consumption in Shanghai"
September 14	Shandong	"Shandong Province's Three-Year Action Plan to Promote New Energy Vehicles in Rural Areas (2023–2025)"
September 18	Guangxi	2023 Guangxi "Love New Cars" Car Purchase Subsidy
September 24	Hainan	"Several Measures to Encourage the Promotion and Application of New Energy Vehicles in Hainan Province from 2023 to 2025"
September 27	Tianjin	"Driving Tianjin" 2023 Tianjin Auto Consumption Carnival
October 23	Shaanxi	Shaanxi Launches "Million Subsidies for Home Charging Piles" Event
October 25	Anhui	"Notice on Accelerating the Promotion of New Energy Vehicles in the Public Domain"
October 30	Gansu	"Several Implementation Measures for Promoting Automobile Consumption in Gansu Province"
November 3	Guizhou	"Several Measures to Accelerate the Promotion and Application of New Energy Trucks"
November 7	Hainan	"Notice on the Application for Subsidies for Encouraging the Promotion and Application of New Energy Vehicles in Hainan Province in 2023"
November 21	Yunnan	"Implementation Measures for Promoting Automobile Consumption"
November 22	Jiangxi	2023 Jiangxi Provincial Car Consumption Coupon
December 15	Shanghai	"Shanghai Municipal Measures for Encouraging the Purchase and Use of New Energy Vehicles"
December 24	Hubei	Hubei Province Issued 200 Million Yuan "Hubei Purchase" Car Consumption Coupons

Data source Provincial government websites

Table 11 Overview of important national new energy vehicle infrastructure policies in 2023

Time	Policy	Core content
June 19	State Council Office "Guidance on Further Building a High-Quality Charging Infrastructure System"	By 2030, a high-quality charging infrastructure system that is widely covered, appropriately scaled, rationally structured, and fully functional will be basically established, strongly supporting the development of the new energy vehicle industry and effectively meeting the charging needs of the public
June 28	"Implementation Opinions on Accelerating the Construction of Charging Infrastructure to Better Support the Rural Promotion and Revitalization of New Energy Vehicles"	Innovate the construction and operation maintenance mode of charging infrastructure in rural areas, including strengthening the layout construction of public charging infrastructure, promoting the sharing of community charging infrastructure construction, increasing the support for charging network construction and operation, promoting new modes such as intelligent orderly charging, and improving the service experience of charging infrastructure operation and maintenance. Support the purchase and use of new energy vehicles in rural areas, and strengthen the promotion, service, and management of new energy vehicles in rural areas
July 20	"Several Measures to Promote Automobile Consumption"	Strengthen the construction of new energy vehicle supporting facilities. Continue to promote the formulation of related standards for battery swap infrastructure, enhance compatibility and universality. Accelerate the promotion and application of the battery swap model, actively carry out pilot projects for vehicle battery swap in the public domain, support the construction of charging and battery swap infrastructure at urban bus stations. Encourage cities and major traffic arteries with conditions to accelerate the construction of battery swap stations

Data source Various ministries' websites

are encouraged to choose new energy vehicles by strengthening the emission and energy consumption standards of fuel vehicles. Encourage technology research and development and innovation, improve the performance and competitiveness of new energy vehicles. Regulate industry supervision, build a national-level operation platform, increase the scale of infrastructure, protect consumer rights, and promote the high-quality development of the new energy vehicle industry.[1]

(This section is written by: Xiang Zhengjian, reviewed by: Gao Hui)

[1] Calculated based on the company's oil and gas production combined in the first half of 2023.

Table 12 Overview of important local new energy vehicle infrastructure policies in 2023

Date	Province	Policy name
March 13	Chongqing	"Notice on Chongqing's 2023 Fiscal Subsidy Policy for Charging and Battery Swap Infrastructure (Draft for Soliciting Opinions)"
March 17	Fujian	"Fujian Province 'Light Storage Charging Inspection' Charging Infrastructure Construction Management Guide (Trial)"
March 31	Sichuan	"Sichuan Province Charging Infrastructure Construction Operation Management Measures"
April 13	Shanghai	"Shanghai Residential Community Electric Vehicle Charging Facility Construction Management Measures"
May 16	Shanxi	"Shanxi Province Electric Vehicle Charging (Swap) Infrastructure Construction Operation Management Measures"
May 17	Inner Mongolia	"Six Policy Measures to Promote the Construction of New Energy Vehicle Charging Facilities in Urban Areas"
June 2	Beijing	"Beijing Residential Area New Energy Vehicle Charging 'Unified Construction and Service' Pilot Work Plan"
June 30	Hunan	"Notice on the Trial Implementation of Time-of-Use Electricity Pricing for Residential Electric Vehicle Charging Facilities"
July 6	Tianjin	"Several Measures to Accelerate the Construction of Charging Infrastructure to Better Support the Rural Promotion and Revitalization of New Energy Vehicles"
July 12	Chongqing	"Notice on the Issuance of Chongqing's 2023 Fiscal Subsidy Policy for Charging and Swapping Infrastructure"
July 18	Jiangxi	"Notice on Further Promoting the Construction of Electric Vehicle Charging Infrastructure"
July 21	Hainan	"Notice on Further Optimizing the Peak-Valley Time-of-Use Electricity Pricing Policy for Electric Vehicles (Draft for Soliciting Opinions)"
July 24	Zhejiang	"Improving the High-Quality Charging Infrastructure Network System and Promoting the Action Plan for New Energy Vehicles to Go to the Countryside (2023–2025)"
July 31	Shandong	"Notice on Further Improving the Time-of-Use Electricity Pricing Policy for Residential Electric Vehicle Charging Piles"
August 4	Henan	"Henan Province's Major New Infrastructure Construction Acceleration Action Plan (2023–2025)"

(continued)

Table 12 (continued)

Date	Province	Policy name
August 11	Hubei	"Implementation Opinions of the Provincial People's Government Office on Accelerating the Construction of a High-Quality Charging Infrastructure System in Hubei Province"
August 11	Chongqing	"Special Plan for Charging and Swapping Infrastructure in Chongqing's Central Urban Area (2023–2025)"
August 15	Henan	"Henan Province's Three-Year Action Plan for the Construction of Electric Vehicle Charging Infrastructure (2023–2025)"
August 15	Henan	"Henan Province's Three-Year Action Plan for the Construction of Electric Vehicle Charging Infrastructure (2023–2025)"
August 16	Gansu	"Several Measures to Accelerate the Construction of Charging Infrastructure and Better Support the Going to the Countryside of New Energy Vehicles and Rural Revitalization"
August 31	Shaanxi	"Notice on Matters Related to the Time-of-Use Electricity Pricing Policy for Residential Electric Vehicle Charging Piles"
September 7	Hubei	"Notice on Implementing Time-of-Use Electricity Pricing for Residential Electric Vehicle Charging Facilities (Draft for Soliciting Opinions)"
September 8	Ningxia	"Implementation Plan for Further Enhancing the Service Guarantee Capacity of Charging Infrastructure in the Autonomous Region"
September 19	Tianjin	"Implementation Plan for Further Building a High-Quality Charging Infrastructure System in Tianjin"
September 20	Beijing	"Detailed Rules for the Implementation of Awards for the Construction and Operation of Electric Vehicle Charging and Swapping Facilities in Beijing in 2023 (Draft for Soliciting Opinions)"
October 10	Jilin	"Notice on Improving Matters Related to Electricity Prices and Service Fees for Electric Vehicles"
October 20	Beijing	"Detailed Rules for the Implementation of Awards for the Construction and Operation of Electric Vehicle Charging and Swapping Facilities in Beijing in 2023"
Time	Province	Policy name
October 30	Shanghai	"Shanghai Public Charging Facilities 'Slow to Fast' 'Vehicle-Network Interaction' and 'Unified Construction and Operation' Demonstration Community Construction Plan for 2023"
November 2	Chongqing	"Guide for the Record of Public Charging and Battery Swap Infrastructure Construction Projects in the City (2023 Edition)"

(continued)

Table 12 (continued)

Time	Province	Policy name
November 9	Sichuan	"Sichuan Province's Plan to Accelerate the Construction of Charging Infrastructure, Support New Energy Vehicles to Go to the Countryside and Rural Revitalization"
November 15	Yunnan	"Notice on the Time-of-Use Electricity Price Policy for Residential Electric Vehicle Charging Piles (Trial)"
December 1	Hainan	"Notice on Publicly Soliciting Opinions on Adjusting the Peak-Valley Time-of-Use Electricity Price Policy for Electric Vehicles (Draft for Soliciting Opinions)"
December 6	Fujian	"Plan to Accelerate the Construction of the Charging Infrastructure System in Fujian Province (Draft for Soliciting Opinions)"
December 13	Anhui	"Anhui Province New Energy Vehicle Charging and Battery Swap Infrastructure Construction and Operation Management Measures (Interim)"
December 19	Zhejiang	"On Further Optimizing and Regulating the Electricity Price for Electric Vehicle Charging and Battery Swap Facilities"

Data source Provincial government websites

Development of the CCUS Industry

New Situations and Requirements for the Development of CCUS Technology in China

The Development of CCUS Technology is Urgent, and the High-Speed Growth Period of the Industry is Coming

In China's energy consumption, the total consumption of fossil energy is large and the proportion is high. CCUS still has a large growth space in the future to support carbon reduction contributions. According to estimates, under the carbon neutrality target, China's CCUS emission reduction demand is about 24 million tons/year in 2025, and will increase to the level of billions of tons per year in 2030. The current carbon reduction capacity of CCUS projects under construction and in operation is less than 10 million tons per year, the target gap is huge, and the demand for industrial growth is high.

High-carbon emission industries urgently need to accelerate the low-carbon transformation, and the window for CCUS facility transformation is relatively tight. The service time of equipment in industries such as China's active coal-fired power plants (40%), cement plants (55%), and steel plants (15%) is relatively short. Forced retirement will cause a large amount of assets to be stranded, amounting to 3.1 trillion to 7.2 trillion yuan (according to the "China Carbon Capture, Utilization and Storage Annual Report (2023)"). Taking coal-fired power plants as an example, installing CCUS can capture 90% of carbon emissions, but installing CCUS facilities will reduce the power generation efficiency of power plants by 20–30%, and increase the power generation cost by about 61% (Table 13). Considering that the payback period of CCUS technology investment is at least 10 to 15 years, the best transformation period for infrastructure is 2030–2035, and it is imperative and urgent to complete low-cost, low-energy consumption CCUS supporting transformation in the future.

The number of global CCUS projects has been growing for six consecutive years. Driven by policies in various countries around the world, in 2023, more than 198 new projects were announced globally, an increase of 102% compared to the same period last year. It is estimated that after all the newly planned projects are completed globally, the carbon dioxide capture capacity will be about 361 million tons/year (Fig. 15).

Table 13 Changes in the cost of installing CCUS in some industries (estimated)

Project	Coal power plant [yuan/(kW h)]	Steel (yuan/ton)	Cement (yuan/ton)
Original cost	About 0.325	About 5500	About 450
Cost after installing CCUS	About 0.525	About 0.525	About 705
Increase	About 61%	About 15%	About 57%

Data source PetroChina Economic Research Institute

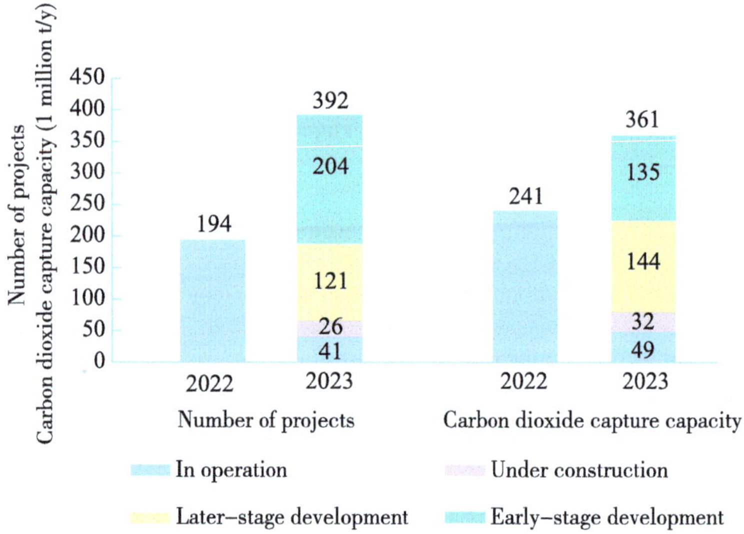

Fig. 15 Number and capture capacity of global CCUS projects in 2022 and 2023. *Data source* GCCSI, 2023

Progress Has Been Made in Technology Research and Development Capabilities, and Some Key Technologies Need to Be Broken Through

On the carbon capture technology side, China is gradually maturing in the field of first-generation carbon capture technology, which covers pre-combustion, post-combustion and oxygen-enriched combustion capture of coal-fired power plants, post-combustion capture of gas-fired power plants, carbon dioxide capture of coal chemical industry, and post-combustion capture of cement kiln tail gas and other technologies. The physical absorption method for decarbonization transformation of thermal power plants is already at the commercial application stage. The second-generation carbon capture technology (such as new membrane separation technology, new absorption technology, new adsorption technology, pressurized oxygen-enriched combustion technology, etc.) is still in the laboratory research or pilot stage.

On the carbon utilization technology side, China is at the demonstration project stage in terms of carbon dioxide enhanced oil recovery (CCUS-EOR), carbon dioxide fertilizer production, and food-grade applications. It is estimated that there are about 14 billion tons of geological reserves suitable for carbon dioxide enhanced oil recovery nationwide. CCUS-EOR technology can increase the recovery rate by about 15%, and the carbon dioxide sequestration rate is about 65%. In August 2023, Jilin Oilfield completed the first full-industry chain, full-process CCUS-EOR demonstration project in China. China's carbon dioxide flooding mining technology has reached the commercial application level, and it is still in the basic research or pilot stage in the field of carbon dioxide enhanced natural gas, shale gas, and coalbed methane extraction. In the emerging carbon utilization directions such as carbon dioxide hydrogenation to methanol and carbon dioxide methanation reforming, they are still in the theoretical research of catalysts or the pilot stage of projects.

On the carbon transport technology side, China has reached the commercial application stage in CO_2 transportation technology for road tankers and ships, mainly used for CO_2 transportation of less than 100,000 tons per year. In July 2023, the CO_2 transportation pipeline of the Qilu Petrochemical–Shengli Oilfield million-ton CCUS demonstration project was put into operation, marking the accelerated move towards commercial application of land pipeline transportation. Submarine pipeline transportation is still in the research stage.

In terms of carbon sequestration technology, China's CO_2 sequestration mainly includes onshore saline aquifer sequestration, seabed saline aquifer sequestration, and depleted oil and gas field sequestration, with a geological sequestration capacity of 1.21 trillion to 4.13 trillion tons. Considering the maturity of technology and operating costs, oil and gas reservoirs are the most suitable sequestration sites, with a CO_2 sequestration capacity of up to 5 billion tons. In addition, saline aquifer sequestration has low cost and high sequestration potential, and will become the main method for China's medium and long-term deep CO_2 emission reduction. In June 2023, China's first offshore carbon sequestration demonstration project (Enping

15-1 oilfield) was officially put into operation, achieving a breakthrough in offshore CO_2 sequestration.

The Policy and Standard System Continues to Improve, and the Development of Large-Scale Clusters Faces Challenges

China is the first country in the world to fund CCUS-related research through national science and technology plans. Since 2003, China has successively introduced a series of CCUS development policies, which have effectively promoted the construction of CCUS demonstration projects and the perfection of the industrial chain. The CCUS policy and regulatory system has initially formed. Compared with foreign CCUS policies such as the US 45Q amendment, Canada's carbon tax policy, and the UK's differentiated grid-connected electricity price contract, China's policies and regulations are mostly scientific planning and guiding policy documents, and there is no special policy for CCUS subsidy incentives and promoting CCUS industrial development. The development of CCUS clusters involves technical breakthroughs and demonstration project construction in carbon sink source matching, capture, transportation, sequestration, monitoring, utilization, etc., and needs to be further strengthened in terms of industrial integrated layout planning, incentive policies, industrial deployment and management systems.

Current Status of China's CCUS Demonstration Project Development

Overview of the Diversified Capture, Utilization, and Sequestration of China's CCUS

According to incomplete statistics, in 2023, China has 91 CCUS demonstration projects in operation or under construction (excluding those in planning), spread across 24 provinces, with a CO_2 capture capacity of 8.74 million tons per year. Among them, there are more than 30 coal-fired power plant CCUS demonstration projects, with a total capture capacity of about 4.7 million tons per year. From the perspective of regional distribution, CCUS demonstration projects are mainly concentrated in the eastern provinces where high-energy-consuming industries are relatively concentrated, and CCUS projects in the central and western regions are accelerating. China has 8 million-ton-level CCUS projects in planning, construction, or operation. The Qilu Petrochemical—Shengli Oilfield CCUS project of Sinopec, as China's first million-ton-level CCUS project, has been fully completed and put into operation, marking the entry of China's CCUS industry into the mature commercial operation stage. According to statistics, China currently has about 10 million tons per year of projects in planning that have signed cooperation agreements and are waiting to start construction.

Technological Progress and Existing Gaps

China attaches great importance to tackling key problems and innovation in CCUS technology. In 2023, Southwest Petroleum University initiated the establishment of the CCUS Science and Technology Innovation International Cooperation Alliance to accelerate the cooperation of CCUS industry, academia, and research.

In terms of application in the coal power industry, in June 2023, the Jiangsu Taizhou Power Plant project of the National Energy Group was officially put into operation. The project applies a new generation of CO_2 absorbents, new fillers and efficient amine absorption and other series of independent technologies, reaching the industry-leading level.

In terms of carbon emission reduction industry, decarbonization practice promotes technological innovation. Hebei Iron and Steel Group launched the CCUS industrial demonstration project in the steel industry in August 2023, and completed the process scheme demonstration, indicating that the "reserve technology" has officially transformed into "application technology".

In the field of CO_2 chemical utilization, as of November 2023, about 40% of the CCUS technology demonstration projects that have been put into operation and planned in China are high-efficiency CO_2 conversion projects. Scientific research teams from the University of Electronic Science and Technology and the Chinese Academy of Sciences have made breakthroughs in technologies for converting CO_2 into ethanol, gasoline, glucose, fatty acids, etc.

In terms of carbon capture technology, The cost of carbon capture accounts for 60% to 80% of the total CCUS project, and cost and economic constraints are key to commercial promotion. Seeking better performance CO_2 absorbents and exploring lower-cost second and third-generation carbon capture technologies are crucial for promoting the comprehensive use of CCUS technology. China's ionic liquid, electrochemistry, membrane separation technology, etc. are at the laboratory stage, and DAC technology is still in its infancy.

In terms of carbon utilization technology, The main oil regions around the four major basins of Songliao, Ordos, Junggar, and Bohai Bay in China have a high degree of source-sink matching for CCUS, and CCUS-EOR has been scaled up. Other forms of utilization, such as seaweed cultivation, carbon dioxide polymerization, carbon dioxide methanation reforming and other technologies are rapidly developing. Green hydrogen and carbon dioxide production of green methanol technology is accelerating the demonstration application, and carbon dioxide carbon fiber, acetic acid and other material technologies still need to be tackled.

On the carbon transport technology side, China's tank truck transport and ship transport have reached the stage of commercial application, and it has the ability to design large-scale pipelines in land pipeline transport technology, but lacks experience in supercritical state transport and ultra-long distance transport scenarios.

On the carbon sequestration technology side, China has huge potential for geological sequestration, with various methods. It now has experience in sequestering 100,000 tons/year of carbon dioxide in the Ordos saline aquifer and the 150,000 tons/year post-combustion CCUS demonstration project at the Guohua Jinjie Power Plant. However, in the carbon sequestration link, exploration site selection technology is the key to carbon dioxide sequestration. The strengthening of deep saline water extraction and sequestration technology has reached the industrial demonstration stage, while heat extraction utilization, enhanced shale gas and natural gas extraction are still in the research stage.

Progress of CCUS Demonstration Projects

In 2023, the progress of China's CCUS demonstration projects accelerated, with a total of 2 million-ton CCUS demonstration projects starting construction, and 6 demonstration projects and engineering projects announced completion or operation, with projects spread across the Northeast, North China, and Northwest regions (Table 14). CCUS technology is widely adopted in high-carbon emission industries such as power, oil, steel, and cement. For example, power companies are promoting large-scale, low-cost carbon capture demonstration projects; steel companies are conducting technical research and demonstration applications around CCUS steel slag mineralization utilization, CCUS coupled with hydrogen metallurgy, etc.; printing and dyeing companies have achieved the first Chinese ten-thousand-ton carbon capture and utilization project. The development of CCUS-related equipment manufacturing and materials has accelerated, and key equipment such as compressors and membrane components have been nationalized. Breakthroughs have also been made in pipeline and monitoring application links.

Policy on CCUS Technology Application Fields

International CCUS Policy

The United States is a global leader in CCUS demonstration projects. It encourages the development of CCUS technology, promotes the commercialization of projects, and increases green financial support. The passed "Inflation Reduction Act" and "Clean Competition Act" both strengthen the "45Q Act" tax credit plan to further promote the commercial development of CCUS.

The EU's "2009/31/EC" is the world's first detailed CCS legislation, which stipulates the responsibilities for carbon dioxide transportation, storage, operation, and closure. In October 2023, the Carbon Border Adjustment Mechanism

Table 14 Situation of China's CCUS project construction (partial) in 2023

Type	Serial number	Project name	Location	Project introduction
Power Industry	1	State Energy Group's 500,000 tons/year CCUS project	Taizhou, Jiangsu	Currently the largest coal-fired CCUS project in Asia
	2	Changxing Island Thermal Power Co., Ltd.'s 100,000 tons/year CCUS project	Chongming, Shanghai	The first Chinese low-concentration CCUS project for gas combustion
	3	Huaneng Yangpu Cogeneration Project's 2000 tons/year CCUS project	Yangpu, Hainan	The first Chinese gas flue gas CCUS project
Petrochemical Industry	4	CNOOC Shenzhen Branch's CCUS project	Shenzhen, Guangdong	The first Chinese offshore CCUS project
	5	PetroChina Xinjiang Oilfield's 2 × 660,000 kW CCUS project (2 million tons)	Xinjiang Karamay	The largest coal-fired CCUS Integrated project in the country
	6	Huaneng Longdong Energy Base's million-ton carbon dioxide capture, utilization and sequestration research and demonstration project (1.5 million tons)	Qingyang, Gansu	The world's largest and lowest energy-consuming coal-fired carbon dioxide capture project
Steel Industry	7	Hebei Iron and Steel Group and BHP Billiton CCUS Project	Handan, Hebei	Steel Industry CCUS Industrial Demonstration Project
	8	Baogang Group and Chengda Company 2 million tons/year CCUS Project	Inner Mongolia Baotou	The first full-industry chain CCUS demonstration project in the steel industry
Cement Industry	9	Qingzhou Zhonglian Cement Co., Ltd. 200,000 tons/year CCUS Project	Qingzhou, Shandong	The first and largest CCUS project in the Chinese cement industry
Chemical Industry	10	Xinjiang Jinjiang Chemical 100,000 tons/year CCUS Project	Kuitun, Xinjiang	Synthetic ammonia urea device tail gas CCUS project

Data source PetroChina Economic Research Institute

(CBAM) entered the transition implementation period, levying carbon taxes on high-carbon import products, which is conducive to promoting the development of CCUS technology.

Australia insists on both positive and reverse incentive policies, advocating for increased supervision of major carbon-emitting companies in the form of compensation costs, forcing companies to reduce carbon emissions year by year. At the same time, the government promises to allocate special funds to the CCS center every year.

Japan has long been committed to a low-carbon development strategy and actively participates in overseas CCUS project investments internationally. Documents such as the "Carbon Cycle Utilization Technology Roadmap" and the "Green Growth Strategy for Achieving Carbon Neutrality by 2050" clearly define the status of the CCUS technology route.

Chinese CCUS Policy

Under the framework of the "$1 + N$" policy system for carbon peak and carbon neutrality, China's CCUS policy is becoming more and more perfect. According to incomplete statistics, as of November 2023, a total of 77 CCUS-related policy documents have been issued, involving planning, standards, roadmaps, technology catalogs, etc. Local governments in Beijing, Shanghai, Guangdong, Jiangsu, and other places have also issued more than 130 path planning and implementation plans to promote the innovation of CCUS technology and the construction of demonstration projects.

In 2023, under the existing policy system, China continued to enhance the guidance of CCUS technology research and development and industrial demonstration, and the layout of the entire industry chain was significantly strengthened (Table 15). The policies show the following characteristics.

First, the strategic orientation of technology is more explicit. The National Development and Reform Commission and other departments issued the "Notice on Promoting the Healthy Development of the Modern Coal Chemical Industry", and the National Energy Administration issued the "Blue Book on the Development of New Power Systems", which deployed the coupled innovative development of modern coal chemical industry with renewable energy, green hydrogen, and carbon capture, utilization, and storage (CCUS). The trend of coupled development of CCUS and new energy is more explicit.

Second, the industrial practice policy emphasizes cluster development. The National Development and Reform Commission and other departments issued the "Guiding Opinions on Promoting the Green Innovation and High-Quality Development of the Refining Industry" and the "Implementation Plan for the Green Low-Carbon Advanced Technology Demonstration Project", strengthening the layout of the entire CCUS industry chain. The Anhui Provincial Development and Reform

Table 15 2023 China CCUS policy situation table (partial)

Year	Department	Policy and Related Content
National Ministries	National Energy Administration	The "Blue Book on the Development of New Energy Systems" requires accelerating the research and development of new CCUS technologies and the integration and demonstration application of the whole process system
	National Development and Reform Commission and other departments	The "Guiding Opinions on Promoting Green Innovation and High-quality Development of the Refining Industry" supports refining enterprises to accelerate the demonstration application of CCUS, effectively reducing carbon emissions
	National Development and Reform Commission and other departments	The "Implementation Plan for Green Low-Carbon Advanced Technology Demonstration Projects" proposes three key directions: full-process scale CCUS demonstration, advanced and efficient carbon dioxide capture demonstration, and carbon dioxide resource utilization and carbon fixation demonstration
	Ministry of Ecology and Environment	The "National Low-Carbon City Pilot Work Progress Assessment Report" sets up the supply of green low-carbon technology, incorporating the CCUS field into evaluation indicators
Local Government	Anhui Provincial Development and Reform Commission	The "Notice on Organizing the Application for Green Low-Carbon Advanced Technology Demonstration Project and Low-Carbon Application Scenarios" stipulates that for the petrochemical, coal chemical, and oil extraction industries, the annual capture, utilization, and storage volume of the project should not be less than 500,000 tons, and a carbon dioxide transportation pipeline should be built in principle; for the coal power, steel, non-ferrous, and building materials industries, the annual capture, utilization, and storage volume of the project should not be less than 100,000 tons
	Hebei Provincial Department of Ecology and Environment	The "Hebei Province Carbon Capture Project Emission Reduction Accounting Methodology" stipulates the accounting process and method for net reduction of carbon dioxide emissions through carbon capture projects in the steel, building materials, power generation, oil, and chemical industries within the scope of Hebei Province
	Shanxi Provincial Development and Reform Commission, etc	The "Guidelines on Promoting the Integrated Development of the Coal Industry and Carbon Reduction Technology" emphasizes the promotion of carbon reduction technology throughout the entire process of carbon capture, utilization, and storage (CCUS)

(continued)

Table 15 (continued)

Year	Department	Policy and Related Content
	Tianjin Municipal Science and Technology Bureau	The "Tianjin Science and Technology Support Carbon Peak and Carbon Neutrality Implementation Plan (2022–2030)" focuses on breakthroughs in carbon capture, utilization, and storage (CCUS), ecosystem carbon sequestration, carbon sink monitoring, etc., accelerates the continuous tackling of artificial synthesis of starch from carbon dioxide, and promotes the research and development of negative carbon technology
	Inner Mongolia Ordos	The "Ordos City Carbon Asset System Construction Plan", developing carbon assets exceeding 50 million tons
	Shandong Dongying	The "Dongying City Carbon Peak Work Plan" proposes to build the largest million-ton CCUS full-industry chain demonstration base in the country

Data source PetroChina Economic Research Institute

Commission issued the "Notice on Organizing the Application for Green Low-Carbon Advanced Technology Demonstration Project and Low-Carbon Application Scenarios", emphasizing that CCUS projects should be equipped with carbon dioxide transportation pipelines in principle, and proposed lower limit standards for the scale of CCUS equipped in different industries. Guangdong Province has initiated the preliminary planning research of a ten-million-ton scale offshore CCS/CCUS cluster, accelerating the development of industrial chain aggregation.

Third, it highlights the construction of basic capabilities and facilities. The National Energy Administration issued the "Several Opinions on Accelerating the Development of Energy Digitalization and Intelligence", proposing digital and intelligent transformation requirements from the perspective of industrial transformation and upgrading, and carrying out energy use monitoring and evaluation of key links such as CCUS. The Ministry of Ecology and Environment issued the "National Low-Carbon City Pilot Work Progress Assessment Report", incorporating the CCUS field into evaluation indicators, and forcing the application of green low-carbon technology.

Fourth, the outlook for the application and industrial layout development of CCUS technology in China.

Innovation in the Field of Carbon Capture Technology Will Accelerate, Playing a Key Role in Reducing CCUS Costs

The cost of carbon capture is a key link in reducing the cost of the entire CCUS chain and is also a key focus of CCUS technology research in China. The 2023 "bp World Energy Outlook" rapid transformation scenario and net-zero scenario predict that by 2050, more than 70% of global CCUS will be deployed in emerging economies led by China and India. With the acceleration of policy support and demonstration project construction for CCUS in China, China is expected to make breakthroughs in key technologies for carbon capture, including new absorbents, key materials, and equipment, driving a reduction of more than 30% in energy consumption and cost. A series of technologies are expected to be widely applied on a large scale around 2035.

CCUS Projects Are Showing a Trend of Cluster Development, and the Corresponding System and Policy Incentives Need to Be Further Explored

The cluster construction of CCUS is the key path to reducing the cost of CCUS in the future. China has already planned CCUS industrial clusters around oil and gas basins, such as the Junggar Basin, Bohai Bay Basin, Yangtze River Delta, and Daya Bay. The scale advantage of developing CCUS industrial clusters is obvious. In the future, when the carbon price or policy subsidy level reaches 100 yuan/ton, 150 yuan/ton, and 170 yuan/ton, petrochemical companies, power companies, and some steel companies will respectively have the potential to carry out full-process CCUS commercial projects. It is predicted that China will continue to improve the CCUS industry norms and standard system, and provide incentive policies that fit the development of CCUS projects from multiple dimensions such as financial support, technology research and development, and project support.

"Cooperative Competition" Has Become the New Normal in the Development of the CCUS Field, and We Should Seize the Key Window Period to Promote the Integrated Development of Industry, Academia, and Research

On November 15, 2023, China and the United States issued the "Sunshine Town Declaration on Strengthening Cooperation to Address the Climate Crisis", proposing that by 2030, the two countries will strive to promote at least five large-scale cooperation projects in the fields of industry and energy, such as carbon capture, utilization, and storage (CCUS). In 2023, the investment in global modular carbon capture systems, small carbon dioxide (CO_2) recovery devices, and direct air carbon capture (DAC) significantly increased, and the cooperation in carbon capture in the power, chemical, and transportation fields deepened. China will continue to seize the strategic opportunity period for the high-quality development of global CCUS, continue to explore the joint research and development and technology sharing system, and promote the progress of CCUS technology and the construction of talent teams through international cooperation platforms.

(This section was written by: Dong Yu, Reviewed by: Gao Hui)

The manufacturer's authorised representative in the EU is Springer Nature Customer Service Centre GmbH, Europaplatz 3, 69115 Heidelberg, Germany. If you have any concerns regarding our products, please contact ProductSafety@springernature.com

Printed and bound by CPI Group (UK) Ltd, Croydon, CR0 4YY
07/10/2026
02229350-0014